ADVANCES IN FUNCTIONAL HETEROCHAIN POLYMERS

ADVANCES IN FUNCTIONAL HETEROCHAIN POLYMERS

MARIA CAZACU
EDITOR

Nova Science Publishers, Inc.
New York

For permission to use material from this book please contact us:
Telephone 631-231-7269; Fax 631-231-8175
Web Site: http://www.novapublishers.com

LIBRARY OF CONGRESS CATALOGING-IN-PUBLICATION DATA

Advances in functional heterochain polymers / Maria Cazacu (editor).
p. cm.
ISBN 978-1-60456-598-0 (hardcover)
1. Heterochain polymers. I. Cazacu, Maria.
QD382.H48A38 2008
547'.7--dc22 *2008023159*

Published by Nova Science Publishers, Inc. ✠ New York

CONTENTS

PREFACE

The aim of this book was to provide some useful information concerning multiple aspects of certain polymers types. The approached polymers are heterochain ones that can develop special or high performance properties of interest for advanced fields: electronics, optic, nanoscience, environmental protection, coatings, etc. Both literature and own authors data are systematized. The advances in synthesis and characterization techniques were emphasized, as well as the evaluation of their functionalities and claimed the main application directions.

The first chapter of the book approaches a special class of heterochain polymers based on the silicones. The most important representative of this polymer family is the polysiloxane - an inorganic backbone consisting in repeating Si-O units with organic side groups attached to the silicon atoms. Although their synthesis and study began many years ago, they continue to stimulate the interest of scientists because of their versatile chemistry that permits obtaining new generations of silicone polymers. Thus, polysiloxane, heterochain one itself, can be involved in more sophisticated structures by combination with different organic or inorganic partners. A large range of commercially available or easy to prepare siloxane precursors can be reacted with organic or inorganic partners resulting in hybrid structures with various internal functions as linkages between sequences. The chapter points out some key aspects concerning the synthesis procedures, properties and main application fields of the polysiloxanes and polymers contained in the backbone, beside siloxane sequences, and other moieties based on elements such as nitrogen, phosphorous, boron, and carbon.

The second chapter is focused on the polymeric liquid crystals (LCPs) obtained by siloxane involving. Polymeric liquid crystals can be tailored in different ways: by choosing adequate mesogenic groups or spacers, synthesis methods and conditions. Siloxane moieties in LCPs are a very interesting alternative for obtaining desired properties. For example, they decrease the transition temperatures compared to organic spacers, and influence the nature of the mesophases and their stability, as well as other properties. The most recent developments (generally after year 2000) in LCPs of various architectures, containing siloxane segments in their structure are reviewed, emphasizing the role of Si-O bonds on their thermotropic and overall properties.

Aromatic polyazomethines are the most available thermotropic liquid crystalline (TLC) materials, which can be used in optics and optoelectronics. From a fundamental point of view, polyazomethines represent today an important class of polymers because (as it was shown by Millaud and Strazielle in 1979) they "are the first macromolecules to present liquid crystal structure in the absence of solvent". One can add also their excellent thermal, mechanical, electronic, optical, optoelectronic and fiber forming properties. The chapter three is focused especially on the authors' own contributions in thermotropic poly(azomethine-ether)s and poly(azomethine-sulfone)s.

Chapter four presents an up-to-date review of the latest research into three heteroatom containing conducting polymer classes, i.e. polyaniline (PANI), polypyrrole (PPy) and polythiophene (PT). The first one contains the heteroatom (nitrogen) placed into the main polymer chain as imine and amine nitrogen, the structure and properties of the polyaniline being strongly influenced by the presence of this atom. The existence of imine sites confers a specific feature for polyaniline unlike other semiconducting polymers, namely the possibility to be doped with protonic acids. The imine nitrogen is more basic and more easily to protonate compared with the amine atom that can be protonated only at high acid concentrations. The last two polymers have a polyenic structure remembering of *trans – cisoidal* polyacetylene but the –NH- or –S- atom bridges stabilize the *cisoidal* conformation and the polymers are very thermal and environmentally stable, their properties being closer of arylene polymers. In the polyaniline class, the emphasis will be centered on synthesis of nanostructures obtained by polymerization of aniline in different conditions and in the presence of organic macrocyclics, organic polymers and inorganic materials. For polyheteroarylenes, the new exciting structures of oligomers and polymers, such as tubular, macrocyclic, star-shape, hyperbranched and dendrimer are briefly presented, as well as their self-assembly and the applications of these polymer classes.

The main issues of polyurethanes containing different chromophores chemically incorporated in the macromolecular backbone, with a focus on the ionic polymers with photoisomerizable or photolabile structures, are presented in chapter five. First, aspects of chemistry and photochemistry of the polyurethanes are presented, such as isocyanate, polyols, chromophoric derivatives, prepolymers and segmented elastomers as well as their properties induced by the presence of chromophore (for instance, azobenzene). Then, some specific applications of the photopolyurethanes and the corresponding ionic forms are described.

Among thermostable polymers, aromatic polyimides represent an important class of heat resistant polymeric materials. However, polyimides are often insoluble and intractable in their fully imidized form, presenting serious processing difficulties. Poly(ether imide)s have received much attention because they can provide a compromise between good thermal stability and processability when compared to aromatic polyimides. Chapter six is a review concerning the progress on the synthesis, characterization and potential applications of poly(ether imide)s. It is focused on the recent development in this field. Different types of structurally modified poly(ether imide)s having flexible units, bulky groups, fluorine atoms, siloxane sequences, as well as hyperbranced compounds or poly(ether imide)s containing other heterocycles are presented. The properties of these polymers like solubility, film forming ability, thermal stability, glass transition temperature, mechanical and dielectric characteristics are discussed in correlation with their structures. Future perspectives of the polymers for use as advanced materials in high performance applications are also given.

Another polyimide category, addition polyimides, were approached in chapter seven with emphasize on the most important chemical concepts for bismaleimide-based addition polyimides. The attention is focused on the classical approaches and also to more recent developments. The developed properties are frequently mentioned, the most important being the inherent thermal stability with applications derived from herein.

Every polymer type was approached by authors with a large experience on the topic. Both the literature and their own data was reviewed.

In: Advances in Functional Heterochain Polymers
Editor: Maria Cazacu
ISBN: 978-1-60456-598-0

Chapter 1

POLYMERS CONTAINING SI, O, AND OTHER ELEMENTS WITHIN BACKBONE

Maria Cazacu [1]
"Petru Poni" Institute of Macromolecular Chemistry, Iasi, Aleea Gr. Ghica Voda 41A, 700487, Romania

ABSTRACT

A special class of heterochain polymers is formed by mixed organic-inorganic polymers. The most important representatives of this polymer family are the silicones that are typically represented by polysiloxane - an inorganic backbone consisting in repeating Si-O units with organic side groups attached to the silicon atoms. Although their synthesis and study began many years ago, they continue to stimulate the interest of scientists because of their versatile chemistry that permits to obtain new generations of silicone polymers. Thus, polysiloxane, itself heterochain one, can be involved in more sophisticated structures by combination with different partners. A large range of commercially available or easy to prepare siloxane precursors can be reacted with organic or inorganic compounds resulting in hybrid structures with various internal functions as linkages between sequences. The chapter is an overview of the synthesis procedures, properties, and main application fields of the polysiloxanes and polymers containing in the backbone, besides siloxane sequences, other moieties based on elements such as: oxygen, nitrogen, phosphorous, boron, carbon.

1. INTRODUCTION

Most of the polymers that have been synthesized and studied along the time are organic macromolecules [1] characterized in principal by a backbone that contains either only carbon or carbon in combination with a selection of non-metallic heteroatoms such as oxygen, nitrogen, sulfur or phosphorus [2]. Even though, as is well known, the elements from the periods in the lower table have tendency to form element-element bonds with decreasing

[1] e-mail: mcazacu@icmpp.ro.

strength, a huge potential exists for the development of polymers with inorganic backbones that challenged the researchers still at the beginning macromolecular science [1].

The main alternative to the hydrocarbons is the polysilanes family. Silicon an element belonging to Group 4 (IVA) of the periodic table is situated under carbon, and belongs, with germanium, to the metalloids. Some of its characteristics are similar to carbon, but overall it can be seen as a completely different element. Silicon has the same oxidation state as carbon, but has the ability to use 3d orbitals for bonding by expanding its valence sphere. Silicon has semi-metallic properties. Silicon and carbon bonding can be compared in many ways. Silane, SiH_4, and methane, CH_4, are both very stable tetrahedral compounds, but while the carbon chain is stable one (carbon is the backbone of life and can form chains of infinite length), the silane chains stability decreases with length, the Si-Si bond being slightly weaker than the C-C bond. By replacing half the silicon units in a polysilane with suitable heteroatoms, thermally stable and non-light-sensitive silicon-based polymers resulted. The best-known examples are the polysiloxanes, whose backbones consist of alternating atoms of silicon and oxygen, commonly known by their trade name of silicones. Siloxanes were first characterized as macromolecules by the English chemist Frederic Stanley Kipping in 1927. The name of silicones come from the similarity between the repeating units in siloxanes and ketone. However, there are differences between these two types of compounds as a result of the basic differences between silicon and carbon chemistries. Carbon shows a strong tendency to form π bonds, and ketones have the structure RR'C=O, with a double bond between the carbon and oxygen. They are generally simple molecular compounds. Silicon, on the other hand, does not show this tendency, and $p\pi$-$p\pi$ double bonds between silicon and itself or other elements are relatively unstable [3].

Polysiloxanes being the main class of inorganic backbone polymers constitute the subject of broad industrial and fundamental interest [1]. Due to their special position occupied in chemistry there are a high number of books and reviews on silicones [3-19]. The structural differences within the polysiloxane class can be done by the length of the chain, the type of the groups attached to the silicon atoms within the chain, end groups nature, by branching or crosslinking. Thus, there are α,ω-bifunctionalized polysiloxanes, side-functionalized polysiloxanes, and siloxane-siloxane copolymers containing siloxane units with different substituents to the silicon atom. Depending on the nature of the substituents, these can be or no reactive or functionally. In the world of silicones many chemistries are now available to make various linear siloxane and organofunctional siloxane polymers [20]. Silicones are known as possessing a unique properties combination and have a wide range of applications.

Although the polysiloxanes are itself heterochain polymers, it can be co-chained with different other elements within the backbone in various combinations and successions that can induce a large range of properties combinations, the possibilities on this direction theoretically being unlimited.

By incorporating other elements or elements groups as sequences or blocks within the siloxane backbone it would be lead to: poly(carbosiloxane)s, poly(siloxazane)s, poly(siloxaphosphazane)s, poly(siloxacarborane)s, polysiloxanes containing metals. A very vast class of polymers containing other sequences beside siloxane within the backbone is siloxane-organic copolymers. Copolymers containing different functional or non-functional bridges between siloxane and organic sequences: ester, silyl ester, ether, silyl ether, anhydride, amide, imide, urethane, alkylene, etc. were already described elsewhere [21]. Different techniques (hydrosilylation [22-24] or polycondensation of appropriate functionalized precursors, transition metals mediated living radical polymerization [25-28], atom transfer radical polymerization, ATRP, [29], in supercritical carbon dioxide [30-32] or in miniemulsion [33]) were approached to obtain such copolymers.

Some of the main groups of heterochain polymers derived from siloxanes will be reviewed in the following.

2. POLYSILOXANES

Polysiloxanes are heterochain polymers whose backbones consist of alternating atoms of silicon and oxygen having organic substituents attached to the silicon atoms (Scheme 1).

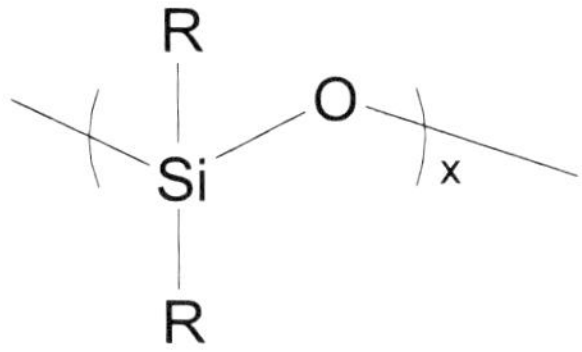

Scheme 1 (R= organic group)

Polysiloxanes (silicones) are ones of the most studied and useful polymers in the modern world, they exhibiting a variety of properties combinations not common to non-metal-containing macromolecules [34]. The presence of certain organic groups attached to the silicon atom in polysiloxane influences the physical, chemical and mechanical properties, typically in a desired way.

Polydimethylsiloxane, PDMS, backbone, the most representative for silicones, consists of alternating Si and O atoms with two methyl groups attached to each Si atom, as seen in Scheme 2.

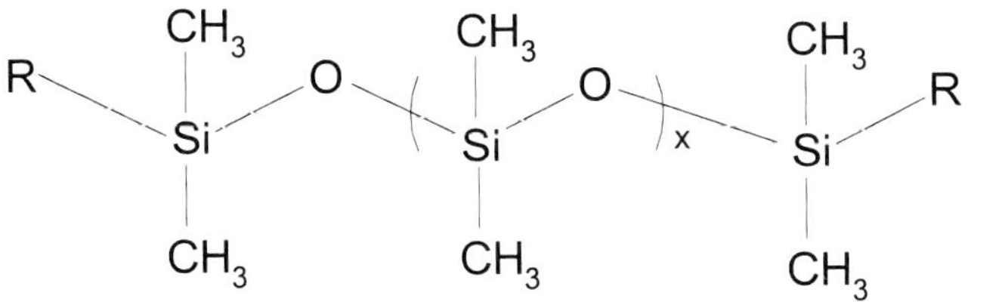

with R: H, organic group (CH_3, C_2H_3), functional group

Scheme 2.

Silicone polymers containing organic groups other than methyl received considerable attention from the beginning of the silicones. These groups incorporated as substituents, are in principal long alkyl, phenyl, fluoro- or cyano-alkyl, vinyl, or H (Table 1). Such substituents modify the properties of the parent PDMS, improving certain properties as mechanical, thermal, solvents, and/or radiation resistance, etc. Siloxane-siloxane copolymers containing units with different substituents at the silicon atoms have been synthesized many years ago.

The researches developed last time followed to obtain copolymers having tailored microstructure.

Table 1. Other common polysiloxanes types

Structure	Particularities	Ref.
Polydiethylsiloxane 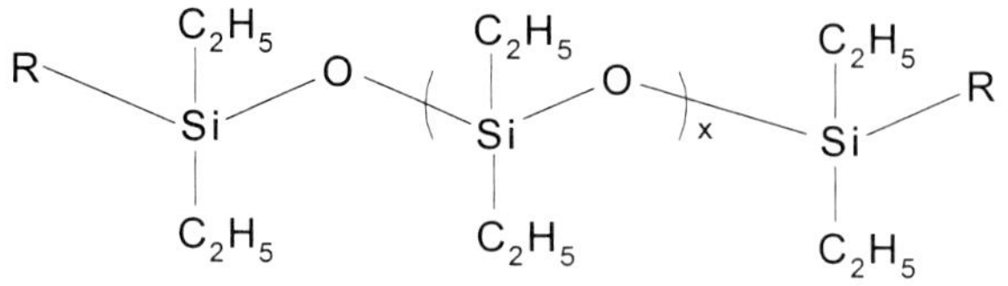	Stress-induced columnar mesophase formation, greatly improved lubrication characteristics, greater compatibility with organic materials, higher viscosity-temperature coefficients, lower compressibility and decreased oxidation stability.	[35-42]
Poly(dimethylmethyl-H-siloxane) 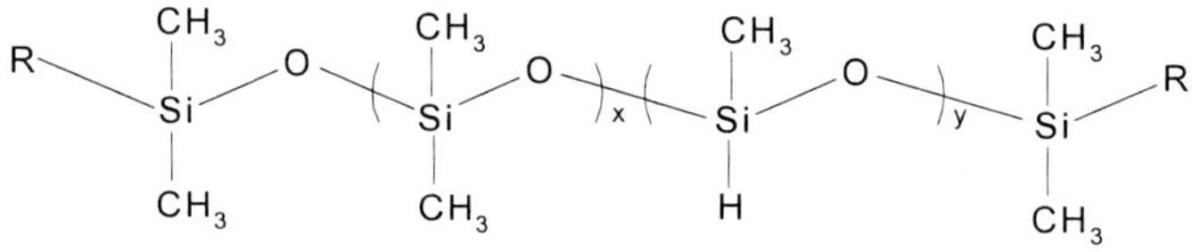	Chemical reactivity, lower Tg	[43,44]
Poly(dimethylmethylvinylsiloxane) 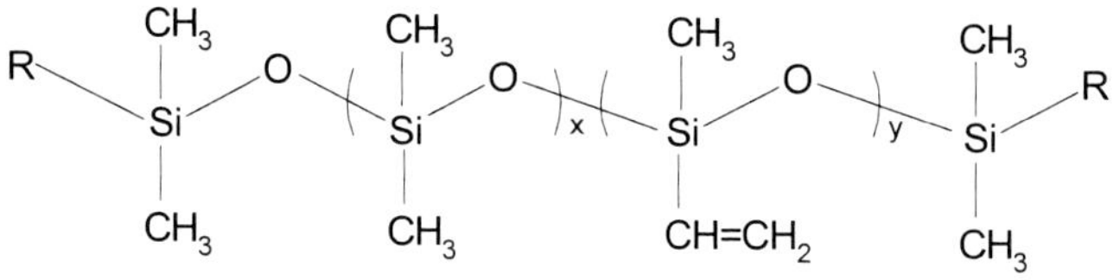	Chemical reactivity, compression resistance	[45-48]
Poly(dimethyldiphenylsiloxane) 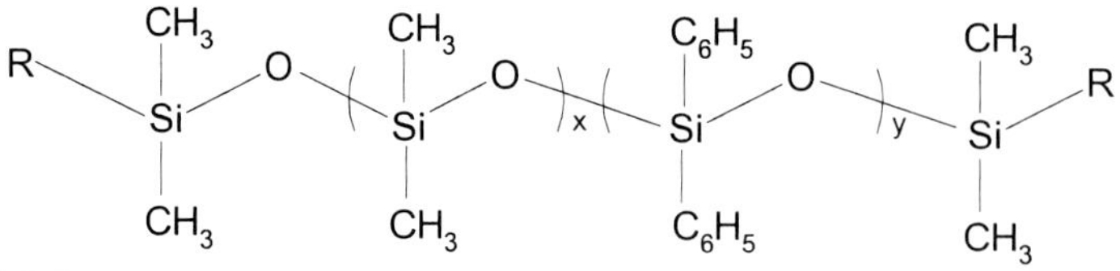 Poly(dimethylmethyphenylsiloxane) 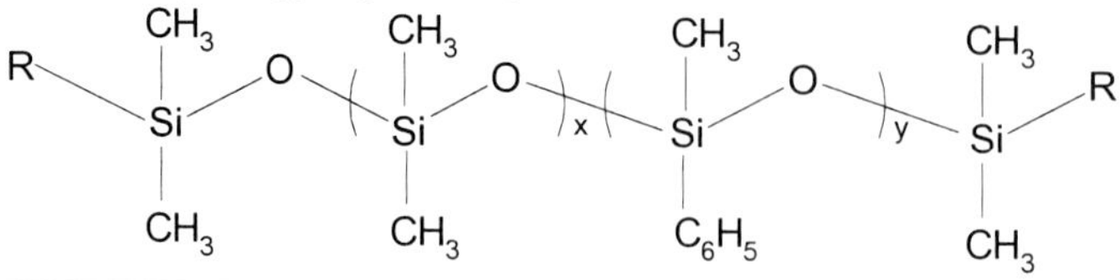	Enhanced radiation and oxidation resistance, thermal stability, lubricity, and shear resistance, reduced compressibility, increased rigidity, increased refractive index	[42,49-55]
Poly(dimethylmethylcyanpropylsiloxane) 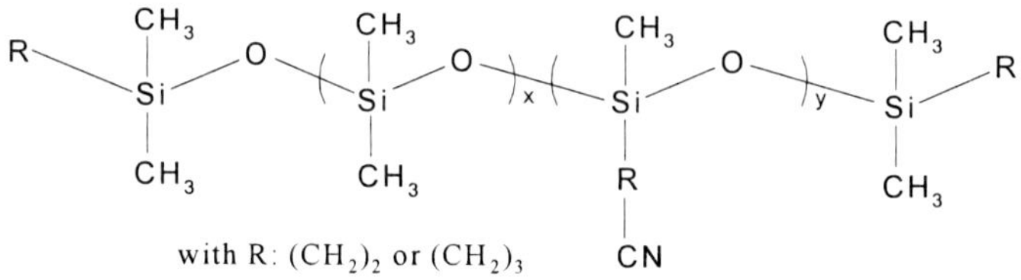 with R: $(CH_2)_2$ or $(CH_2)_3$	Polarity, solvent resistance, high dielectric, antistaticity.	[56-58]

Structure	Particularities	Ref.
Poly(dimethylmethyltrifluoropropylsiloxane) 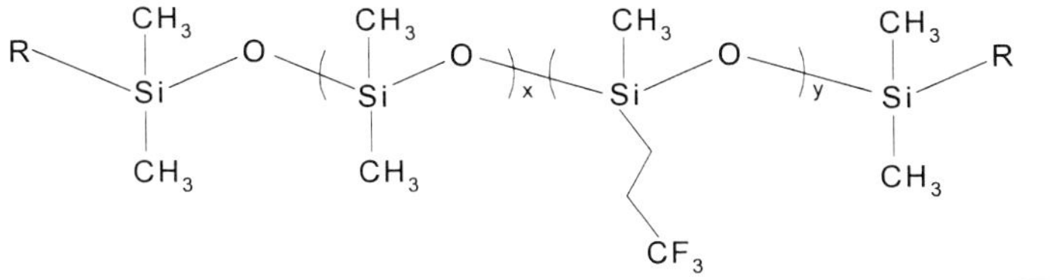	Chemical and solvent resistance, lubricity and thermal stability	[54, 59,60]
Side-functionalized polysiloxanes 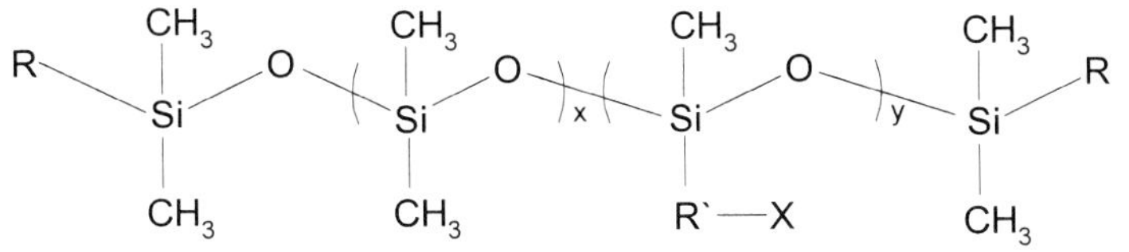 with R`-organic moiety; X – functional group	Reactive and functional polymers	[61-94]

With R: H, organic group (CH_3, C_2H_3) or functional groups

Some recent studies were focused on the synthesis, structure and properties of siloxanes having silyl [95] or trimethylsiloxy [96] substituents at the silicon.

2.1. Polysiloxanes Synthesis

The linear polysiloxanes are prepared by the hydrolysis of the dihalo- or dialkoxydiorganosilanes when corresponding silanols are formed. The outstanding characteristic of the silanols is the ease in which they condense to yield siloxane polymers as is shown in Scheme 3 for the $(CH_3)_2SiCl_2$ case.

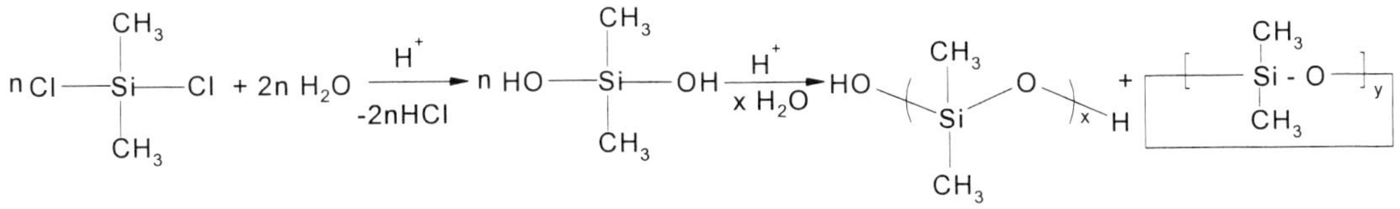

y = 3 (hexamethylcyclotrisiloxane, D_3); 4 (octamethylcyclotetrasiloxane, D_4); 5 (decamethylcyclopentasiloxane, D_5)

Scheme 3.

Thus, the hydrolysis reaction product consists in a mixture of cyclic compounds (where y = 3, 4, 5, etc.) and open chained compounds having hydroxyl end groups. The cyclic species are isolated and purified by fractional distillation and then polymerized to the linear polymer by a process called equilibration (Scheme 4) [3,5-7]. Equilibration of cyclic siloxanes is the process by which the -Si-O- linkages are continuously broken and reformed until the system reaches an equilibrium condition at the most thermodynamically stable state. The chain length of the polymer can be controlled by adding a monofunctional compound, usually $(CH_3)_3SiCl$ (but silanes of the type X-R$(CH_3)_2$SiCl with X – function can also be used) called endblocker, to the reaction mixture.

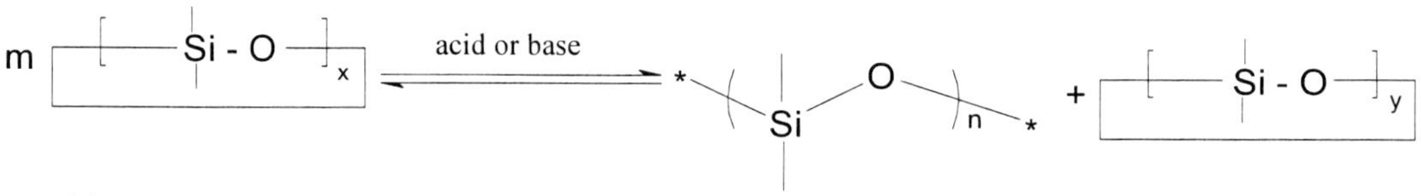

with: x = 3 or 4, y = mixture of 3,4, 5......

Scheme 4.

The equilibrium position depends on the starting cycle size and the silicon substituent nature and also on the reaction conditions (concentration of cyclosiloxane units, solvent type, initiating/catalyzing system, temperature) [3,5-7,97].

By the choosing the proper monomer/catalyst/solvent system a higher propagation rate toward the depolymerization and chain scrambling ones can be reached, thus the reaction being kinetically controlled. There is an excellent review on this subject [6]. A suggestive representation of the two types of processes is shown in Figure 1 adapted from [4].

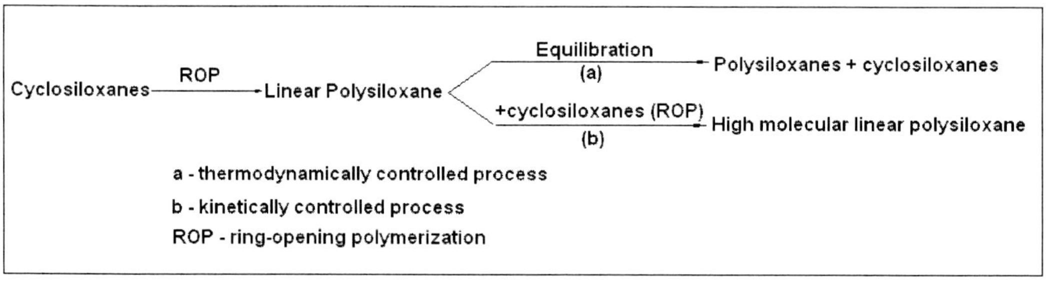

Figure 1 [4].

2.1.1. Catalysts for Ring-Opening Polymerization of the Cyclosiloxanes

Due to its polar character, the siloxane bond can be broken only by ionic mechanism, either anionic or cationic. As a result, either acid or base catalyst systems are usually used [5].

A large range of catalysts was approached in the synthesis of polysiloxanes.

In general, the acid catalysts include powerful acids such as H_2SO_4, CF_3SO_3H or $HClO_4$, and also Lewis-Brönsted acids, such as $SnCl_4/H_2O$. The basic catalysts for the ring-opening polymerization are alkaline metals hydroxide and quaternary alkyl-ammonium hydroxides. Some reports are referring to the opening of the siloxane rings on the surface of some metals and dispersed metallic oxides, ion exchanging resins [98], compounds with platinum and palladium [99], and graphite intercalation compounds [100,101].

Basic Catalysts

Anionic polymerization is the main method used for preparing high molecular mass polysiloxanes. The catalysts for this type of polymerization are usually hydroxyl and silanolate anions that are generated through the reaction of the cyclic siloxanes with different strong bases such as hydroxides, oxides or silanolates of the alkaline metals, ammonium and phosphonium hydroxides. Organo-metallic compounds such as *n*-butyl lithium, electron-transfer complexes (naphthalene sodium and naphthalene potassium) and metal alkoxides can also be used in the cyclosiloxanes polymerization processes [4]. Anion exchangers containing ammonium or phosphonium groups [102-105] can also be used as catalysts in siloxane synthesis.

For the polymerization reactions initiated with the alkaline metal hydroxides, it has been shown that the efficiency of the catalysts increases with the growth in the degree of ionization of their salts or that of the silanolates, in other words it grows when the size of the cation increases. The environment in which the reaction takes place is nonpolar, whether it is in solution or in mass. The concentration of the free ions will therefore be insignificant, due to the increased affinity of the silanolate ions for the cation. The active propagation center is in the form of ion pairs. Even more, it is supposed that these in turn form associations in which the ions are in a latent state. Only a small proportion of the free, no associated ions exist in equilibrium with these aggregates. The aggregation process strongly reduces the reaction rate. It was found that the most powerful aggregates are formed by lithium, while the ions that are larger in size, such as Cs^+ form only traces. The ammonium or phosphonium ions do not form such complexes. That is why the efficiency of these alkalies varies in this order [5]:

$$LiOH < NaOH < KOH < RbOH < CsOH = R_4NOH = R_4POH$$

The potassium silanolate has proved to be more efficient than the potassium hydroxide, while the potassium amide (KNH_2) and potassium oxide (K_2O) are even more active than the silanolate. Tetramethylammonium methoxide is almost as efficient as the cesium hydroxide. The rate of the polymerization process grows as the value of the dielectric constant of the environment [4].

Some pathways approached to reduce the complexation degree of the ionic pairs in anionic polymerization are reported in literature:

- The polymerization in solution, in a solvent that strongly interacts with the cation, the most common system being $THF^{-}Li^{+}$ [6].
- The polymerization in mass or in solution in presence of hexamethylphosphortriamide (HMPT) or DMSO [6] as nucleophile additions.
- The polymerization with inclusion complexes of the counter ion [6,8,36, 106] by using some macrocyclic ligands (cryptands or crown ethers).
- The polymerization in the presence of counter ions that weakly interact with other ions. It is known that some organic cations interact very weak with other ions. Most often the tetraalkyl ammonium or phosphonium ions, R_4N^+ and R_4P^+, are used, these additionally possessing the advantage to be termolabile and can be deactivated at the end of the reaction by raising the temperature at > 100 °C for the phosphonium and at > 135 °C for the ammonium ones [6].

Acid catalysts

The polymerization process initiated by acids resembles that initiated by bases in what regards the efficiency in polymer and other reaction products that form at equilibrium. Yet the way to the final product is entirely different, because the global equations used to describe the reaction rate and the relative reactivities of the monomers differ considerably from those of the processes initiated with bases. The most used catalysts for the polymerization through cationic mechanism are:

Brönsted acids: H_2SO_4 and his derivatives (alkyl and aryl sulphonic acids, chloro sulphonic acid, diorganylsilyl sulphates, aluminum sulphate) and other mineral and organic acids (HCl, $HClO_4$, CF_3COOH, CF_3SO_3H, etc).

Strong Lewis acids that includes a variety of electrophil agents:

- inorganic or organic cations: $AlCl_2^+$, $C_6H_5CO^+$, $(C_6H_5)_3C^+$;
- neutral molecules: PF_5, Tf_2O, TfOEt, CH_3I, with Tf = $F_3CSO_2^-$;
- mixtures: MX_n + RX (M-metal, R-organic radical, X-halogen).

Some studies have proved that the Lewis acids do not initiate alone the cationic polymerization. There were even prepared stable systems such as: D^* + MX_n [107]. Efficient systems are obtained in the presence of a cocatalyst (Brönsted acids) that provides the proton, which is the actual initiator in the polymerization, obtained through the formation and the splitting of the complexes: $H^+[AlCl_4]^-$, $H^+[FeCl_4]^-$, $H^+[FeCl_3OH]^-$, $H^+[SnCl_4OH]^-$.

However, it has been found [106] that Lewis super acids such as ethylborontriflate ($Et_3B_2(OTf)_3$) initiate alone the polymerization of D_3, in a strong enough polar solvent (70% CH_2Cl_2 + 30% CH_3CN). But the achieved results in what regards the reaction rate and the polymer yields were modest. Along the years there were reported results on the polymerization of the cyclosiloxanes with nonprotic cationic initiation systems: silyltriflate-esters or $R_3SiH/(C_6H_5)_3C^+B(C_6H_5)_4^-$ [108]. In 1998 there were reported [109] very interesting results, specifically the achievement of *live cationic polymerization systems* of the cyclosiloxanes, including those that are differently substituted, in bulk or in solution (methylene chloride), at room temperature, in the presence of trimethylsilyltriflate as a catalyst. Addition of 2,6-ditertbutyl pyridine was used as proton acceptor. It results in siloxane polymers with molecular mass value pre-established through the initial monomer/catalyst ratio, having one trimethylsilyl group at one end of the chain and in which the monomer's microstructure is re-found. The cyclic oligomers quantity was minimum. Thus, if at the polymerization of the methyltrifluoropropyl substituted cyclosiloxane with sulphuric acid the yield in polymer at equilibrium is 17%, for the polymerization of the same monomer with trimethylsilyltriflate, in the conditions shown above, it gets a polymer yield of 87%.

These results open the way for solving a problem of great importance that for years has preoccupied many groups of researchers, and that is to create conditions for the cationic polymerization under kinetic control of the cyclosiloxanes, that could avoid excessive cyclization. Also, the efficient synthesis of many polysiloxane structures that have on the silicon atom electrophil substituents or other groups reactive towards bases the becomes possible. The live character of the process also offers the possibility to obtain some siloxane-organic blockcopolymers through cationic mechanism.

Other sources of changeable protons, with peculiar importance in the industrial systems, as catalysts for siloxane polymerization, are the compounds of intercalation, activated clays, ion-exchanging resins.

Intercalation compounds Lewis acid - graphite [107,110,111], with the structure $C_x(M^{n+}X_n)_y$, where M = Fe, Al, Sb, etc. and X = Cl, Br, etc., such as $C_{30}SbCl_5$, $C_{77}FeCl_3$,

$C_{64}AlBr_3$, where x – the number of carbon atoms from the graphite layers intercalated with the molecules of Lewis acid were tested in siloxane polymerization.

It is supposed that these acids posses an ionic character due to the transfer of electrons from graphite to the acid participating at the process of intercalation. The resulted compounds can be represented with the formulas:

$[graphite]^+[Fe^{2+}Cl_2Cl^-]^-.xFeCl_3$

$[graphite]^+[SbCl_6]^-.ySbCl_5.zSbCl_3$

If one considers that these compounds are structurally analogous with the complexes of the Lewis acids with the proton donors having the structure: $H^+[M^{n+}X_nY]^-$, it may be assumed that the layers of graphite act like as the protons from the mentioned complex.

These catalysts present a series of advantages:

- they are active in polymerization;
- in some reactions they posses greater activity than the Lewis acids – cocatalyst complex;
- it is possible to work in bulk;
- they are easy to separate.

Although clear results exist about them catalysing the polymerization of the cyclosiloxanes, there are relatively few data in the literature in this field regarding their use as catalysts, as they present some disadvantages:

- the resulted polymers have relatively low molecular mass;
- the resulted polymers are slightly colored due to the extraction of the metallic halide in the homogenous phase;
- there is the suspicion that the proton donor impurities interfere in the reaction.

Activated Clays

The use of inorganic ion exchangers has appeared as an alternate way for the polymerization with sulphuric acid. The catalytic properties of the mineral clays and polyelectrolytes have been known for a long time [112]. The using ion exchangers in catalysis is an industrial viable way and depends on the availability of robust, durable materials, capable of performing well in extreme conditions of temperature, strong ionic, radioactive, oxidant and reducing concentrations [113]. In order to be separated easier from the reaction environment, the acid catalysts were laid on support materials such as bentonite and absorbing clays.

The first reports on the use of the inorganic ion exchangers as catalysts for the polymerization of the cyclosiloxanes dates from 1949 [114] and refers to the use of the acidic clays for the polymerization of dimethylsiloxanes. Later, other few studies [115,116] have brought their contribution in the field of catalysis with acid activated clays applied at the rearranging of the siloxanes. Thus, "terra alba" clay treated with sulphuric acid, washed and dried, was used as a catalyst for the polymerization of octamethylcyclotetrasiloxane (D_4) in the presence of the hexamethyldisiloxane, $(CH_3)_2SiOSi(CH_3)_2$, at the temperature of 80 °C.

Following the change of the viscosity, the same variation as in the case of using sulphuric acid in similar conditions was found. A correlation between the reaction rate and the catalyst quantity has been established, and at a constant ratio between the D_4 and hexamethyldisiloxane there have not been noticed changes in the equilibrium viscosity with the change of the quantity of catalyst or of the temperature. The Si-H bond from organo-H-siloxanes, which is extremely sensitive to the alkaline catalysts, is very well preserved at the polymerization with sulphuric acid. But the best catalyst for these siloxanes has proved to be the activated clay, which attacks the Si-O-Si bond, leaving intact the Si-H bond [117]. During the years the use of different types of acidic activated clays has been intensively studied [118-126].

Ion Exchangers

However, in time, the inorganic ion exchangers were slightly replaced by the organic, synthetic ion exchanging resins [113]. The organic ion exchangers were used for the first time to obtain siloxane polymers in 1958 [127], and then, many papers and especially patents have appeared on the use in this purpose of the ion exchangers, including sulphonated phenol-formaldehyde resins or sulphonated styrene-divinylbenzene copolymers [128-134]. The use of other ion exchangers (having acid or alkaline active groups on different supports (polystyrene, cellulose, etc.) was also suggested [129]. Starting with 1970, many applicative studies have been done on the polymerization of cyclosiloxanes with sulphonated macroporous styrene - divinylbenzene ions exchangers [117]. Such catalyst type proved to be very effective in the synthesis of the siloxane polymers [135-139] and copolymers [140-145] with low molecular mass (<100000) as well as for the synthesis of functional siloxane oligomers [146-150].

Anion exchangers based on groups of the type $-C_6H_4-N^+(CH_3)_3$ and $-CONH(CH_2)_3N^+(CH_3)_2CH_2C_6H_5$ attached on a macroporous styrene – divinylbenzene support were also tested as catalyst for the synthesis od siloxane polymers and copolymers [102,151].

The ion exchangers-catalyzed polymerization has proved to be highly efficient and, at least of the point of view of the ease of catalyst separation, this procedure offers a net advantage. Other advantages of using the ions exchangers, that are not to be neglected, are:

- obtaining polymers and copolymers without rests of catalyst on their ends;
- the possibility of multiple use and therefore of making the process continuous;
- high selectivity;
- the possibility of using in the same system of both cations and anions exchangers, each with his own function [55];
- moderate and noncorossive action;
- non toxicity;
- easy handling.

Even more, taking in consideration just a few of these characteristics, the ions exchanger catalysis can be considered eco-friendly.

Other Catalyst Systems

Some catalytic systems, such as: ≡SiOLi/THF, ≡SiOLi/HMPA, ≡SiOK/18Crown6, $\equiv SiONMe_4$, Schwesinger base (peralkylated polyammino-phosphazenes)/MeOH, CF_3SO_3H, $CF_3SO_3SiMe_3$/ CF_3SO_3H, $K^+B(C_6F_5)^-_4$/ ≡SiCl were recently tested for the chemo- and regioselective anionic or cationic ring-opening polymerization of some model cyclotrisiloxanes having dimethyldiphenyl- or dimethylmethylvinylsiloxane mixt units [152,153].

The "silicateine" proteins proved to catalyze the synthesis of silica and organic modified silsesquioxanes starting from the etoxysilanes at pH neutral and low temperature [154]. This creates the premise for the synthesis of silicon-based polymers and materials in friendly environment conditions.

2.1.2. Catalysts for Siloxane Polycondensation

The polycondensation of silanol groups is the most important reaction leading to the formation of the siloxane bond [155]. This reaction is largely used for the synthesis of different siloxane molecular architectures, such as linear siloxane polymers and copolymers, silsesquioxanes, silica networks and glasses. That is why the mechanism of this reaction was intensely studied [155]. In the same time, this reaction type is suitable for the preparation of well-defined siloxane copolymers [156]. Therefore, there is a great interest in new effective and selective catalysts for this process.

The strong protic acids, such as H_2SO_4, CF_3SO_3H, HCl and CH_3SO_3H induce the polycondensation of the functional silanes in mild conditions and are easily tolerated by some functional groups such as SiH or $C_6H_5CH_2Cl$ that are reactive towards bases. As a result, the acid catalysts are the most used in the silanol polycondensation. The tris(pentafluorophenyl)boron, $[B(C_6F_5)_3]$ is a remarkable stable and as strong as BF_3 Lewis acid that has found application as catalyst for the condensation of the polysiloxanediols with organic-H-silanes, and the reaction takes place in mild conditions [156].

Because the catalytic activity in the polycondensation of the silanols grows with the strength of the acid used as catalyst, some superacids have been tested for this purpose. Thus, the protonated tetrakis(pentafluorophenyl)borate complex, $B(C_6F_5)_4^-H^+$, which is a superacid, was found to be a very efficient catalyst in the condensation of the silanols [157,158]. Compared with the conventional acids this complex presents the advantages that it is more efficient and more soluble in nonpolar solvents. Its large anion has an unusually low nucleophilic character and exceptional inertness. In the presence of a very small amount of such acid the transformation of the oligodimethylsiloxanediol in a high molecular mass polymer takes place very rapidly at room temperature in a nonpolar solvent, such as toluene, without the formation of cyclic compounds.

Strong bases, such as hydroxides or silanolates are known to be also very efficient catalysts for the condensation processes. However, the polycondensation driven in the presence of strong bases is dominated by disproportionation processes that imply the breaking of the siloxane bond and leads to oligomers and statisical polymers. Just like in polymerization, it was proved that an important role in disproportionation processes is played by the counter ion. Thus, disproportionation is favoured by the presence of small cations, such as Li^+, that interact with the silanolate anion. That is why the use of good solvents for the base, capable to weaken the interactions between the pairs of ions, is preferred. Most of the studies were lead by using the bases of the alkaline metals, in solvents such as dioxane-water, methanol, deuterated water and isopropanol. But these solvents are known to be very active in breaking the Si-O bond, leading to a very complex reaction system. An alternative is represented by the bases with very large cations such as ammonium and phosphonium. Recently, perammino-oligo-phosphazenium hydroxide, a very strong base, soluble in aromatic hydrocarbons, has proved to be a very efficient catalyst, both for the polymerization of the siloxanes and for the polycondensation of the silanols [159].

The cyclosiloxane polymerization can be conducted in bulk (in fluid state) or on surface (in solid-state [160,161]), in solution (in non-polar solvents such as toluene, xylene, isooctane, n-heptane, methylene chloride, chloroform, chlorobenzene, ethylic ether, etc.) in principal and, according to more recently reports, in miniemulsion [162-171].

Silicones can be prepared having a wide range of compositions, architectures, viscosities, and reactivities and processed as fluids (oils), adhesives, foams, coatings, greases, rubbers (room/low/high temperature), etc.

2.2. Properties of the Polysiloxanes as a Consequence of the Si-O-Si Bond Particularity

Polysiloxanes are characterized by combinations of chemical, mechanical, electrical, and other properties that, when taken together, are not found in any other commercially available class of materials.

The siloxane bond particularity is considered responsible to induce such special properties combinations to silicones: high flexibility, high permeability to gases, low glass transition temperature, T_g, low surface energy or surface tension, good resistance to UV radiation, excellent release properties and surface activity, good damping behavior, antifriction and lubricity, hydrophobic and physiological inertness, shear stability, weak intermolecular forces, and excellent dielectric strength, low volatility at high molecular weight, and high volatility at low molecular weight, good performance at extreme (high/low) temperatures and minimal temperature effect on the properties [172]. PDMS is homogeneous, isotropic, and optically transparent down to about 300 nm [173]. Silicones are odorless, colorless and water resistant.

Some key properties of PDMS arise from the high and flexible bond angle around the O atom in the siloxane backbone, ranging between 135 and 180° with an energy minimum at 145°. This facilitates rotation of $Si(CH_3)_2$- group around the backbone axis. The siloxane backbone unit Si-O-Si has a bond length of 1.63 Å, where as a C-C-C backbone unit has a bond length of 1.54 Å and a bond angle of 112°. Because the bond lengths are longer in siloxanes, they can move further and change conformation easily, making for a flexible material. Thus, polysiloxanes show two types of flexibility: torsion flexibility and bending flexibility [5,7]. These polymers can be made less flexible or more rigid by varying the structure. For instance, combining two polysiloxane chains into a ladder structure, insertion of rigid groups into the structure, or adding bulky side groups will increase rigidness and will also affect the glass transition temperature; bulky groups that hinder the flexibility will raise the T_g. Thus, while PDMS have Tg=-125°C, by replacing a methyl with another group, Tg value modifies as following: H -138°C, C_2H_5 -135°C, C_3H_7 -120°C, $C_3H_4F_3$ -70°C, C_6H_5 -

90...-75°C. Polydiphenylsiloxanes have T_g between 50-100°C depending on the molecular mass [42, 172].

In the same time the siloxane bond length is smaller as compared with sum of ionic radius of the two involved atoms (Si and O). This is assigned to partial double bond character. On the other hand, the difference between the electronegativities of the two atoms of the silxoane bond has as a result a bond having partial ionic character. As a result, the siloxane bond can be broken only heterolytic, by ionic mechanism [5,7].

Polysiloxanes have a much higher permeability to gases than other polymers. Besides this, the representant PDMS has also other unusual properties including: low internal pressure, low bulk viscosity, low temperature viscosity coefficient, low entropy of dilution, and low excess volume upon mixing. The reasons suggested for these behaviors involve low intermolecular interactions and high rotational and oscillatory freedom of the methyl side groups. Another explanation is based on the chain's irregular cross section, which is large at the substituted silicon atom and small at the un-substituted oxygen atom [7].

Low surface energy, or surface tension, is another key feature of polysiloxanes. The low surface tension of PDMS (21-23 mJ m^{-2})[3] is a result of closely packed methyl groups at the surface [174]. In correlation with this, the boiling point for some polysiloxanes is low as well.

However, the mechanical properties of PDMS are usually very poor at room temperature, unless reinforced with silica and vulcanized. In order to develop useful properties, very high molecular weights are required. Even for chemically crosslinked and reinforced PDMS, the tensile strength is still relatively low, compared to other elastomers [173].

2.3. Applications of the Silicones

Due to their properties, silicone-based materials have a long history of excellent performance in the fields ranging from medical to aerospatial: industrial chemistry, automobiles, cosmetics, pharmaceutic, electronic and energetic industries, in environment protection, etc., where they can fulfield different functions. As a a result, the polysiloxanes are extensively produced and currently sold as high-performance sealants, lubricants, release agents, surfactants, water repellents, adhesives, foam control agents, biomedical devices, personal care and cosmetics, dielectric encapsulation and high-temperature electrical insulation, high-temperature paints, window gaskets, O-rings, contact lenses, and numerous and variable human biological implants and prosthetics, glass sizing agents, greases, hydraulic fuids, heat transfer fuids, lubricants, fuser oil, masonry protectants, process aids to mention just a few of their applications, rubber molds, etc. [34,172].

The world-wide sales of polysiloxanes or silicones at the beginning of this new millennium are approximately $10 billion per year. Commercial products range from those entirely composed of silicone to products where the silicone is a low level but key component [15], the silicone term used herein in its broadest sense, that is, any polymeric structure that contains repeating silicon-oxygen groups in the backbone, side chains or cross links regardless of the substitution on the silicon atom.

Silicones are currently used in many different pharmaceutical products (than 350 registered drug products) as actives or excipients or as process aids to manufacture such products due to their unique set of physico-chemical properties. Recent application examples indicate that silicones can contribute to drug delivery system innovations, novel treatments and manufacturing process improvements [175].

Siloxane polymers have numerous medical applications such as: prostheses, artificial organs, facial reconstruction and tubing and catheters taking advantage of the inertness, stability, and pliability of these polymers; artificial skin, contact lenses, and drug delivery systems utilize their high permeability as well. A wide range of medical-purpose items was manufactured using vulcanizates of unsaturated siloxane rubbers with silicon hydrides of various structures. These materials are biologically-compatible with a high complex of medical and engineering properties [176]. Silicone implants are used in reconstructive and cosmetic breast surgery.

Silicone polymers are recognized as good emollients, not so much for their lubricity but for their high capability to spread due to low surface tension, which is also lower than their critical surface tension of wetting [175], being used in cosmetic products.

Due to the low surface energy and inherent hydrophobicity, PDMS surfaces inhibit cell adhesion being thus a useful substrate for the fabrication of cell arrays because it is biocompatible, inexpensive, and durable and can be readily integrated with microsystems. The surface chemistry of PDMS must be tailored to improve its bioactivity and enable patterned cell proliferation [177].

PDMS as a soft polymer has been widely used in microfluidic devices and microcontact printing technology [178], soft lithography applications and nanoindentation, where their attractive physical and chemical properties (elasticity, optical transparency, flexible surface chemistry, low permeability to water, low toxicity, and low electrical conductivity) are of interest [179].

PDMS is the most commonly used membrane material for separation of organic vapors and gases [180].

Adhesives based on silicones can bond together materials as diverse as concrete, glass, granite, steel and plastics, allowing for innovative architecture and engineering. Certain silicone adhesives are widely used in pressure-sensitive tapes and labels when application conditions or the nature of substrate surfaces surpass the performance boundaries of organic-based adhesives. Since the year 2000, there has been continuing interest in and pursuit of new uses for silicone pressure-sensitive adhesives, especially in applications such as medical and industrial tapes. Driven by needs, silicone adhesives based on new silicone chemistry and cure mechanisms have also emerged [181].

The high transparency in the UV-visible region, controlled refractive index, and stable thermo-opto-mechanical properties are the key attributes of silicones that make them attractive materials for high-brightness LEDs.

PDMS-rubber has a number of properties that make the material suitable for the production of high-voltage *insulators* [182]: good dielectric properties, high surface mobility, excellent resistance to weathering, low surface free energy, hydrophobic surface properties, non-adhering surface, insolubility in water, low glass transition temperature, high thermal and oxidative stability, low reactivity, toxicity and combustibility. In addition, it has been proved that the silicone rubber has the potential to recover its hydrophobicity after exposure to ageing factors in the environment. Several plausible mechanisms for the hydrophobic recovery are suggested in the literature. Most researchers believe that the migration of low molar mass PDMS is the dominant mechanism for the hydrophobic recovery [183].

Silicone-based masonry water repellents provide excellent protection against water, chloride ion intrusion and extend the life of parking decks. Unlike other coatings, silicone water repellents do not seal the pores but form a very thin layer on the pore walls, which reduces water intake without compromising the material's vapor permeability or breathability. Water is prevented from penetrating the material from the outside while moisture that has accumulated inside can escape. Silicones are used in a wide range of applications for bonding components and sealing against moisture or environmental contaminants.

Polysiloxanes are generally recognized as the newest generic class of high performance protective coatings and include coating types based on inorganic siloxane and organic-inorganic siloxane hybrids. Polysiloxane coatings based on pure inorganic siloxane binder are curable at ambient temperature and have high solids and low volatile organic compounds (VOC), excellent thermal stability and good resistance to certain acids and solvents.

Silicones can be applied in a thin protective layer on a substrate by methods such as dip, flow, spray, and selective robotic coating. Selection of a silicone coating material for a particular application involves the consideration of various performance and processing criteria.

Silicone rubbers and gels are widely used in electronics to ensure mechanical and environmental protection. A full range of products can be found in various cure speeds, viscosities, and performances, many of them improving the thermal cycling protection, stress relief, material strength, flame retardancy, or optical clarity.

Silicones are found in board-level adhesion, coating, and encapsulation applications, and contribute to the long-term, reliable performance of many components and assemblies. They are also commonly used in a variety of electronics applications to provide heat resistance, flame retardancy, low volatility for sensitive components, and moisture protection.

With its range of adhesives, coating, sealants, encapsulation and potting materials, able to work in harsh conditions, silicones are used in solar energy, automotive, aviation and aerospace industry.

In the search for materials for new applications, the design of polymers with new architectures to give specific properties is required. Some recent advances regarding siloxane materials with non-linear architectures are well discussed in literature [20].

3. Polymers Containing other Elements within the Chain Beside Siloxane Bond

Polysiloxanes can be also co-chained with different other elements in structures like those simplified presented in Table 2. Different linkages between siloxane and the others moieties, as well as various combinations and successions of the sequences can led to a large diversity of siloxane-containing heterochain polymers.

Table 2. Some types of backbones containing other moieties beside siloxane

Poly(siloxazane)s	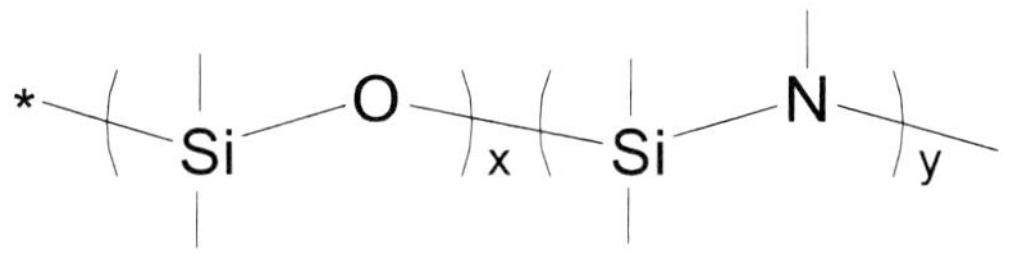
Poly(siloxaphosphazene)s	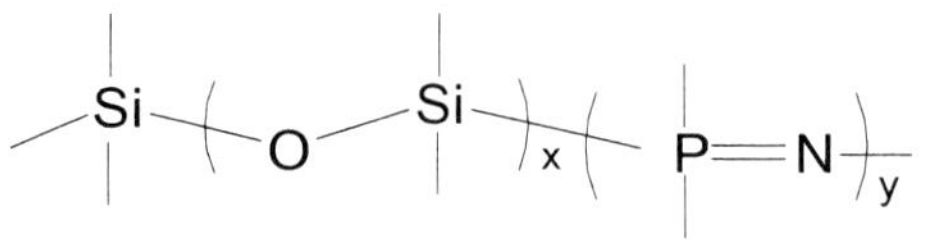
Poly(carboranylenesiloxane)s	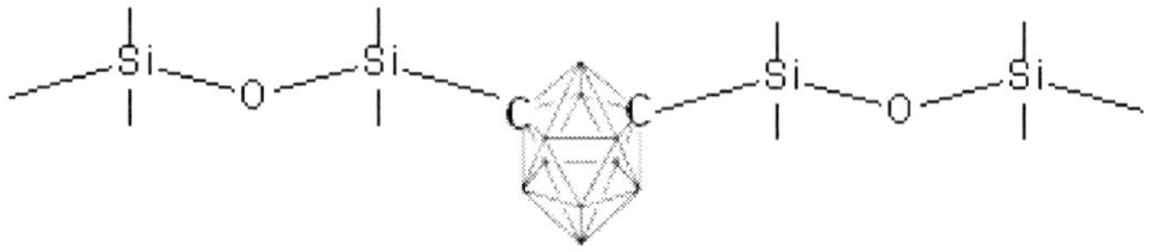
Poly(carbosiloxane)s	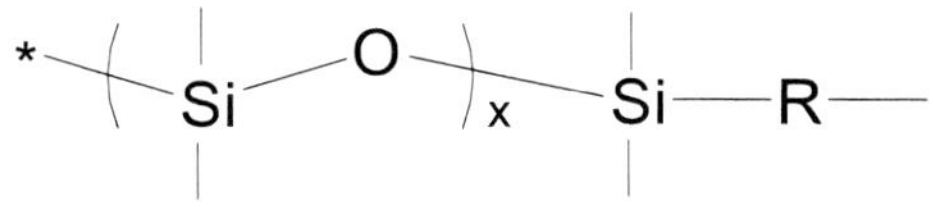 R - aliphatic or aromatic sequence
Siloxane-organic copolymers	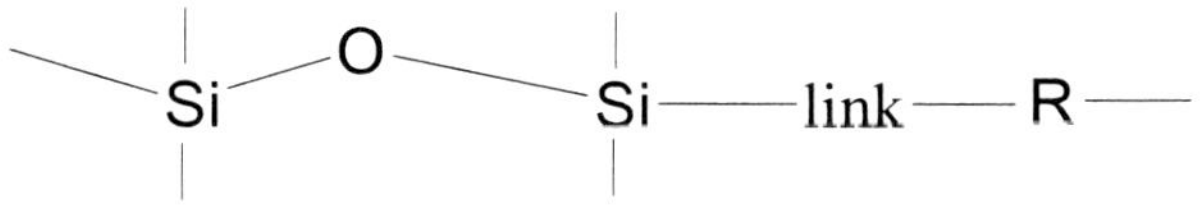 R - organic sequence link - linking between siloxane and organic moieties
Poly(metallosiloxane)s	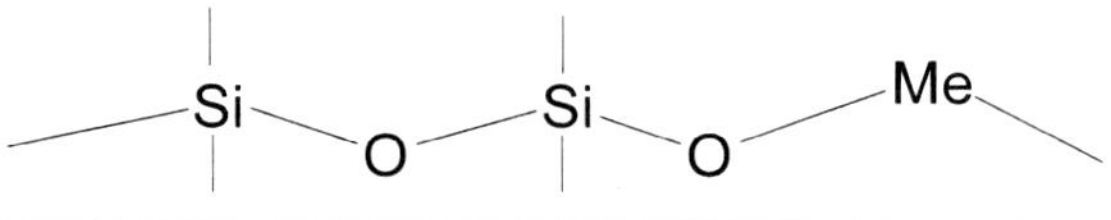

The last two types of the heterosiloxanes were reviewed elsewhere [21,184].

3.1. Poly(siloxazane)s

Polysilazanes are polymers with backbones consisting of silicon and organic or hydrogen substituted nitrogen [185]. Polysilazanes appeared as alternatives for siloxane compounds and many studies have been made on possible improvements, which could be attained through substitution of the Si-O linkage of silicones by the isosteric Si-NH linkage [186,187]. A particular point of interest in this area is the possible crosslinking of polymers through the trifunctional nitrogen atom, giving networks of another type then those obtained by crosslinking through the trifunctional or tetrafunctional silicon atoms. This particularity combined with the very good thermal stability of the Si-N bond made polysilazanes to become valuable materials, especially as ceramic precursors. Such polymers can be considered either inorganic (perhydropolysilazanes) or organic (organopolysilazanes or poly(carbosilazanes)) and have both silicon and nitrogen atoms in their backbones [185].

Polysilazanes are the answer for a broad range of challenging applications that cannot be satisfied by conventional materials. These materials enable long-term, cost-effective and "user-friendly" solutions in applications where high temperature stability, corrosion resistance or long term durability are critical factors.

Principally due to their high pyrolytic mass yields of silicon-based ceramic material (often > 80 wt%), polysilazanes have been extensively used in the past as precursors to silicon dioxide, silicon nitride and silicon carbide ceramics. They offer a convenient route for preparing ceramic materials such as fibers, coatings or 3-D continuous fiber-reinforced ceramic matrix composites, which often cannot be made through traditional ceramic processing methods [185].

Various organic polysilazanes have been proved to be able to form amorphous, non-glassy silicon carbonitride (SiCN) ceramics. These ceramics are resistant to chemical degradation, creep, oxidation, thermal shock, decomposition and softening. Thus, they are quite attractive for a variety of structural ceramic and electronic applications. A viable technique to obtain ceramic devices for micro-electro mechanical systems by microcasting UV-curable organopolysilazanes, for example, has recently been developed. A microfabrication technique for constructing 3-D photonic crystals by deep X-ray lithography using organopolysilazanes has also been demonstrated [185].

Co-existence of SiO and SiN bonds in the same macromolecular backbone permits to obtain polymers, poly(siloxazane)s, which exhibit intermediate properties between those assigned to the two compounds types. Such compounds can be useful precursors for coating and ceramic materials (i.e, SixNyCzOw). Poly(siloxazane)s have been synthesized either by copolymerization of cyclosiloxanes and cyclosilazanes or by homopolymerization of the cyclosiloxazanes. Like in the case of the polysiloxanes synthesis, in general a mixture of linear and cyclic compounds is obtained. The using of controlled anionic ring-opening polymerization of heptamethyl-1,3-dioxa-5-aza-2,4,6-trisilacyclohexane have recently been reported [188] that results in polymers with a regular structure and a limited amount of by-products. n-Butyllithium was used as initiator and DMF as a promoter. Dimethylvinylchlorosilane was used as quenching agent, resulting in structure like that presented in Scheme 5.

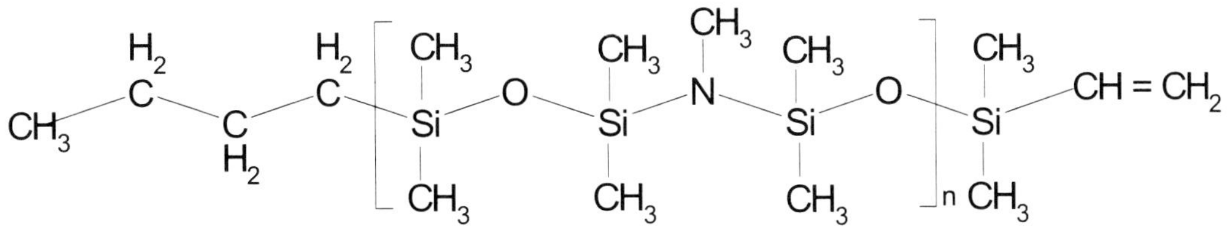

Scheme 5.

The poly(siloxazane)s comprising $-(SiH_2)_n$-NH- and $-(SiH_2)_n$-O- sequences as the main repeating units were synthesized by reacting a dihalosilane or an adduct thereof with a Lewis base, with ammonia and water vapor or oxygen. Novel silicon oxynitride shapes essentially composed of silicon, nitride and oxygen were obtained from the poly(siloxazane)s [189].

High molecular weight poly(siloxazane)s as silicon oxynitride ceramic precursors have been obtained by the ammonolysis of isocyanatodisiloxanes [190].

According to another source [191], poly(siloxazane) was prepared by the ammonolysis of cyclic and bicyclic oligosiloxazanes, 1,3-diamino-1,3-di-t-butoxy-1,3-dimethylsiloxane with 1,1,3,3-tetraisocyanato-1,3-dimethyl-1,3-disiloxane in 1:1 and 1:2 molar ratios, respectively in THF at 213K. Silicon oxynitride was prepared by the pyrolysis of the synthesized poly(siloxazane) under an ammonia atmosphere. The ceramization process was monitored by IR and NMR spectra, which showed the replacement of a methyl group with an amino group followed by a condensation reaction to provide silicon oxynitride at 1173 K.

Ring-opening polymerization of nitrogen-containing cyclic organosilicon monomers such as cyclosilazanes, cyclosiloxazanes and cyclocarbosilazanes has also been investigated. Although these monomers can be polymerized through both cationic and anionic processes, only the latter allows high level of control of the polymer architectures. The resulted polymers exhibit promising properties, particularly higher thermal stability and mechanical resistance as compared with polysiloxanes [192].

Polysiloxane containing N,N`-bis(diphenylsilyl)tetraphenylcyclodisilazane (Scheme 6) was prepared by the anionic polymerization with a "seed solution" as initiator. The synthesis of monomers N,N`-bis(hydroxydiphenylsilyl)tetraphenylcyclodisilazane and 1,3-dichloro-1,1,3,3-tetraphenyldisilazane was reported. Such special kind of polysiloxanes containing cyclodisilazane in the main chain was proved to have excellent thermal stability. However, the application of this polysiloxane was limited by the complicated synthesis of the monomer, N,N`-bis(chlorodiphenylsilyl)tetramethylcyclodisilazane [193].

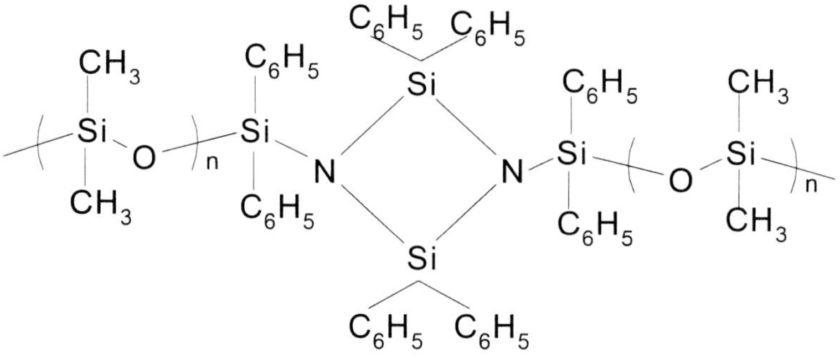

Scheme 6 [193].

3.2. Poly(siloxaphosphazene)s

Polyphosphazenes are a class of inorganic backbone polymers, which compete with many organic systems in their molecular structural diversity and property variations [1,194].

Polyphosphazenes consist of an inorganic phosphorous-nitrogen backbone (repeating unit (NP)n) and side groups that may be organic, organometallic or inorganic in nature (Scheme 7).

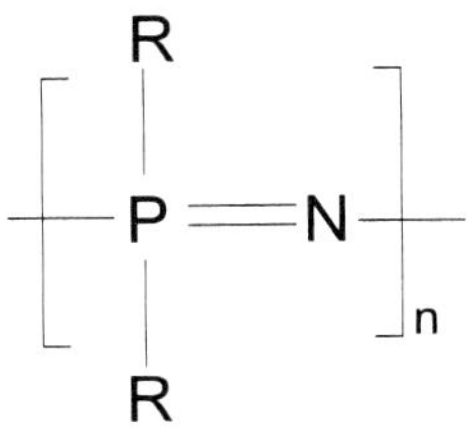

Scheme 7.

Polyphosphazenes are a unique class of polymers, which are differentiated from most conventional polymers by their route of synthesis, their synthetic versatility, tunable hydrolytic degradability, high thermal and oxidative stability and a broad range of physical, chemical and mechanical characteristics. Due to their unique synthetic pathway, polyphosphazenes offer the promise to avoid many of the issues associated with classical polymerization.

They constitute the second group of inorganic-organic polymers to be developed extensively. The properties of these macromolecules vary over a broad range depending on the side group structure. Polyphosphazenes are well suited for the assembly of useful liquid crystal polymer systems because the phosphorous-nitrogen backbone is very flexible, and this can serve to enhance mesophase stability, thus increasing the useful temperature range over which these materials can be used. The polyphosphazene backbone is also optically transparent in the visible region, which is an important feature in the development of optical devices. These properties, combined with the ease with which the macromolecular properties can be modified by choosing the appropriate co-substituents, make polyphosphazenes of considerable interest for both fundamental and applied research. They have potential applications that includes both biomedical and advanced materials (conductive polymers, elastomers, etc) [195].

Polyphosphazene-containing polymers are also of general interest because of their capability to combine with another polymer, either organic or inorganic, incorporating the properties and advantages of the two polymer components without sacrificing the overall chemical, physical, and mechanical properties of each component. As a result, an emerging field of research involves an attempt to create a new area of polymer science at the interface between organic/inorganic, by the synthesis of hybrid organophosphazene-organosilicon systems. There are excellent reviews on the progress in this new field [1,194]. Polyphosphazene-polysiloxane hybrid systems are of particular interest due to their anticipated surface and bulk characteristics. Copolymers that combine polyphosphazenes with polysiloxanes could, in principle, produce tailorable materials with unique combinations of properties. Potential applications include nonburning elastomers, viscosity modifiers, coatings, solid polymer electrolytes, and biomedical materials [196].

However, the incorporation of organosiloxanes and polyphosphazenes into one material, although successful, has not resulted in linear phosphazene-siloxane block copolymers combining the attributes of the two inorganic backbone systems. Moreover, until recently, the preparation of block copolymers containing polyphosphazenes has been limited to species with two phosphazene components. The synthesis of telechelic polyphosphazenes, which may be used to couple with preformed organic polymers, has proved to be difficult [197].

The synthesis of macromolecules with a polyphosphazene backbone and a variety of organosilicon side groups has been reported [196]. Such as, poly(phosphazene-graft-siloxane) materials with 10-20% oligomeric siloxane grafts along the poly(methylphenylphosphazene) chain were prepared. However these polymers do not combine polyphosphazenes and polysiloxanes into one continuous backbone. Recent advances in the living cationic polymerization of polyphosphazenes have allowed the preparation of linear block copolymers of polyphosphazenes with polyorganosiloxanes. Thus, linear A-B, A-B-A, and B-A-B block copolymers containing polyphosphazenes and polysiloxanes were synthesized and characterized (Scheme 8).

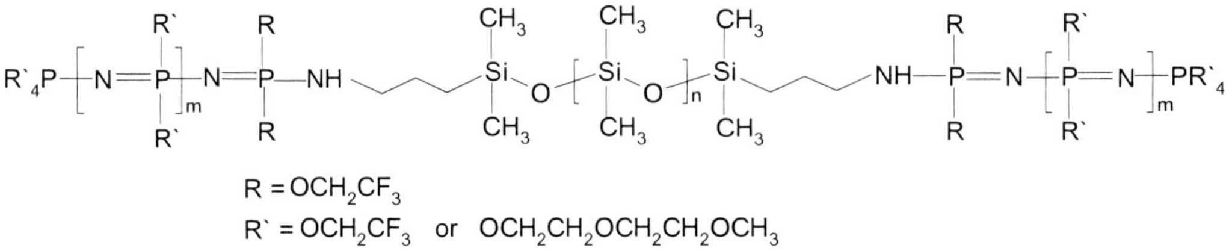

Scheme 8 [196].

Two methods were used in the synthesis of these novel block copolymers. The first used hydrosilylation technique to couple monotelechelic polyphosphazenes with commercially available α,ω-bis(H)-polydimethylsiloxane. The second involved the termination of living polydichlorophosphazene chains with phosphoranimine-terminated PDMS. The semicrystalline, hydrophobic copolymers, $[(CF_3CH_2O)_2P{=}N]_x$-$[(Me)_2Si\text{-}O]_y$, and the water-soluble copolymers, $[(CH_3OCH_2CH_2OCH_2CH_2O)_2P{=}N]_x$-$[(Me)_2Si\text{-}O]_y$, were prepared by the linkage of either hydrophobic or hydrophilic organic groups to the polyphosphazene backbone. These materials with a polyphosphazene content from 19 to 73 mol % were obtained having controlled molecular masses [196].

A new type of siloxane-phosphazene copolymer system has been synthesized via hydrosilylation of a high molecular weight poly(aryloxyphosphazene), with oligo-H-siloxanes of different molecular weight, allowing the preparation of different materials with an increasing siloxane content up to 80 wt%. The starting H-functionalized siloxanes were quantitatively grafted and led to cross-linked structures characterized as a microphase-separated copolymer system. Thin films with good mechanical properties were prepared having up to 60 wt% siloxane. The transport properties of the films for n-butanol, water and their saturated mixture as permeants have been determined by sorption, desorption and pervaporation measurements. It was established that all the tested materials have a stronger affinity for butanol than for water and that the difference in transport properties in the series is mainly based on the type of continuous phase [198].

3.3. Poly(carboranylenesiloxane)s

In the continuing interest in high thermostable polymers and polymeric precursors for ceramics, a special attention was given to the polymers containing boron, silicon, and carbon [199]. The polymers with high percentages of nitrogen and boron are valuable precursors to the boron nitride ceramic and boron nitride fibers. These materials are also of interest for coatings, nanomaterials, catalysts, therapeutic agents, and especially electronic applications such as nonlinear optical and fluorescent light-emitting materials. The development of the polymers containing boron is challenged by the abilities of this element to delocalize electrons due to the presence of an empty p orbital, capacity to capture neutrons and its ability to form protective coatings. The most studied class of boron polymers containing boron and siloxane bonds are the carboranylenesiloxane polymers. More recent advances refereed to the silicon-containing borazine and boroxine polymers [34].

Carborane is a cluster compound containing carbon (C) and boron (B) atoms as well as hydrogen (H) atoms external to the framework of the cluster. The cluster compound is one with insufficient electrons. Carboranes are of interest because of their nonclassical bonding, their relatively high thermal stability, and their ability, when containing the ^{10}B isotope, to capture neutrons efficiently. Carboranes exhibit a high degree of electron density delocalization, and their bonding can be described more accurately by employing molecular orbital theory.

Numerous polymers containing carborane cage structures within the backbone have been reported that proved to exhibit appreciable enhancement in their thermal stability. The development of carborane-siloxane polymers in the early 1960s was a major breakthrough in the search for high-temperature elastomers. These elastomeric materials show superior thermal and oxidative stability at elevated temperatures compared to siloxane-based polymers. The carborane-containing systems retained their elastomeric properties after aging in air at 315°C for 300h. The exceptional thermal and thermo-oxidative properties of the polymers containing siloxane and carborane groups have been of particular interest to aerospace and defense industries especially for use in sealing assemblies for landing gears, flight control, and fuel systems and in coating insulations for cables [4,200-203]. There is also a high demand for such materials in the electronic industry in the manufacture of resist layers of computer chips. However, initially the carboranylenesiloxanes having generally low molecular weight polymers were obtained, a fact that has limited their use in applications requiring high molecular weight materials [202,203]. A solution to this problem was found to consist in the introduction of unsaturated crosslinkable units in such materials, which on curing yielded extended polymer networks of sufficient strength for various applications [199]. In this aim, the diacetylene group can be incorporated into the backbone (Scheme 9) [34]. Otherwise, this cross-linking group has also been introduced into a variety of polymers. Similar diacetylene-containing materials have been reported that contain carborane and disiloxane groups [204].

Because of the superior thermal and oxidative stability of the polycarboranesiloxane elastomers it has been studied the effects of incorporating acetylenic units into the structure for cross-linking purposes. The presence of acetylene units in the backbone of a polymeric material provides many attractive advantages. The acetylene remains dormant during processing under ambient conditions.

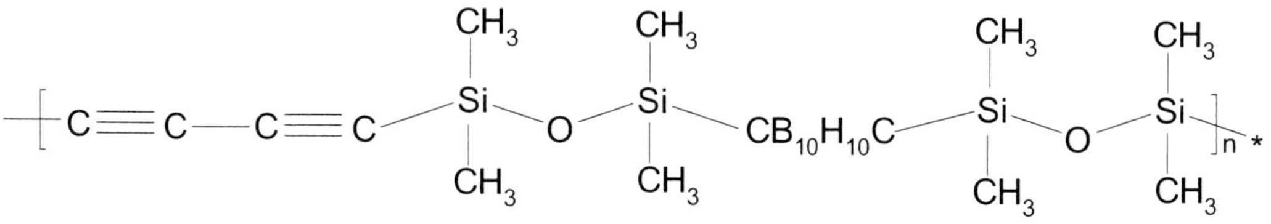

Scheme 9 [34].

During the thermal or photochemical reaction, a conjugated cross-link is formed as a result of the addition polymerization reaction occurrence without the formation of volatile byproducts [199].

The cross-linking can also occur through vinyl groups introduced with methylvinylsiloxane units in the backbone. Therefore, the existing methods for the production of network polymers from a precursor carboranylenesiloxane containing side unsaturated organic functionality are based either on the polymerization at the vinyl groups in presence of organic peroxides in air (at 315° for 300 h) [205] or on the thermal crosslinking of the diacetylene groups in an inert atmosphere (in the range 250-400°C for several hours) [210]. Under such conditions, a loss of the elasticity of the polymeric product has been observed to occur after the network formation [205] such derivatives being converted to plastics by curing. This limits their use in certain domains [206,207]. However, in applications involving high temperature coatings, sealings, composites, etc., there is a need for elastomeric materials. There is also a need for the curing to be effected in an expeditious manner under an ambient or inert atmosphere. Thus, the development of a less severe and more expedient curing protocol for the production of *elastomeric* crosslinked carboranylenesiloxane network systems is desirable. It has been discover that carboranylenesiloxanes having unsaturated internal or terminal crosslinkable groups can be cured by using the chloroplatinic acid (H_2PtCl_6-OH_2O), a hydrosylation catalyst [204, 208-210]. The Karstedt catalyst also catalyzes ambient-condition hydrosilylation reactions in hexane of a monomeric vinyl-containing carboranylenesiloxane and three-branched siloxane crosslinker monomers resulting in elastomeric network polymers at very rapid rates of formation. The flexible and transparent films of the saturated elastomeric network polymers were observed to possess low glass-transition temperatures (below -35 °C). Similar hydrosilylation reactions at two different reactant ratios involving a diethynyl-containing carboranylenesiloxane and the siloxane crosslinkers produced partially or completely hydrosilylated polymeric networked systems, which were transparent and elastomeric at room temperature. The prepared elastomeric polymeric networks were found to have glass-transition below 0 °C and degradation temperatures in the range of 500-550 [211]

The preparation of linear and amorphous D2-meta-carborane-siloxane polymers having molecular weights up to 250,000 by a new condensation process was recently reported [212]. These materials were compounded and vulcanized to give elastomers having outstanding retention of mechanical properties after heat-aging at 300 °C. An investigation of the effects of structural variations in the gum stock, filler, antioxidant, and peroxide curing agent was carried out. In addition, low temperature properties, flammability, solvent resistance, and adhesive properties (lap-shear strength) were determined. The property profile and vulcanization chemistry of the D2-carborane-siloxane system suggest utility in high performance sealant, gasket, O-ring and coating applications [212].

Such polymers were diversified by enclosing silarylene sequences as in Scheme 10 [34]. The resulted copolymers proved to have high char yields (up to 85%) at 1000°C in N_2 and in air, reflecting the thermal stabilizing effects of the carborane and aromatic units in the polymeric backbone [34].

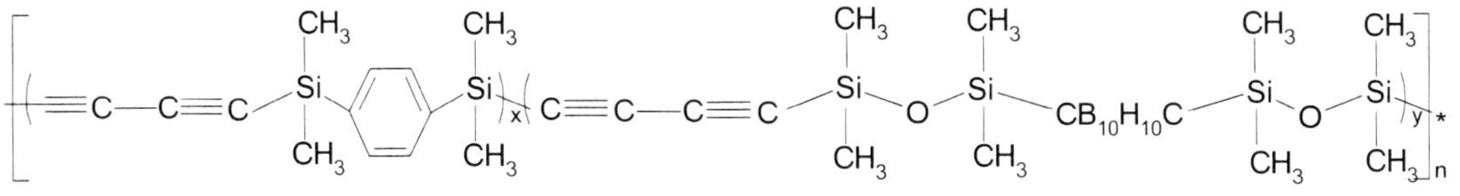

Scheme 10 [34].

Elastomeric high molecular weight alternating and random block copolymers comprising soft carborane-siloxane elastomer and high performance rigid polymer blocks were prepared. Methods for preparing such block copolymers are also disclosed. These copolymers exhibited the advantages of high temperature elastomers while allowing the use of conventional thermoplastic fabrication techniques [213].

Some carborane groups in carboranylenesiloxyl–diacetylene copolymers were replaced with organometallic systems (i.e., ferrocenyl) (Scheme 11) [34] resulting in elastomeric material by thermal treatment [34,214,215].

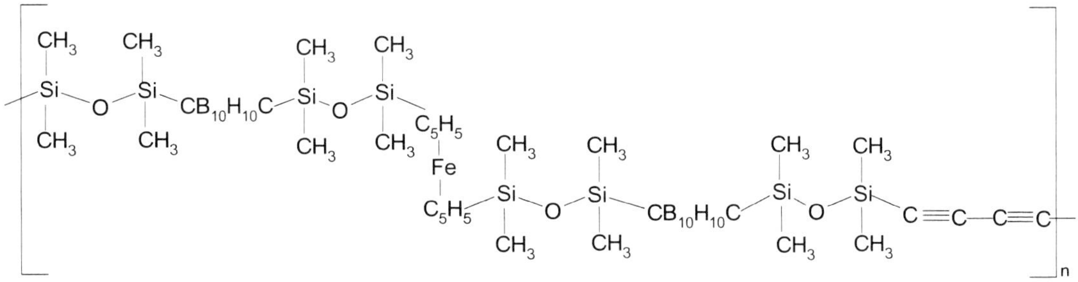

Scheme 11 [34].

It has been observed that these polymers possess exceptional thermooxidative stabilities.

The utility of the thermo-oxidative stable diacetylene-containing carboranylenesiloxane as a viable system for the formation of nanoparticles of the transition metal carbide has also been demonstrated [215].

Other different structures containing carborane and siloxane were approached. A D_2-*m*-dodecacarboranesiloxane containing alternating block copolymer –[A-B-]$_n$ was prepared by the reaction of an ureidosilane-terminated polysulfone hard block with a carboranesilanol-terminated carboranesiloxane oligomer. The resulting block copolymer exhibited thermoplastic elastomeric behavior. This unique polymer had excellent thermal stability up to 400°C [216].

A new synthesis for carboranesiloxane polymers involves the alcoholysis of the bischlorodimethylsilylcarborane monomers and in situ generation of the HCl catalyst.

Alcoholysis is applicable to the synthesis of most carborane-siloxane polymers. The attack of the B-H moieties in the carborane cage can be minimized both by the utilization of a tertiary alcohol and by the addition of excess acid when a primary alcohol is employed [217].

Carborane-siloxane polymers were conveniently prepared by the polymerization of a carborane disilanol and either a silyl diamine, an ureido-silane, or a silyl dicarbamate. The resulting linear polymers, elastomeric in nature, were characterized by relatively high molecular weights. These gumstocks are compounded with suitable fillers and additives and fabricated into cured elastomeric articles [218].

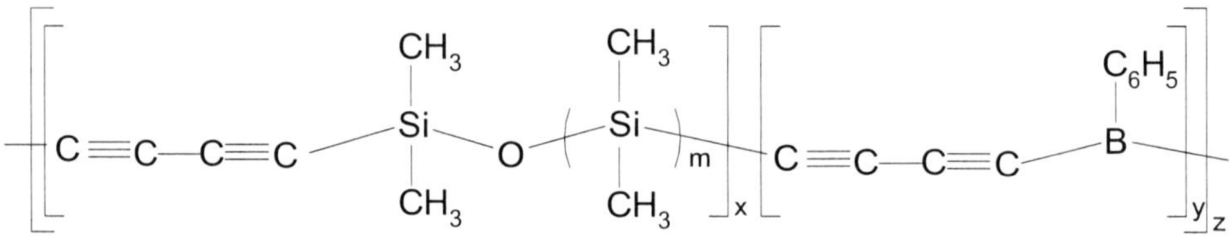

$x = 0, 1, 2; y = 0.8, 0.9; z = 0.2, 0.1$

Scheme 12 [34].

The synthesis of linear boron–silicon–diacetylene polymers (Scheme 12) having different compositions using phenylboron dichloride (PBD) as the source for boron has also been reported. These polymers were observed to possess exceptional thermooxidative stabilities. This topic of the carboranesiloxane copolymers is very actually and complete reviewed in [34].

4. Conclusion

Although became classical many years ago, the polysiloxanes continue to fascinate by their oneness and versatility. The unlimited possibilities in design of the architecture and functionality challenge to new researchers and open new and new perspectives. Siloxanes can be associated in simple or complex structures with both organic and inorganic moieties resulting in a large variety of compounds with wide range of properties. If the field of the siloxane-organic polymeric structures is one very active and vast, that involving inorganic partner appears to be rather scarce. However, in the last time it looks that the interest in such field revives.

Acknowledgements:

The author is grateful for financial support by Projects CEEX-MATNANTECH 10/2005.

REFERENCES

[1] Harry, R.; Kuharcik, A.; Kuharcik, S.E. *J. Inorg. Organometal. Polym.* 1996, 6(1), 1-41.

[2] Rehahn, M. *Acta Polymer.* 1998, 49, 201-224.

[3] Yilgör, I.; McGrath, J. E. *Adv. Polym. Sci.* 1988, 86, 1-86.

[4] Dvornic, P.R.; Lenz, R.W.; High Temperature Siloxane Elastomers; Huthig and Wepf: New York, 1990.

[5] Voronkov, M. G.; Mileshkevich, V. P.; Yuzhelevski, Yu. A. The Siloxane Bond - Physical Properties and Chemical Transformations, Consultants Bureau, New York, 1978, pp. 159-212.

[6] Chojnowski, J. *J. Inorg. Organomet. Polym.* 1991, 1, 299-323.

[7] Noll, W., Chemistry and Technology of Silicones, Noll, W. (Ed), Academic Press Inc. (London) Ltd, New York, 1968.

[8] Boileau, S., Anionic Polymerization of Cyclosiloxanes with Cryptates as Counterions in "Ring-Opening Polymerization Kinetics, Mechanisms and Synthesis", Ed, by J. E. McGrath, ACS Symposium Series 286, Washington DC, p. 23, 1985.

[9] Sobolevskii, M.V., "Oligoorganosiloxani. Svoista, polucenie, primenenie", Moskva, 1985.

[10] Saam J.C., in "Silicone Based Polymer Science", Ziegler, J.M.; Gordon Fear, F.W. (Eds.), Am. Chem. Soc., Washington DC, 1990.

[11] Kendrick, T.C.; Parbhoo, B.; White, J.W., The Silicon -Heteroatom Bond, Patai, S.; Rappoport Z. (Eds.), 1991, John Wiley and Sons Ltd., cap 4.

[12] Torkelson, A., "Developments and trends in silicone fluids", in Silicone Technology, Bruins, P.F. (Ed.), Interscience Publishers, 61, 1970.

[13] Freeman, G.G. " Silicones - An Introduction to their Chemistry and Applications", P. D. Ritchie; H. Barker (Eds.), London-Ilife Books LTD, 1962.

[14] Mark, J.E. in "Silicon-Containing Polymers" in: "Silicon-Based Polymer Science: A Comprehensive Resource", 47, ACS, 1991.

[15] Clarson, S.J.; Fitzgerald, J.J.; Owen, M.J.; Smith, S.S.; Van Dyke M.E. (Eds.), The Science and Technology of Silicones and Silicone-Modified Materials #964, First Edition ACS, 964, 2007

[16] Synthesis and Properties of Silicones and Silicone-Modified Materials, Aldrich, S. Clarson, Oxford University Press, 2003, Z557536.

[17] Jones, R.S.; Ando, W.; Chojnowski, J., Eds. Silicon-Containing Polymers - The Science and Technology of Their Synthesis and Applications; Kluwer Academic Publishers: Dordrecht, 2000.

[18] Clarson, S.J.; Fitzgerald, J J.; Owen, M.J.; Smith, S.D. (Eds.), Silicones and Silicone-Modified Materials., ACS Symp. Ser., 729, Oxford University Press: Washington D.C. 2000; 20-37.

[19] The Analytical Chemistry of Silicones Books, Smith, A.L. (Ed.), John Wiley and Sons Inc, 1991.

[20] Bischoff, R.; Cray, S.E. *Prog. Polym. Sci.* 1999, 24, 185–21.

[21] Cazacu, M.; Racles, C. "Recent developments in siloxane-based polymers and copolymers" in "New Trends in Nonionic (Co)polymers and Hybrids", Dragan, S. (Ed.), Nova Science Publishers, 2006
[22] Chiellini, E.; Laus, M. In Handbook of Liquid Crystals; Demus, D.; Goodby, J.; Gray, G.W.; Spiess, H. -W.; Vill, V.; Ed.; Wiley-VCH: Weinheim, Germany, 1998, vol. 3, pp 28-32.
[23] Chauhan, M.; Chauhan, B. P. S.; Boudjouk, P. *Org. Lett.* 2000, 2, 1027–1029.
[24] Ready, T. E.; Chauhan, B. P. S., Boudjouk, P. *Macromol. Rapid Commun.* 2001, 22, 654-657.
[25] Haddleton, D.M.; Crossman, M.C.; Dana, B.H.; Duncalf, D.J.; Heming, A.M.; Kukulj, D.; Shooter, A. *J. Macromolecules* 1999, 32, 2110-2119.
[26] Haddleton, D.M.; Jasieczek, C.B.; Hannon, M.J.; Shooter, A. *J. Macromolecules* 1997, 30, 2190-2193.
[27] Miller, P.J.; Matyjaszewski, K. *Macromolecules* 1999, 32, 8760-8767.
[28] Huan, K.; Bes, L.; Haddleton, D.M.; Khoshdel, E. *J. Polym. Sci.Part A: Polym. Chem.* 2001, 38, 1833-1842.
[29] Coessens, V.; Pintauer, T.; Matyjaszewski, K. *Prog. Polym. Sci.* 2001, 26, 337-377.
[30] Chambon, P.; Cloutet, E.; Cramail, H.; Tassaing, T.; Besnard, M. *Polymer* 2005, 46, 1057-1066.
[31] Kilic, S.; Michalik, S.; Wang, Y.; Johnson, J.K.; Enick, R.M., Beckman, E. *J. Ind. Eng. Chem. Res.* 2003, 42, 6415-6424.
[32] Akgun, M.; Deniz, S.; Baran, N.; Uzun, N.I.; Akgun, N.A.; Dincer, S. *Polym. Int.* 2005, 54, 374-380.
[33] Stumbé, J.F.; Calderara, F.; Riess, G. *Polym. Bull.* 2001, 47, 277-282.
[34] Kolel-Veetil, M.K.; Keller, T.M. The State of the Art in Boron Polymer Chemistry in: Macromolecules Containing Metal and Metal-Like Elements, Volume 8: Boron-Containing Polymers, Abd-El-Aziz, A.S.;Carraher Jr., C.E.; Pittman Jr., C.U.; Zeldin, M. (Eds.), John Wiley and Sons, Inc. 2007, and reference herein.
[35] Tsuchihara, K.; Fujishige, S. *Polym. Bull.* 1999, 43, 129-134.
[36] Hedden, R.C.; Cohen, C. *Polymer* 2000, 41(18), 6975-6979.
[37] Godovsky, Yu.K.; Valetskaya, L. A. *Polym. Bull.* 1991, 27(2), 221-226.
[38] Koerner, H.; Luo, Y.; Li, X.; Cohen, C.; Hedden, R.C.; Ober C. K. *Macromolecules* 2003, 36 (6), 1975 –1981.
[39] Liu, L.; Yang, S.; Zhang, Z.; Wang, Q.; Xie, Z. *J. Polym. Sci. Part A: Polym. Chem.* 2003, 41(17), 2722-2730.
[40] Kimura, H.; Kanesaka, S.; Kuroki, S.; Ando, I.; Asano, A.; Kurosu, H. *Magn. Reson. Chem.* 2005, 43, 209-216.
[41] Godovsky, Yu. K.; Makarova, N. N.; Fischer, H.; Finkelmann, H.; Keller. A. *Philos. Trans.: Phys. Sci. Engin.* 1994, 348 (1686), 45-57.
[42] Petrarch Catalog 2000.
[43] Saam, J.C. *J. Inorg. Organomet. Polym.* 1999, 9, 3-15.
[44] Yactine, B.; Ganachaud, F.; Senhaji, O.; Boutevin, B. *Macromolecules* 2005, 38, 2230-2236.
[45] Bauer, J.; Nicola, H.; Kickelbick, G. *Chem. Commun.* 2001, 137-138.
[46] Chojnowski, J.; Cyprik, M.; Fortuniak, W.; Rozga-Wijas, K.; Scibiorek, M. *Polymer* 2002, 43, 1993-2001.

[47] Cazacu, M.; Marcu, M.; Petrovan, S.; Holerca, M.; Simionescu, M. *Polym. Plast. Technol. Eng. J.* 1996, 35, 2, 327-347.
[48] Cazacu, M.; Marcu, M.; Holerca, M.; Petrovan, S.; Lazarescu, S. *J. Macromol. Sci.-Pure Appl. Chem.* A 1996, 33, 1, 65-76.
[49] Deshpande, G.; Rezac, M. E. *Polym. Degrad. Stabil.* 2001, 74, 363-370.
[50] Deshpande, G.; Rezac, M. E. *Polym. Degrad. Stabil.* 2002, 76, 17-24.
[51] Yang, M.H.; Huang, W.J.; Chien, T.C.; Chen, C.M.; Chang, H.Y.; Chang, Y.S.; Chou, C. *Polymer* 2001, 42, 8841-8846.
[52] Oishi, M.; Kawakami, Y. *Macromolecules* 2000, 33, 1960-1963.
[53] Van Dyke, M.E.; Clarson, S.J. *J. Inorg. Organomet. Polym.* 1998, 8, 111-117.
[54] Saam, J.C. *J. Inorg. Organomet. Polym.* 1999, 9, 3-15.
[55] Cazacu, M., Marcu, M., Dragan, S., Matricala, C., Simionescu, M., Holerca, M., 1997, *Polymer* 38(15), 3967-3971.
[56] Bacqué, E.; Richard, C.; Pillot, J.P.; Birot, M.; Dunoguès, J.; Pétraud, M.; Gérardin, C.; Taulelle, F. *J. Inorg. Organometal. Polym.* 1995, 5(2), 169-182.
[57] Markides, K.; Blomberg, L.; Hoffmann, S.; Buijte, J.; Wannman, T. *J. Chromatography* 1994, 302, 319-340.
[58] Richter, B.E.; Kuei, J.C.; Castle, L.W.; Jones, B.A.; Brashaw, J.S; Lee, M.L. *Chromatography*, 1983, 17(10), 570-573.
[59] Ameduri, B.; Boutevin, B. *J. Fluorine Chem.* 2005, 126, 221-229.
[60] Kobayashi, H.; Nishiumi, W., *Die Makromol. Chem.* 1993, 194(5), 1403-1410.
[61] Abdellah, L.; Boutevin, B.; Caporiccio, G.; Guida-Pietrasanta, F. *Eur. Polym. J.* 2002, 38, 1515-152.
[62] Abdellah, L.; Boutevin, B.; Caporiccio, G.; Guida-Pietrasanta, F. *Eur. Polym. J.* 2003, 39, 49-56.
[63] Wang, W. *Eur. Polym. J.* 2003, 39, 1117-1123.
[64] Kim, W.G.; Ahn, H.K.; Lee, H.W.; Kim, S.H.; Crivello, J.V. *Opt. Mater.* 2003, 21, 343-347.
[65] Jang, M.; Crivello, J.V. *J. Polym. Sci. Polym. Chem.* 2003, 41, 3056-3073.
[66] Lee, I.J.; Song, G.S.; Lee, W.S.; Suh, D.H. *J. Power Sources*, 2003, 114, 320-329.
[67] Iojoiu, C.; Abadie, M.J.; Harabagiu, V.; Pinteala, M.; Simionescu, B.C. *Eur. Polym. J.* 2000, 36, 2115-2123.
[68] Khosla, S.; Raina, K.K.; Coles, H. *J. Curr. Appl. Phys.* 2003, 3, 135-140.
[69] Yao, D.S.; Zhang, B.Y.; Sun, Q.J.; Zhang, L.F. *J. Appl. Polym. Sci.* 2005, 95, 946-952.
[70] Jia, Y.; Zhang, B.; Feng, Z.; Guan, Y. *Eur. Polym. J.* 2003, 39, 1701-1706.
[71] Chan, W.L.; Yana, H.; Szeto, Y.S. *Mater. Lett.* 2004, 58, 882-884.
[72] Hua, J. ; Li, Z. ; Long, K.; Qin, J.; Li, S.; Ye, C.; Lu, Z. *J. Polym. Sci. Part A Polym. Chem.* 2005, 43, 1317-1324.
[73] Li, Z.; Qin, J.; Yang, Z.; Ye, C. *J. Appl. Polym. Sci.* 2004, 94, 769-774.
[74] Li, Z.; Li, J.; Qin J.; Qin, A; Ye, C. *Polymer* 2005, 46, 363-368.
[75] Lin, C.H; Hsu, C.S. *Polym. Bull.* 2000, 45, 53-59.
[76] Toulokhonova, I; Bjerke-Kroll, B.; West, R. *J. Organomet. Chem.* 2003, 686, 101-104.
[77] Scibiorek, M.; Gladkova, N.K.; Chojnowski, J. *Polym. Bull.* 2000, 44, 377-384.
[78] Bosman, A.W.; Sijbesma, R.P.; Meijer E.W. *Materials Today*, April, 2004, 34-39.
[79] Abed, S.; Boileau, S.; Bouteiller, L. *Polymer* 2001, 42 (2), 8613-8619.
[80] Moriguchi, N.; Tsugaru, T.; Amiya, S. *J. Molec. Struct.* 2000, 523, 93-101.

[81] Wu, Q.; Lee, M.L.; Harrison, R.G. *J. Chromatogr.* A, 2002, 954, 247-258.
[82] Antonisse, M.M.G.; Snellink-Ruel, B.H.; Lugtenberg, R.J.M.; Engbersen, J.F.J.; van den Berg, A.; Reihoudt, D.N. *Anal. Chem.* 2000, 72, 343-348.
[83] Arslan, A.; Kiralp, S.; Toppare, L.; Yagci, Y. *Intern. J. Biol. Macromol.* 2005, 35, 163-167.
[84] Hou, S.S.; Chung, Y.P.; Chan, C.K.; Kuo, P.L. *Polymer* 2000, 41, 3263-4272.
[85] Rutnakornpituk, M. *Eur. Polym. J.* 2005, 41, 1043-1052.
[86] Kuo, P.L.; Hou, S.S.; Teng, C.K.; Liang, W.J. *Colloid Polym. Sci.* 2001, 279, 286-291.
[87] Akgun, M.; Deniz, S.; Baran, N.; Uzun, N.I.; Akgun, N.A.; Dincer, S. *Polym. Int.* 2005, 54, 374-380.
[88] Feng, S.; Cui, M. *React. Funct. Polym.* 2000, 45, 79-83.
[89] Persson, J.C.; Jannasch, P. *Macromolecules* 2005, 38, 3283-3289.
[90] Slade, R.C.T.; Varcoe, J.R. *Solid State Ionics* 2001, 145, 127-133.
[91] Zhou, C.J.; Guan, R.F.; Feng, S.Yu. *Eur. Polym. J.* 2004, 40, 165-170.
[92] Ogawa, T. *J. Polym. Sci. Part A Polym Chem.* 2003, 41, 3336-3345.
[93] Senthilkumar, U.; Rajini, R.; Reddy, B.S.R. *J. Membr. Sci.* 2005, 254, 169-177.
[94] Kazmierski, K.; Hurduc, N.; Sauvet, G.; Chojnowski, J. *J. Polym. Sci. Part A Polym. Chem.* 2004, 42, 1682-1692.
[95] Unno, M.; Matsumoto, T.; Mochizuki, K.; Higuchi, K.; Goto, M.; Matumoto, H. *J. Organometal. Chem.* 2003, 685, 156-161.
[96] Cai, G.P.; Weber, W. *Macromolecules* 2000, 33, 6310-6314.
[97] Vadala, M.L., Rutnakornpituk, M.; Zalich, M.A.; St Piere, T.G.; Riffle, J. S. *Polymer* 2004, 45, 7449-7461.
[98] Kan ,C.Y.; Zhu, X.L.; Yuan Q.; Kong X.Z., *Polym. Adv. Tech.* 1997, 8, 631-633.
[99] He, W.D.; Cao, C.T., *Polym. Int.,* 1996, 39, 31-36.
[100] Kong, X.Z.; Ruckenstein, E., *J. Appl. Polym. Sci.* 1999, 73, 2235-2245.
[101] Wright, P.V., "Cyclic Siloxanes" in Ring-Opening Polymerization, Vol. II, Ivin, K. J.; Saegusa, T. (Eds.), Elsevier Applied Science Publishers, London and New York, 1984.
[102] Cazacu, M.; Marcu, M.; Dragan, S.; Matricala, C. *J. Appl. Polym. Sci.* 1996, 60, 731-734.
[103] Tundo, P.; Venturello, P., *J. Am. Chem. Soc.*, 1979, 101 (22), 6606-6613.
[104] Molinari, H.; Montanari, F.; Quici, S.; Tundo, P., *J. Am. Chem. Soc.* 1979, 101 (14), 3920-3927.
[105] Bernard, M.; Ford, W. T.; Tylor, T. W. *Polym. Prepr.* 1984, 25 (1), 170.
[106] Boileau, S. *Polym. Prepr.* 1984, 25 (1), 204.
[107] Rashkov, I.; Gitsov, I.; Panayotov, I. *Polym. Bull.* 1983, 10, 487-490.
[108] Bischoff, R.; Sigwalt P. *Polym. Int.* 1995, 36 (1), 57-71.
[109] Jallouli, A.; Saam, J.C. *Polym. Prepr.* 1998, 39 (1), 448.
[110] Rashkov, I.; Gitsov, I.; Panayotov, I., Polymer Chemistry, IUPAC-Macro`83, Bucharest, Romania, 1983, 644.
[111] Rashkov, I.; Gitsov, I. *J. Polym. Sci.-Polym. Chem.* 1986, 24, 155-165.
[112] Matyjaszewski, K.; Pintauer, T.; Gaynor, S. *Macromolecules* 2000, 33, 1476-1478.
[113] Streat, M., "Ion Exchange- A Technologist`s Perspective of The 20th Century", in Advances in Ion Exchange for Industry and Research, Williams P. A.; Dyer, A. (Eds.), Royal Society of Chemistry, p. 3, 1999.
[114] Britton, B.C.; White, H.C.; Moyle, C., Patent USA 2 460 805, 1949.

[115] Patent Germany (West), 957 662, 1954.
[116] Ishizuka, T.; Aihara, T, *Kogyo Kagaku Zasshi* 1956, 59, 1198; C. A. 52, 13307, 1958.
[117] Sundermeyer, W.; Noll, W., Patent Germany (West) 922 377, 1960.
[118] Adams, F.; Carmichael, J.B.; Zzeman, R.J. *J. Polym. Sci.* A, 1967, 1, 741-746.
[119] Patent USA 2460805 (1949); C. A. 43, 323C, 1949.
[120] Andrianov, K.A.; Krasovskaya, T.A., *Khim. Prom.* 1956, 462.
[121] Sobolevskii, M.V.; Skorokhodov, I.I.; Ditsent, V.E.; Sobolevskaya, L. V., *Visokomol. Soedin.* A, 1969, 11, 1109.
[122] Sakiyama, M.; Okawara, R. *J. Organomet. Chem.* 1964, 6, 473-479.
[123] Patent USA 2 866 760, 1958 ; C. A. 53, 6679g, 1959.
[124] Patent Germany (East), 25 090, 1963.
[125] Patent Cehoslovaky 108 303, 1963; C. A. 60, 9459c, 1964.
[126] Gates, B.C., "Catalytic Materials", Cap. 12 in "Materials Chemistry - An Emerging Discipline", ACS, p. 301, 1995.
[127] Patent Germany (West), 1 069 388, 1958.
[128] Stewart, H. F. *J. Organomet. Chem.* 1967, 10, 229-236.
[129] Patent Germania (Vest) 2 152 270, 1972.
[130] Schindler, S.; Ruhlemann, K. *Plaste Kautsch.* 1978, 7, 384-387.
[131] Braun, I.; Hamann, H.; Tschernko, G. *Plaste Kautsch.* 1979, 11, 603-605.
[132] Hamann, H.; Michael, K.; Spirin, I.; Tschernko, G. *Plaste Kautsch.* 1978, 7, 380-383.
[133] Sobolevskii, M.V., "Oligoorganosiloxani. Svoista, polucenie, primenenie", Moskva, 1985.
[134] Brak, M.T.; Bagley, N.N.; Kurilenko, O.D. *Visokomol. Soedin.* A, 1975, 17, 5, 1034-1037.
[135] Marcu, M.; Lazarescu, S.; Cazacu, M.; Bancila, M. *Mater. Plast.* 1992, 29(3-4), 180-182.
[136] Marcu, M., Stiubianu, Gh., Lazarescu, S., Streba, E., Stanciu, A., Cazacu, M. *Mater. Plast.* 1990, 27, 4, 211-224.
[137] Marcu, M.; Simionescu, M.; Cazacu, M.; Lazarescu, S.; Ibanescu, C. Iran. *J. Polym. Sci. Techn.* 1994, 3(2), 95-104.
[138] Cazacu, M.; Marcu, M. *Macromol. Rep.* 1995, A 32(7), 1019-1029.
[139] Cazacu, M.; Marcu, M.; Hurduc, N.; Lazarescu, S.; Simionescu, M. Iran. *J. Polym. Sci. Techn.*, 1994, 3(1), 72-78.
[140] Cazacu, M.; Marcu, M.; Holerca, M.; Petrovan, S.; Lazarescu, S. *J. Macromol. Sci.-Pure and Appl. Chem.* 1996, A 33, 1, 65-76.
[141] Cazacu, M.; Marcu, M.; Petrovan, S.; Holerca, M.; Simionescu, M. *Polym. Plast. Technol. Eng. J.* 1996, 35, 2, 327-347.
[142] Cazacu, M.; Marcu, M.; Dragan, S.; Matricala, C.; Simionescu, M.; Holerca, M. *Polymer* 1997, 38(15), 3967-3971.
[143] Cazacu, M.; Vlad, A.; Marcu, M.; Panasenko, A.; Racles, C. *Rev. Roum. Chim.* 2000, 45 (6), 561-566.
[144] Vlad, A.; Cazacu, M.; Marcu, M.; Racles, C. *Rev. Roum. Chim.* 2001, 46(8), 113-118.
[145] Cazacu, M.; Vlad, A. *Rev. Roum. Chim.* 2003, 48 (6), 477-484.
[146] Cazacu, M.; Marcu, M.; Caraiman, D.; Lazarescu, S.; Vlad, A. *Rev. Roum. Chim.* 1998, 43(8), 757-763.

[147] Cazacu, M.; Marcu, M.; Vlad, A.; Caraiman, D.; Racles, C. *Eur. Polym J.* 1999, 35, 1629-1635.
[148] Simionescu, M., Marcu, M., Cazacu, M., Racles, C. *Eur. Polym. J.*, 2002, 38, 229-233.
[149] Rimniceanu, C.; Cazacu, M.; Marcu, M.; Rimniceanu, I. *Rev. Roum. Chim.* 2002, 47(6), 46-54.
[150] Cazacu, M., Airinei, A., Marcu, M. *Appl. Organomet. Chem.* 2002, 16, 643-648.
[151] Cazacu, M.; Marcu, M.; Petrovan, S.; Drăgan, S.; Caraiman, D. *Rev. Roum. Chim.* 1999, 44(1), 77-82.
[152] Cyprik, M.; Kurjata, J.; Chojnowski, J., *J. Organomet. Chem.* 2003, 686, 373.
[153] Chojnowski, J.; Cyprik, M.; Kazmierski, K. *Macromolecules* 2002, 35, 9904-9912.
[154] Zhou, Y.; Shimizu, K.; Cha, J.N.; Stucky, G.D.; Morse, D.E. *Angew. Chem. Intern. Ed.* 1999, 38 (6), 779-796.
[155] Cyprik, M.; Apeloig, Y. *Organometallics* 2002, 21, 2165-2175.
[156] Rubinstajn, S.; Cella, J. A. *Macromolecules* 2005, 38, 1061-1063.
[157] Grzelka, A.; Chojnowski, J.; Cypryk, M.; Fortuniak, W.; Hupfield, P.C.; Taylor, R.G., *J. Organomet. Chem.* 2004, 689, 705-713.
[158] Jutzi, P.; Muller, Ch.; Stammler, A.; Stammler, H.G. *Organometallics* 2000, 19, 1442-1444.
[159] Grzelka, A.; Chojnowski, J.; Cypryk, M.; Fortuniak, W.; Hupfield, P.C.; Taylor, R. G. *J. Organomet. Chem.* 2002, 660, 14-26.
[160] Prut, E.V.; Trofimova, T.M.; Enikolopyan, I.S. *Vysokomol. Soedin.* 1964, 6, 2102.
[161] Cawla, A.S.; St.-Pierre, L.E. *J. Appl. Polym. Sci.* 1975, 19 353.
[162] De Gunzbourg, A.; Favier, J.C.; Hémery, P. *Polym. Int.* 1994, 35, 179-188.
[163] Kan, C.Y.; Zhu, X.L.; Yuan, Q.; Kong X.Z. *Polym. Adv. Tech.* 1997, 8, 631-633.
[164] He, W.D.; Cao, C.T. *Polym. Int.* 1996, 39, 31-36.
[165] Kong, X.Z.; Ruckenstein, E. *J. Appl. Polym. Sci.* 1999, 73, 2235-2246.
[166] De Gunzbourg, A. ; Maisonier, S. ; Favier, J.C. ; Maitre, C. ; Masure, M.; Hemery, P. *Macromol. Symp.* 1998, 132, 359,.
[167] Maitre, C.; Ganachaud, F ; Ferreira, O. ; Lutz J.F.; Paintoux Y., Hemery P. *Macromolecules* 2000, 33, 7734-7737.
[168] Barrere, M.; Maitre, C.; Ganachaud, F.; Hemery, P., *Macromol. Symp.* 2000, 150, 359-364.
[169] Baile, M. ; Chou, Y. J. ; Saam, J.C. *Polym. Bull.* 1990, 23, 251-257.
[170] Landfester, K.; Tiarks, F.; Hentze, H.P.; Antonietti M. *Macromol. Chem. Phys.* 2000, 201, 1-5.
[171] Barrere, M.; Ganachaud, F.; Bendejacq, D.; Dourges, M.A.; Maitre, C.; Hemery, P. *Polymer* 42, 7239-7246, 2001.
[172] Kuo, A.C.M., Polymer Data Handbook, Oxford University Press, Inc. 1999, 411.
[173] Xia, Y., Whitesides, . M., *Angew. Chem. Int. Ed.* 1998, 37, 550 – 575.
[174] Hillborg, H., Tomczak, N., Ola`h, A., Scho¨nherr, H., Vancso, G. J. *Langmuir* 2004, 20, 785-794.
[175] Colas, A.; Rafidison, P. *Pharma Chem.* 2005, Oct., 46-49.
[176] Rahimi, A. *Iran. Polym. J.* 2004, 13 (2), 149-164.
[177] Patrito, N.; McCague, C.; Norton, P.R.; Petersen, N.O. *Langmuir* 2007, 23, 715-719.
[178] Qin, M.; Wang, L.K.; Feng X.Z. *Langmuir* 2007, 23, 4465-4471.

[179] Buzio, R.; Bosca, A.; Krol, S.; Marchetto, D.; Valeri, S.; Valbusa, U. *Langmuir* 2007, 23, 9293-9302.
[180] Makrodimitri, Z.A.; Dohrn, R.; Economou, I.G. *Macromolecules* 2007, 40, 1720-1729.
[181] Lin, S.B.; Durfee, L.D.; Ekeland, R.A.; McVie, J.; Schalau, G.K. *J. Adh. Sci. Technol.* 2007, 21(7), 605-623.
[182] Kim, J.; Chaudhury, M.K.; Owen, M.J. *IEEE T. Dielect. El. Ins.* 1999, 6(5), 695-702.
[183] Hillborg, H.; Gedde, U.W. IEEE T. Dielect. El. Ins. 1999, 6(5), 703-717.
[184] Cazacu, M. "Siloxane based Polymeric Structures containing Complexed Metals" in "Advances in Organometallic Chemistry Research", Yamamoto, K. (Ed.), Nova Science Publishers, 2007
[185] Lukacs, A. *Ph. D. Thesis 2006, Polysilazane Precursors* 2006
[186] Caraiman, D. ; Cazacu, M. ; Marcu, M. ; Racles, C. *Rev. Roum. Chim.* 1999, 44(5), 455-460.
[187] Kruger, C.; Rochow, E. G. *J. Polym. Sci. Part A: Polym. Chem.* 1964, 2, 3179-3189.
[188] Brochon, C.; Mingotaund, A.F.; Schappacher, M.; Soum, A. *Macromolecules* 2007, 40, 3547-3553.
[189] Polysiloxazanes, silicon oxynitride fibers and processes for producing same United States Patent 5079323, http://www.freepatentsonline.com/5079323.html
[190] Yoshinori, T.; Takahiro, G.; Yoshimoto, A. *Nippon Kagakkai Koen Yokoshu,* 2001, 79 (2), 797.
[191] Takahiro, G.; Yoshinori, T.; Yoshimoto, A. *J. Ceramic Soc. Japan* 2006, 114 (1330), 492-496.
[192] Bouquey, M.; Brochon, C.; Bruzaud, S.; Mingotaud, A.F.; Schappacher, M.; Soum, A. *J. Organometal. Chem.* 1996, 521 (1-2), 21-27.
[193] Zhou, A.; Zhang, Z.J. ; Xu, C.H. ; Xie, Z.M. *J. Appl. Polym. Sci.* 2001, 82, 929-933.
[194] Allcock, H R.; Kuharcik, S.E. *J. Inorg. Organometal. Polym.* 1995, 5(4), 307-342.
[195] Allcock, H.R.; Klingenberg, E.H. *Macromolecules* 1995, 28(13), 4351-4360.
[196] Allcock, H.R.; Prange, R. *Macromolecules* 2001, 34 (20), 6858 -6865 and references herein.
[197] Prange, R.; de Denus, C.R.; Nelson, J.M.; Crane, C.A.; Laredo, W.R.; Reeves, S.D.; Allcock, H.R. US Patent 6392008/2002, http://www.patentstorm.us/patents/6392008-description.html
[198] Roizard D.; Clement R.; Lochon P.; Kerres J.; Eigenberger G. *J. Membr. Sci.* 1996, 113(1) 151-160.
[199] Henderson, L.J.; Keller, T.M. *Macromolecules* 1994, 27, 1660-1961.
[200] Papetti, S.; Schaeffer, B.B.; Gray, A.P.; Heying, T.L. *J. Polym. Sci.: Part A-1* 1966, 4, 1623-1636.
[201] Schwam, D.; Litt, M.H. *Adv. Perform. Mater.* 1996, 3(2), 153-169.
[202] Papetti, S.; Schaeffer, B.B.; Gray A.P.; Heying, T.L. *J. Polym. Sci. Part A* 1966, 4, 1623-1636.
[203] Mayes, N.; Green, J.; Cohen, M.S. *J. Polym. Sci. Part A* 1967, 5, 365-379.
[204] Kolel-Veetil, M.K.; Beckham, H.W.; Keller, T.M. *Chem. Mater.* 2004, 16 (16), 3162 – 3167.
[205] Henderson, L.J.; Keller, T.M. *Polym. Prep.* 1993, 34(1), 345.
[206] Bucca, D.; Keller, T. *J. Polym. Sci. Part A: Polym. Chem.* 1999, 37(23), 4356-4359;
[207] Homrighausen, C.L.; Keller, T. *Polymer* 2002, 43(9), 2619-2623.

[208] Chalk, A.J.; Harrod, J.F. *J. Am. Chem. Soc.* 1965, 87, 16-21
[209] Spier, J.L. *Adv. Org. Chem.* 1979, 17, 407-413.
[210] Houser, E.J.; Keller, T.M. *J. Polym. Sci.: Part A: Polym. Chem.* 1998, 36, 1969-1972.
[211] Kolel-Veetil , M.K.; Keller, T.M. *J. Polym. Sci. Part A: Polym. Chem.* 2006, 44(1), 147 – 155.
[212] Peters, E.N. ; Bohan, J.J. ; Hedaya, E.; Kawakami, J.H. ; Kwiatkowski, G.T., The Preparation and Properties of D2-Carborane-Siloxanes, Technical Report, 17 JUL 1974, http://stinet.dtic.mil/oai/oai?verb=getRecordandmetadataPrefix=htmlandidentifier=AD A003544
[213] United States Patent 4208492, http://www.freepatentsonline.com/4208492.html,
[214] Houser, E.J.; Keller, T.M. *Macromolecules* 1998, 31, 4038-4039.
[215] Kolel-Veetil, M.K.; Qadri, S.B.; Osofsky, M.; Keller, T.M. *Chem. Mater.* 2005, 17 (24), 6101 -6107.
[216] Stewart, D.D.; Peters, E.N.; Beard, C.D.; Moffitt, R. B.; Kwiatkowski, G.T.; Bohan, J.J.; Hedaya, E. *J. Appl. Polym. Sci.* 1979, 24(1), 115-123.
[217] Kesting, R.E.; Jackson, K.F.; Newman, J.M. *J. Appl. Polym. Sci.* 1971, 15(11), 2645-2650.
[218] Hedaya, E.; Kawakami, J.H. ; Kwiatkowski, G.T., US Patent 4145504 http://www.wikipatents.com/4145504.html

In: Advances in Functional Heterochain Polymers
Editor: Maria Cazacu
ISBN: 978-1-60456-598-0

Chapter 2

SILOXANE-CONTAINING LIQUID CRYSTALLINE POLYMERS (LCPS)

Carmen Racles[1]
"Petru Poni" Institute of Macromolecular Chemistry, Iasi, Aleea Gr. Ghica Voda 41A, 700487, Romania

ABSTRACT

Polymeric liquid crystals (LCP) can be tailored in different ways: by choosing adequate mesogenic groups or spacers, synthesis methods and conditions. Siloxane moieties in LCPs are a very interesting alternative for obtaining desired properties. For example, they decrease the transition temperatures compared to organic spacers, and influence the nature of the mesophases and their stability, as well as other properties. The most recent developments in LCPs of various architectures, containing siloxane segments in their structure are reviewed, emphasizing the role of Si-O bonds on their thermotropic and overall properties.

Keywords: *siloxane, liquid crystals, thermotropic behavior, microphase segregation*

INTRODUCTION

Over the years, a great amount of information was accumulated on liquid crystalline (LC) structures. Siloxane-organic compounds, containing different mesogenic groups linked either directly or through a spacer to the siloxane backbone, into various architectures, have been synthesized and studied and their properties have been exploited in high technology applications. Siloxane –containing liquid crystals are outstanding materials, either we talk about linear polymers having mesogenic groups in the main chain or side chain, crosslinked structures like liquid crystalline elastomers (LCE), dendrimeric LCs or polymer-dispersed liquid crystals (PDLC) with siloxane matrix.

---

[1] raclesc@icmpp.ro.

Aside their structure and properties, or their synthetic paths that are sometimes predictable and sometimes unusual, some questions arise from the specificity of the siloxane-containing materials. It is known and thus predictable that siloxane moieties decrease the transition temperatures in LCPs, due to their flexibility and bulkiness. Polysiloxane backbone is used in order to obtain high mobility of the mesophase and mesogenic properties at moderate temperatures.

As it is largely accepted, the incompatibility of siloxane moieties with most of the organic structures has tremendous consequences over the properties of hybrid compounds. The phase separation in siloxane-organic LCs is considered responsible for many of their unpredictable properties, as well as for their mesophase behavior. For example, as a consequence of the microphase separation, the more ordered mesophases may be favored and sometimes the mesophase range is enlarged due to the increase of the isotropization temperature.

Here we resume some of the most recently reported data (generally after year 2000) on different kinds of LC systems containing polysiloxane moieties, their synthesis, LC characteristics and special properties and applications. Low molecular mass LCs are not a subject of this review, and they will not be treated, with very few exceptions.

Earlier reported (before 2000) polysiloxane LCs, and especially ferroelectric LCPs have been reviewed by Teyssier and Boileau [1]. Some of the trends that had been observed there are still actual, and the continuous development of modern investigation techniques, as well as the pemanent seek of better performances in areas like optoelectronics, sensors and actuators, microelectronics, medical science etc. maintain the interest of the scientific comunity for this subject.

The LC polysiloxanes remain an extremely active field of research, especially from a physical-chemistry point of view. There are a lot of studies and new or newly employed methods of investigation serving this purpose. From a chemist's perspective, there is also increasing information on new structures and materials. In this context, the purpose of this review is not to exhaustively present the most important achievements in this area, but rather to point out its actuality and importance. Especially the synthesis aspects and phase behavior of different types of siloxane-containing LC polymers are reviewed.

1. Main Chain and Side Chain Siloxane-Containing Liquid Crystals

The new trends in main chain and side chain LC polysiloxanes are related to structural variations and special properties of siloxanes:

- Use of new mesogens - as much as their novelty can be proven given the huge number of such structures;
- Combinations of siloxane and other flexible units;
- Investigations of the influence of siloxane moiety on the thermotropic properties;
- Studies or at least anticipation of microphase separation in such systems, with its consequences on the thermotropic and other properties.

1.1. The Effect of Microphase Segregation

One important characteristic observed in many LC systems is that the siloxane moiety trends to suppress the nematic phase. The segmental incompatibility of the siloxane and organic parts of the molecules induces microphase segregation; the siloxane groups have a tendency to aggregate in layers, thus strongly promoting smectic phases and depressing the nematic phase.

Besides this aspect, the influence of microphase segregation in siloxane-containing LCs and other copolymers over different properties is more and more taken into account and investigated by direct or indirect methods.

The microphase separation between the siloxane backbone and the mesogenic groups in a side-chain LC siloxane copolymer is considered responsible for the formation of extremely ordered layers in the polymer Langmuir-Blodgett (LB) film [2].

The LC behavior of ferroelectric siloxane homo- and copolymers has been found to differ in its nature, even if similar phase sequence was observed [3]. The SmC* to SmA transition was second order, de Vries type for the copolymers [4], while it was first order and not of de Vries type for the homopolymers. This was explained by the microphase separation in the copolymers, which has consequences for the mechanical properties of the corresponding elastomers. Such microphase separation occurs into sublayers of the mesogens, which are separated by sublayers of unsubstituted dimethylsiloxane units. This theory is sustained by the occurrence of a diffraction halo at 2θ of approximately 11°, which is assigned to the liquid-like packing in the dimethylsiloxane sublayer [5] and is typical for dimethylsiloxane rubbers. It was also found that weak crosslinking has no influence on the layer spacing and its temperature dependence, while a stronger crosslinking affects them.

An unusual anisotropic conductivity, with the value perpendicular to the director being 200 times higher than that parallel to the director was found for siloxane-containing LCs with cyanobiphenyl mesogen and alkyloxy spacers [6]. On the other hand, a reduced dielectric anisotropy was observed, compared to non-siloxane analog structures. The hypothetical explanation was the microphase separation of the siloxane moieties into "liquid-like regions", which inhibits the mobility across the smectic layers, while increasing the antiparallel correlations of molecular dipoles. These materials, as well as their mixtures [7] may be very interesting for use in bistable, electro-optic storage and display devices.

Well-defined diblock copolymers of polystyrene and side chain liquid crystal (SCLC) polysiloxanes of general structure shown in Scheme 1 have been synthesized by anionic polymerization followed by hydrosilylation [8].

Such copolymers are phase segregated systems, since the siloxane block has Tg under room temperature and polystyrene blocks are glassy till about $60^{\circ}C$. Thus, a mobile LC phase exists within a rigid supporting matrix. The copolymers show very interesting morphologies depending on the LCP weight fraction and mesogen structure.

The covalent linking of segments which have a tendency to phase separate can be used to tailor the stability range and the properties (eg viscosity) of LC materials. Aside from largely studied combinations of calamitic mesogens and siloxanes, the introduction of a third phase separating segment, like perfluorinated carbon chains, can bring new information. Two types of calamitic mesogens were either terminally [9] or laterally [10] attached to flexible linear chains containing alkyl and siloxane units linked to a central perfluorinated chain.

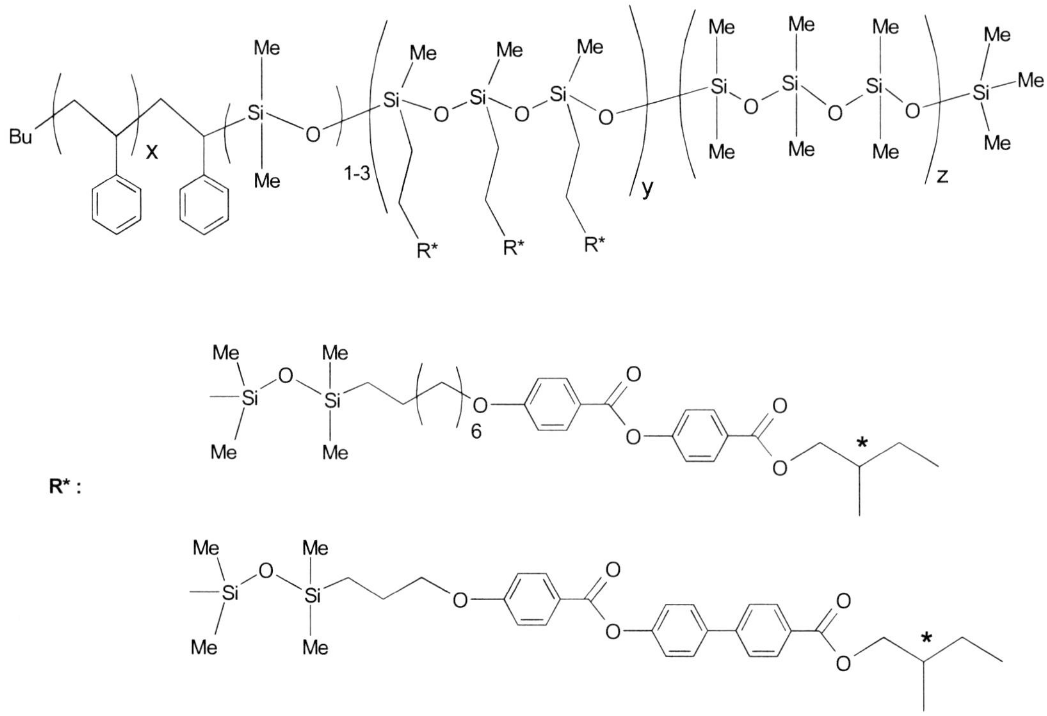

Scheme 1.

The mesogens were bonded to a disiloxane, trisiloxane or tetrasiloxane moiety in a hydrosilylation reaction and a perfluorinated chain was included between the siloxane segments, as shown in Scheme 2.

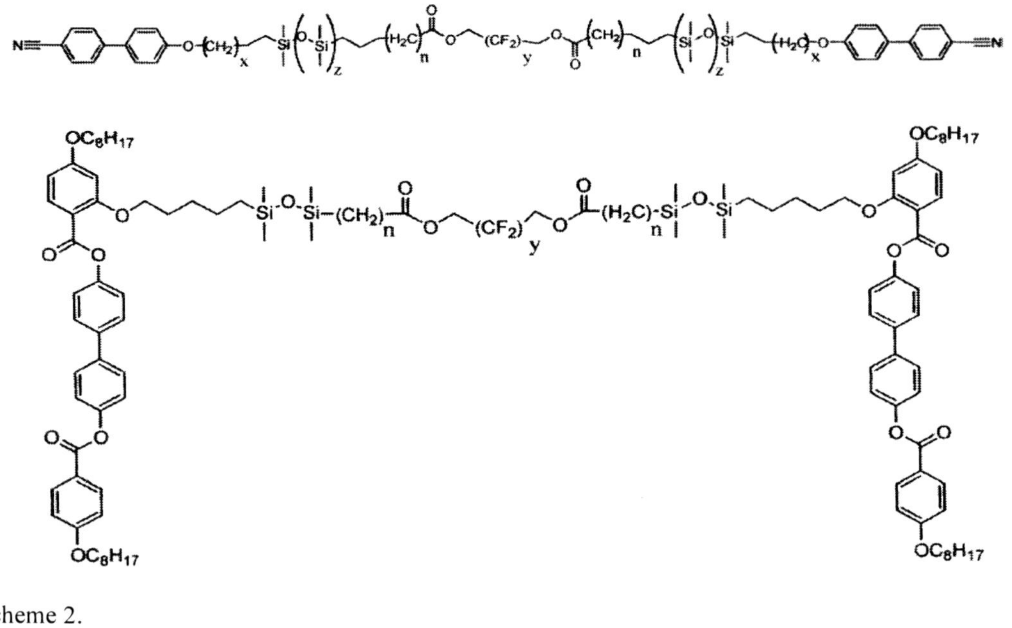

Scheme 2.

These materials exhibited nematic and smectic phases with transition temperatures close to ambient. When comparing the effect of siloxane and perfluorinated groups on the liquid crystalline phase behavior, it was observed that the enlargement of siloxane moiety was considerably more effective in lowering the transition temperatures than a decrease in the size of the perfluorinated unit. The transition temperatures of the side-on mesogens were decreased by the perfluorinated siloxane core and increased by the lengthening of the alkyl spacer, but the effect was less dramatic than observed in the linear materials.

Polysiloxanes containing fluorinated chains linked to the mesogenic groups (Scheme 3) were obtained by hydrosilylation [11, 12]. Smectic mesophases were observed and discussed in relation with the mesogen structure and fluorinated chain length [11]. While in monomers the biphenyl core leads to polymorphism, and the phenyl benzoate group induces a monomorphic LC behavior, this effect was not observed in polysiloxanes. An increasing of the number of the fluoromethylene units increased the isotropization temperatures of polysiloxanes.

Polysiloxanes with 4-(4-cyanophenoxycarbonyl)phenyl 4-(2-perfluoroalkylethoxy)-2-propyleneoxybenzoate side groups were compared to a hydrocarbon analogue [12]. The liquid crystal properties and the monolayer formation on the water-air interface were discussed in correlation with the presence of the perfluoroalkyl groups. The nature of the mesophase depends on the structure of the side chain. Smectic A phase was observed for medium length perflourinated groups (perflourohexyl and –octyl), while the same mesogenic core linked to shorter fluoroalkyl, as well as to longer hydrocarbon "tail" lead to nematic phase.

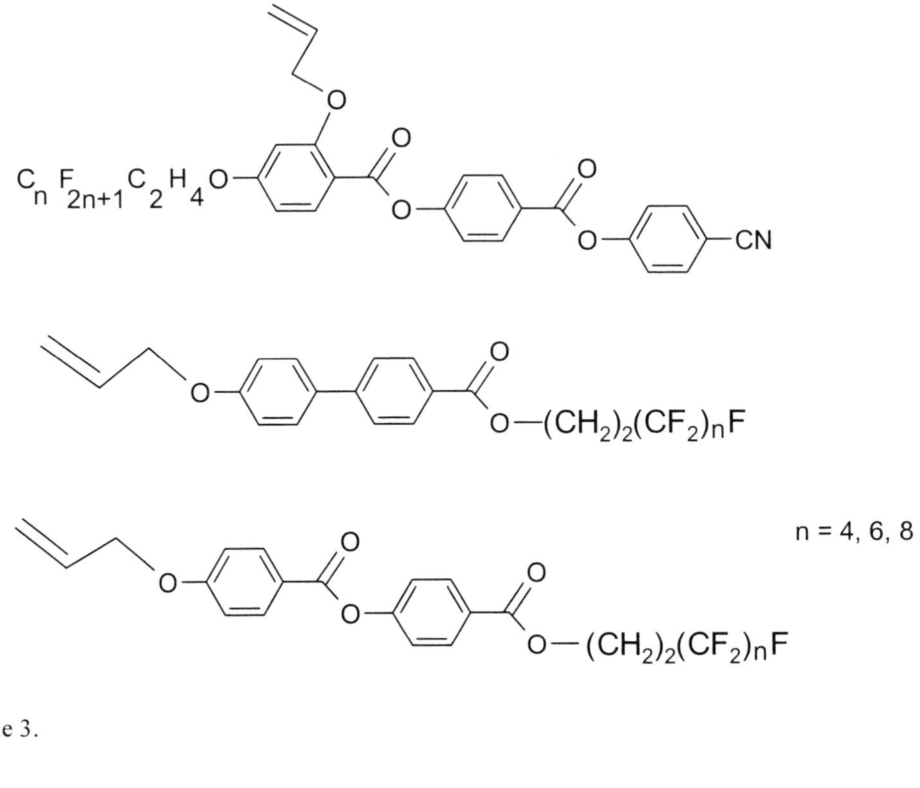

Scheme 3.

1.2. Ferroelectric Liquid Crystals

Ferroelectric liquid crystals (FLC) have been intensively studied, especially for electro-optical applications. Ferroelectric side chain LC polymers combine processing properties of the polymers (as the ability of forming thin films) with valuable properties of the mesogenic groups, like spontaneous polarization, high mobility and memory effect. Although they have higher viscosity than the low molecular weight FLC, they are used for memory devices or light valves [13].

Siloxane dimers generally have a considerably stable anti-ferroelectric (AF) phase due to the siloxane unit [14-17], over a broad temperature interval. Their molecular tilt is usually large, sometimes very close to 45°; a value of 53° was reported [18]. The temperature dependence of the molecular tilt of the siloxane dimers in the AF as well as in the ferroelectric (F) phase is generally minor, compared to monomeric compounds, which is very useful for display applications.

Siloxane-containing dimers with the structure shown in Scheme 4 were studied for optical and electro-optical properties [19]. These materials exhibited a broad temperature range AF phase and large molecular tilt, close to 45°, which was almost temperature-independent. They also showed very low in-plane birefringence, and consequently an extremely dark state in the AF phase, with contrast ratios as high as 1000:1. This study concluded that the optically isotropic siloxane moiety did not contribute to the optical properties, but served to enhance the promotion of the AF phase and the temperature independent properties.

Both ferro- and antiferroelectric mesophases were observed in dimers containing siloxane moiety (Scheme 5) [20]. The ferroelectric phase was present in molecules having an elongated conformation, whereas the antiferroelectric phase appeared in molecules with a V-shaped conformation. Thus these compounds may develop two distinct smectic structures as a function of temperature, synclinic at high temperature and anticlinic at lower temperature.

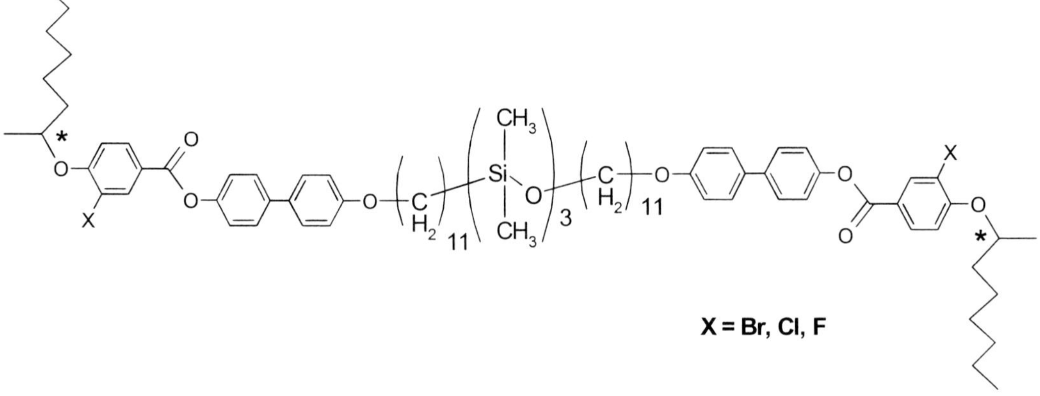

Scheme 4.

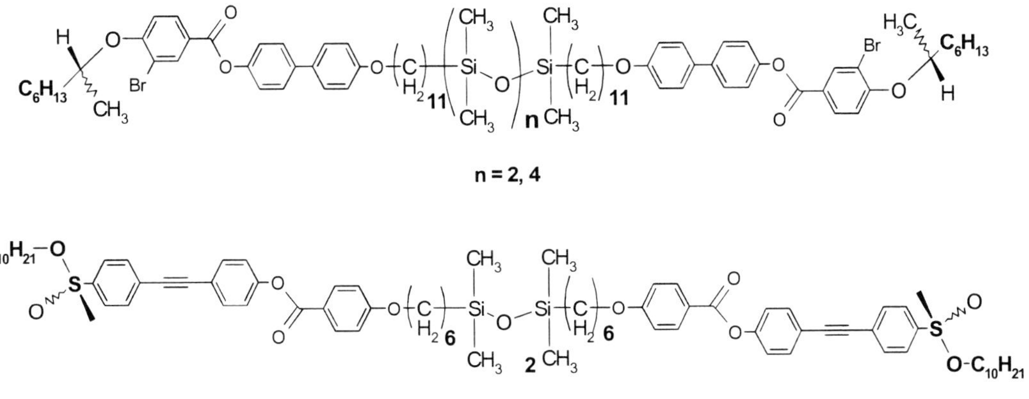

Scheme 5.

The hydrosilylation reaction was used to attach halogen-containing chiral mesogens to the polysiloxane chain [13, 21] (Scheme 6). The side-chain copolymers exhibited SmC* mesophase over a wide temperature range which included room temperature.

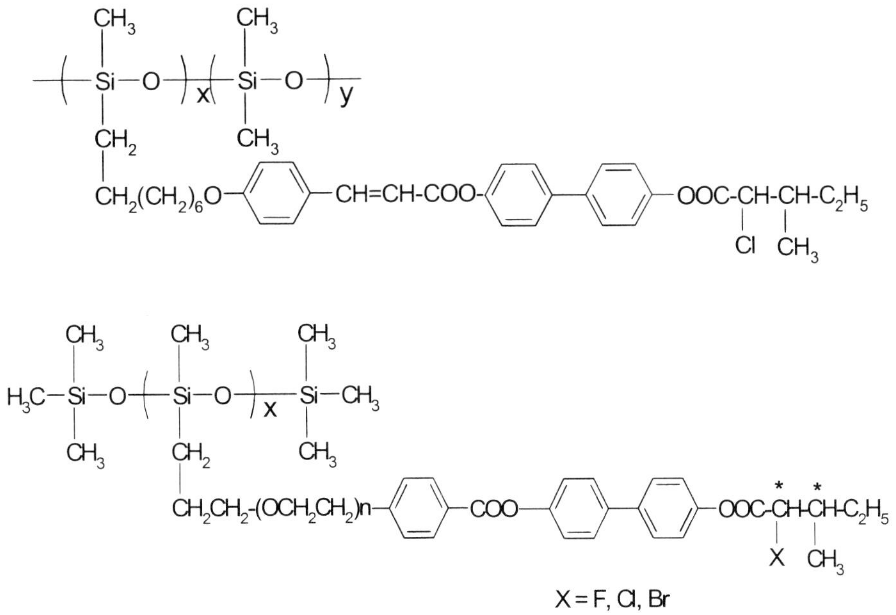

Scheme 6.

The flexible siloxane chain is thought to be responsible for the enhancement of decoupling of the motions of side chain and main chain and thus for better thermal stability of the mesophase. Evidence for microphase separation in copolymers was found by XRD [13].

1. 3. Cholesteric LC Polymers

Cholesteric LC polymers are very attractive, especially due to their unique optical properties (thermochromism, selective reflection, circular dichroism) and they have applications in optical fibers, thermal imaging, paint technologies, flat-panel displays etc.

LC cyclosiloxanes containing a cholesteric and a nematic mesogen showed cholesteric and blue phases [22]. The temperature range of the blue phase was about 20°C, which is broader than the previously reported ones.

A side chain LCP with polysiloxane backbone (homopolymer) containing a diosgeninyl derivative mesogen [23] had a completely different mesomorphic behavior compared to its copolymer homologues containing a menthyl derivative: it exhibited smectic mesophase versus cholesteric Grandjean texture observed in the copolymers. The given explanation is that the polymeric chains hindered the formation of a helical supramolecular structure of the mesogens.

An important challenge is to synthesize polymers with large temperature range for applications, which means a Tg as low as possible and a clearing (isotropization) temperature (Ti) as high as possible, but inferior to the decomposition temperature. One approach was the introduction of chiral nonmesogenic moieties together with cholesteric mesogens, in polysiloxane copolymers [24] (Scheme 7 (a)). Once again, the well known hydrosilylation reaction was the synthetic pathway. The thermotropic properties were influenced by the copolymer's composition. It seems that a small amount of nonmesogenic lateral groups stabilizes the cholesteric phase, due to the interaction of the cholesteryl and menthyl groups, which strengthen the helicity. For large content of nonmesogenic moiety, the copolymers exhibited cholesteric LC properties, but within a narrow temperature range.

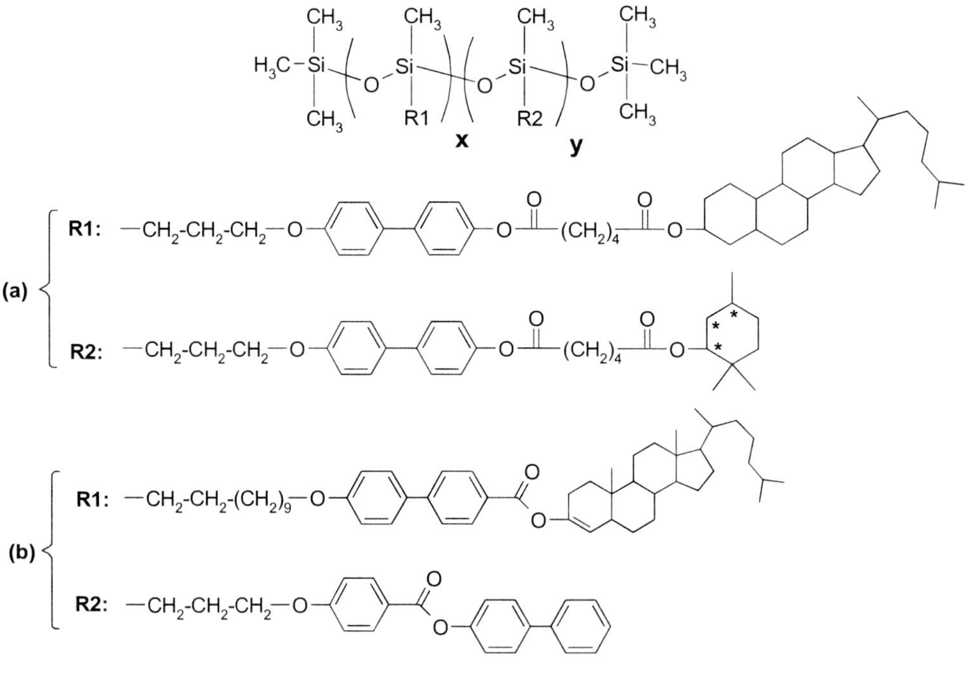

Scheme 7.

Side-chain liquid crystalline copolysiloxanes containing cholesteryl4-(10-undecen-1-yloxy)biphenyl-4'-carboxylate and 4-biphenyl 4-allyloxybenzoate mesogenic side groups (Scheme 7 (b)) showed cholesteric phase [25]. The temperature range of the mesophase increased with the content of cholesteryl derivative. The copolysiloxane containing 50/50 % of the two mesogens was used as stationary phase in gas chromatography capillary columns. It showed chiral resolution properties for three enantiomers, i.e., linalool, menthol and menthone. The chiral resolution ability was attributed to the helical supermolecular structure of cholesteric liquid crystalline polymers, since liquid crystal stationary phases yield separation based upon differences in solute molecular shape.

1.4. LC Poly(Azomethine)s Containing Siloxane Segments

Poly(azomethine)s are known as thermally stable and environmentally resistant polymers [26] and as thermotropics with potential aplications for high-performance fibers or in opto-electronics [27-30].

The wholly aromatic poly(azomethine)s have high melting or softening temperatures and low solubility. One way to reduce the transition temperatures and to improve the solubility is the introduction of siloxane segments as flexible spacers between the rigid conjugated aromatic azomethine moieties. In our group a rather large variety of poly(siloxane-azomethine) main chain LCs, having ether or ester links have been synthesized [31-37], from low molecular mass models and dimesogenic oligomers, to segmented copolymers. Basically polycondensation reactions were used as the synthetic pathway. In table 1 the chemical structure of some poly(siloxane-azomethine)s and their mesophase behavior are presented.

When comparing with the starting mesogenic compounds or with similar polymers having aliphatic spacers instead of siloxane, a few general conclusions were obvious:

- the introduction of the siloxane moieties led to an important decrease of the melting temperature, more pronounced with increasing siloxane chain length;
- a broadening of the mesophase domain was observed (especially in dimers);
- lower isotropization temperatures were also obtained in most of the cases, thus avoiding the thermal degradation within the mesophase;
- mostly smectic mesophases were observed for polymers containing disiloxane moiety and nematic for longer siloxane segments (this is probably due to the polydispersity of the starting siloxanes);
- for long siloxane segments the thermotropic behavior was suppressed, probably due to „dilution" of the mesogen;
- good solubility was generally achieved using siloxane moieties as soft segment.

Table 1. Chemical structures and phase behavior of polyazomethines containing siloxane moieties

Chemical structure	Mesophase range	Ref.
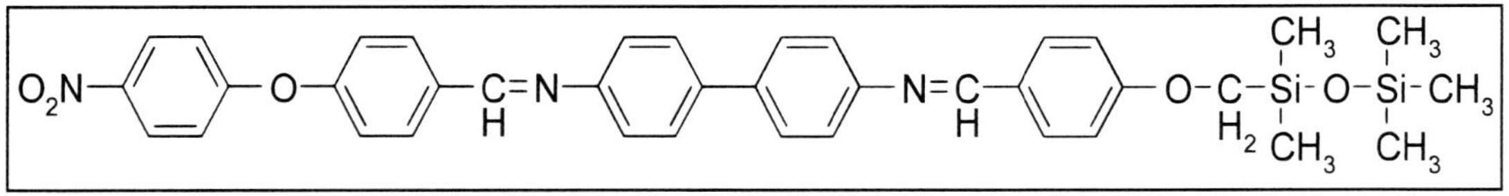	Cr 189 N 357 I	[31]
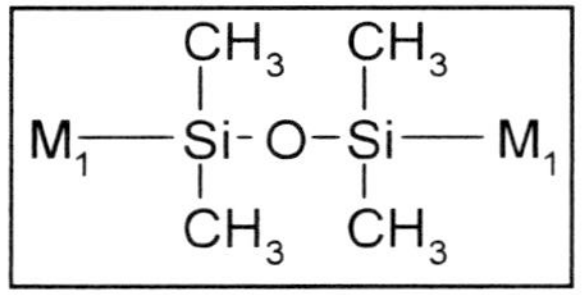 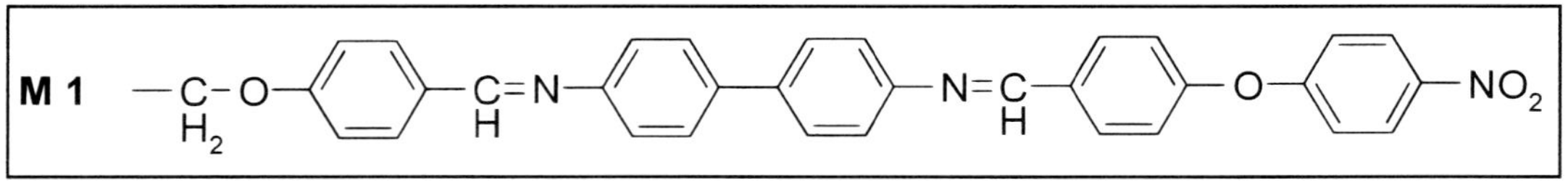	Cr 180 N 275 I	[31]
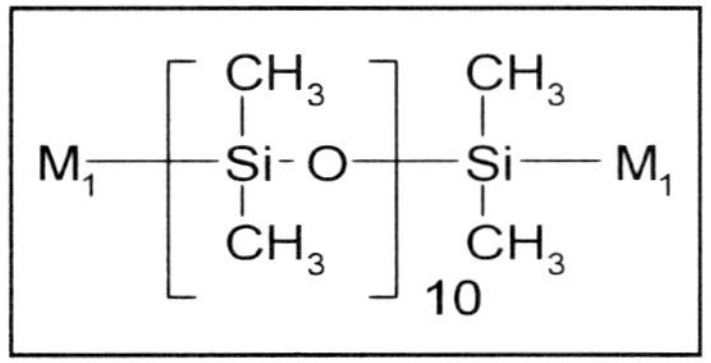	Cr 170 N >300 I	[31]

Chemical structure	Mesophase range	Ref.
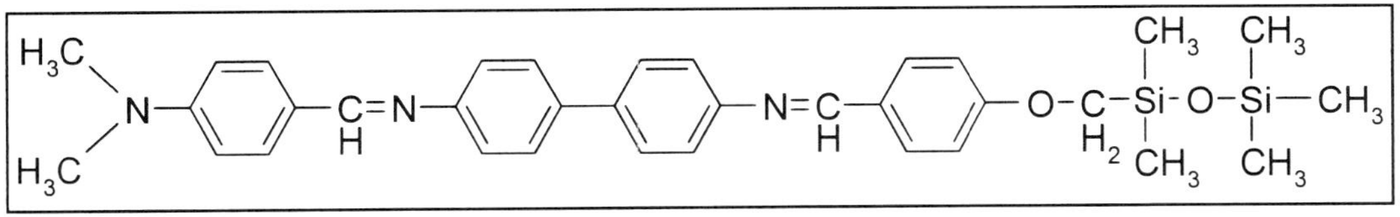	Cr 174 N 362 I	[31]
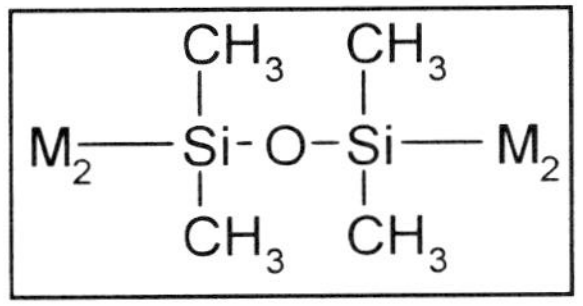 **M 2** 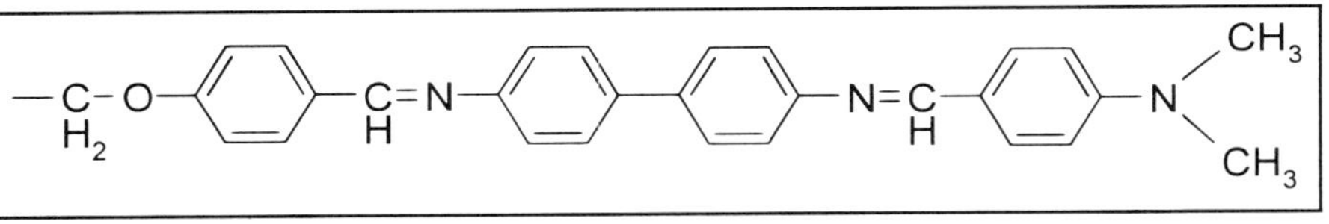	Cr 170 N >300 I	[31]
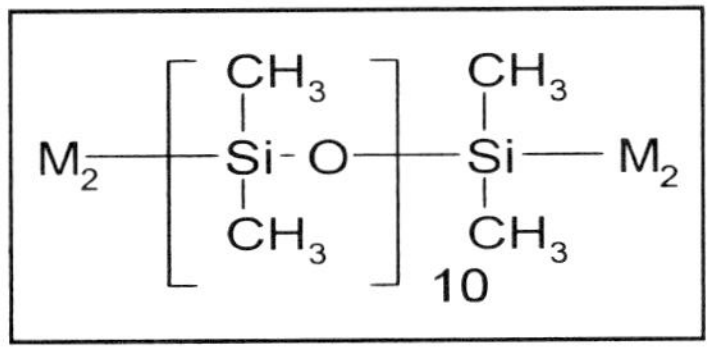	Cr 175 N >400 I	[31]

Table 1. (Continued)

Chemical structure	Mesophase range	Ref.
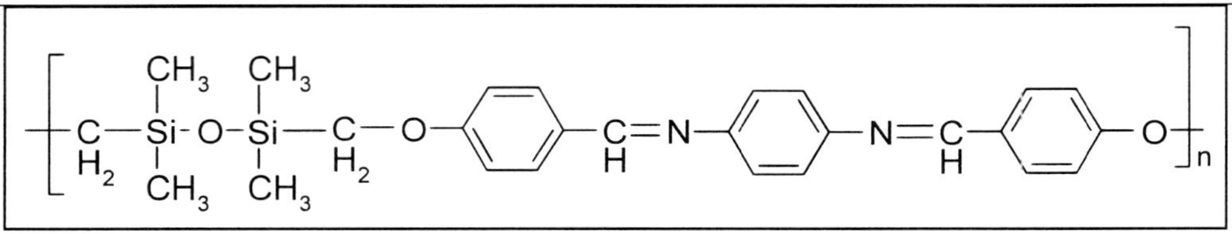	g 60 Cr 140 Sm 195 I	[36]
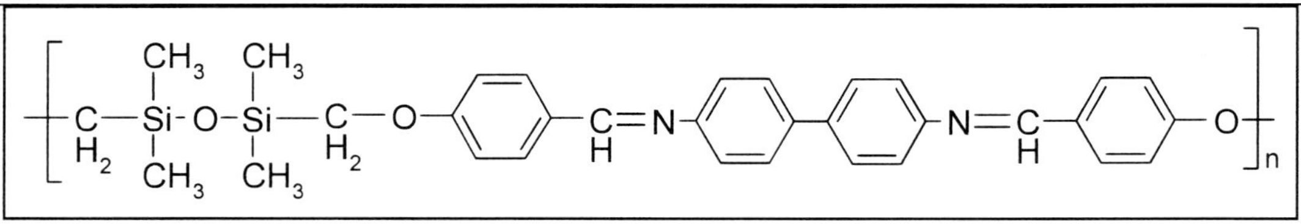	g 80 Cr 140 Sm 200 I	[36]
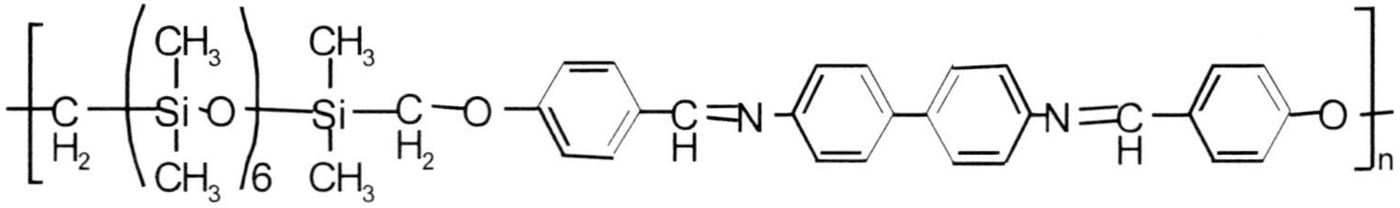	LC behavior not observed	[36]

Chemical structure	Mesophase range	Ref.
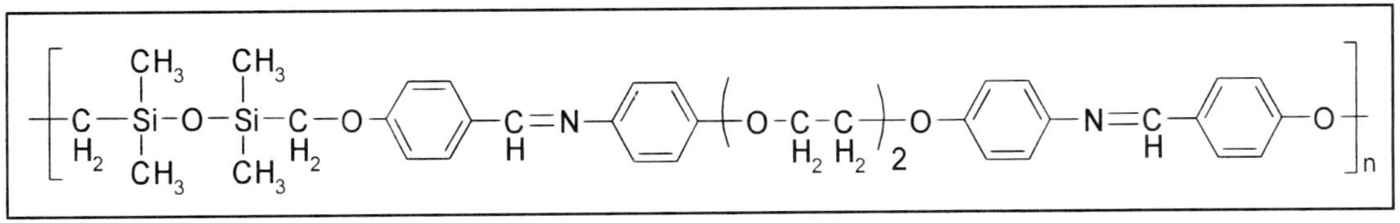	g 83 Cr 147 Sm 162 I	[36]
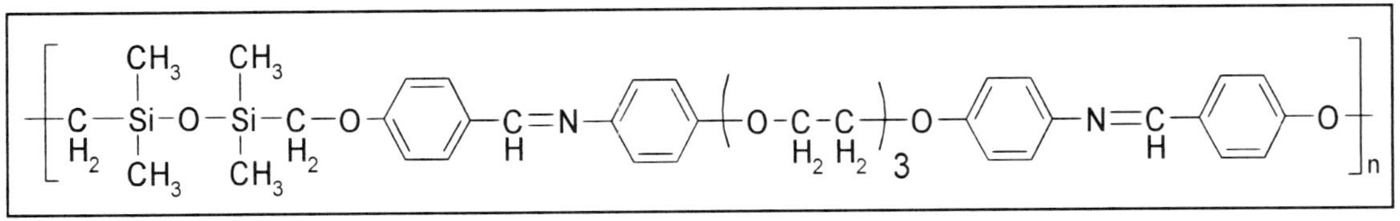	g 66 Cr 113 Sm 135 I	[36]
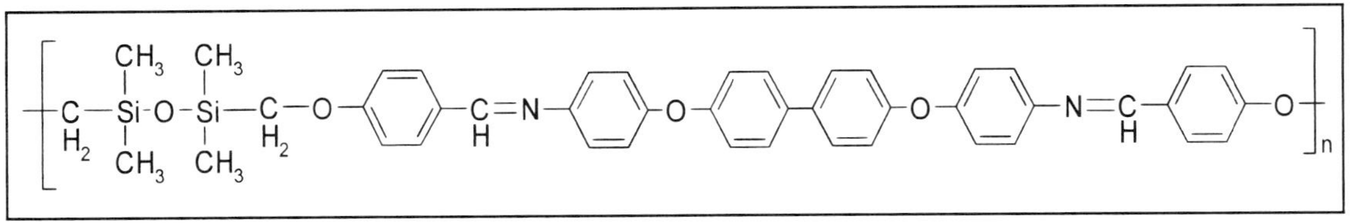	g 95 Cr 245 Sm 270 I	[36]

Table 1. (Continued)

Chemical structure	Mesophase range	Ref.
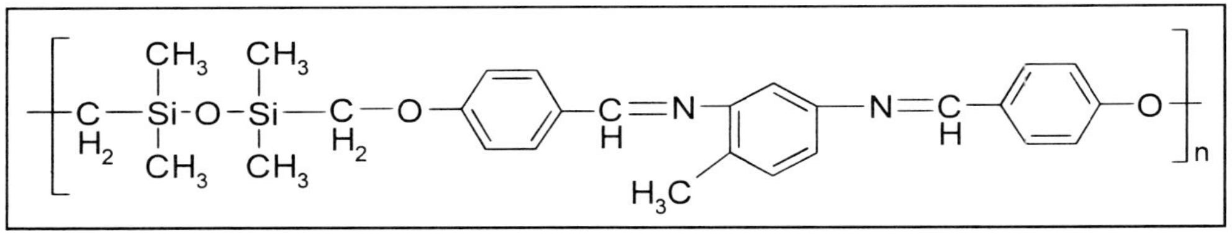	LC behavior not observed	[36]
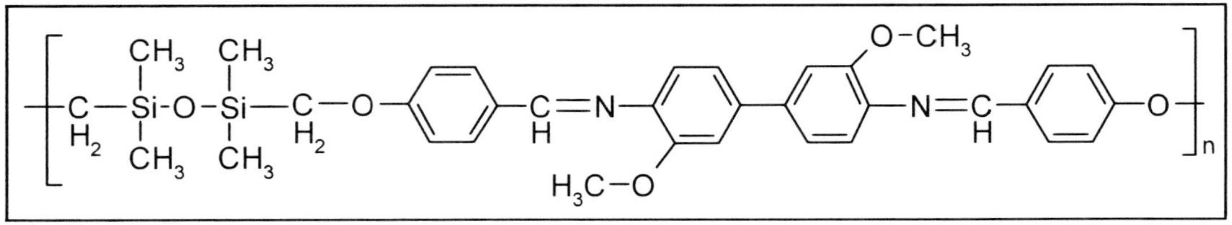	Cr 141 S 165N 245 I	[34, 37]
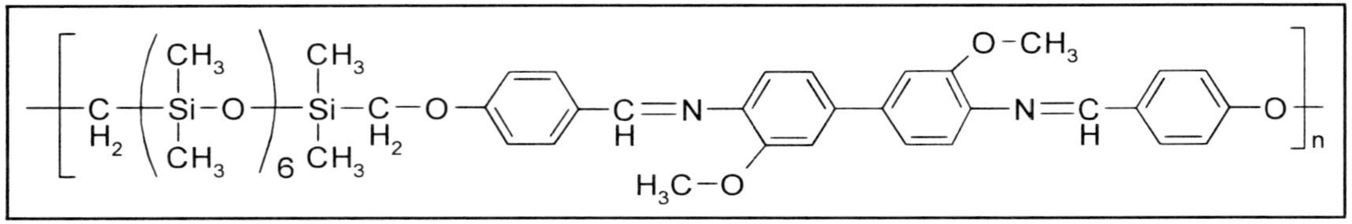	LC behavior not observed	[34]

Chemical structure	Mesophase range	Ref.
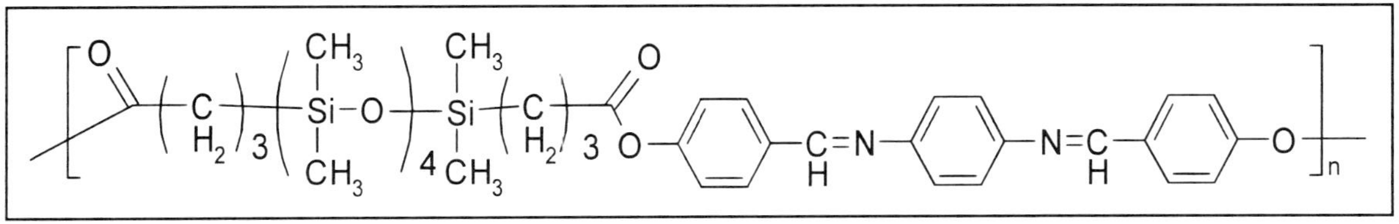	Cr 60 N 168 I	[32]
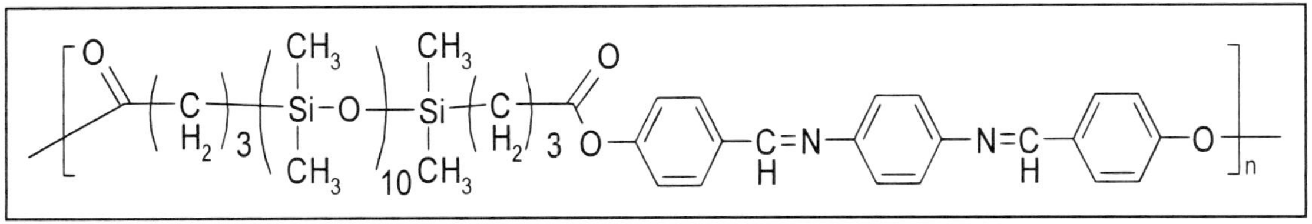	LC behavior not observed	[32]
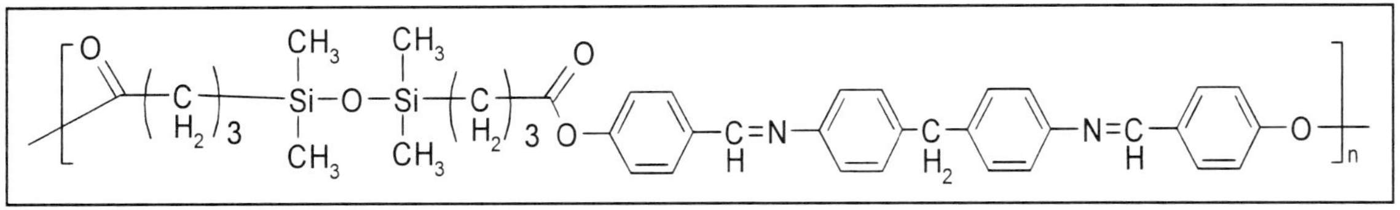	Cr 104 Sm 129 I	[32]

Table 1. (Continued)

Chemical structure	Mesophase range	Ref.
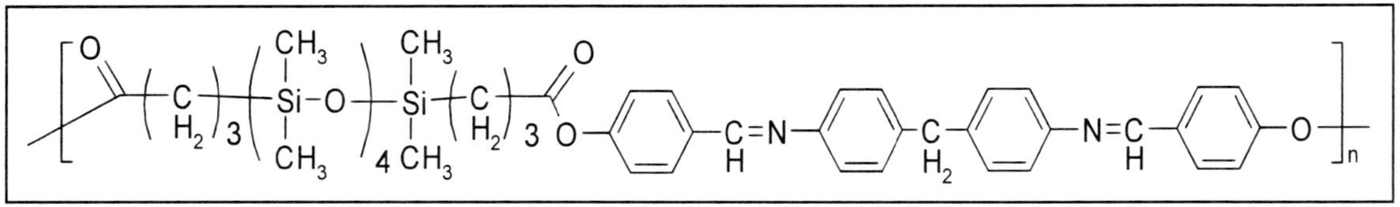	Cr 69 N 102 I	[32]
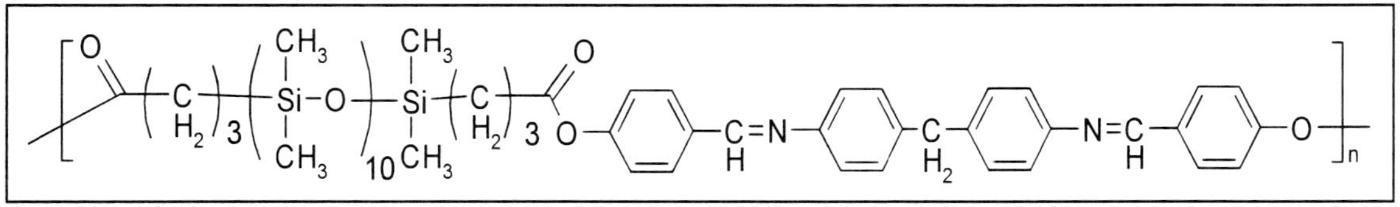	LC behavior not observed	[32]
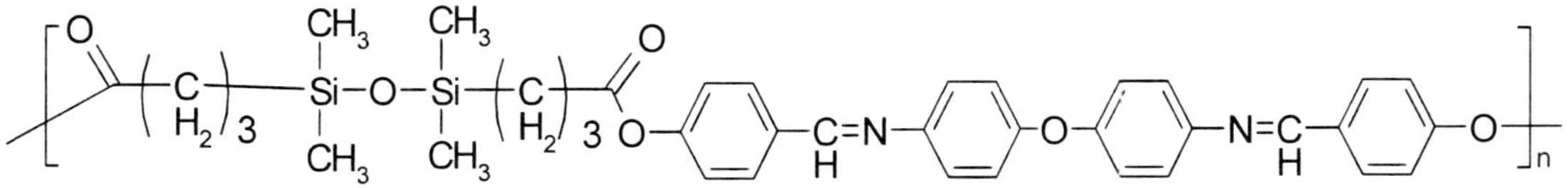	Cr 166 Sm 188 I	[32]

Chemical structure	Mesophase range	Ref.
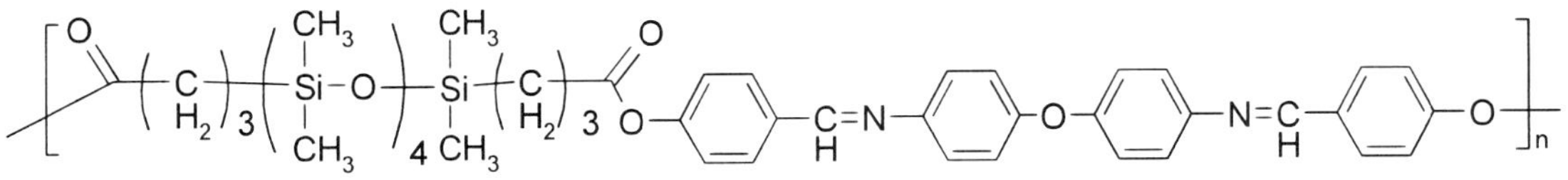	Cr 130 N 174 I	[32]
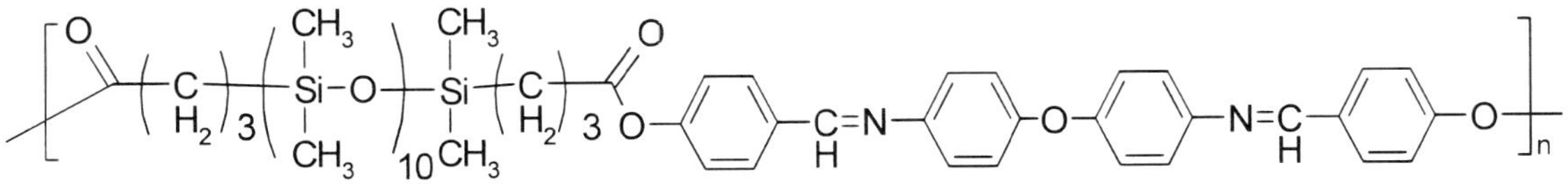	LC behavior not observed	[32]

1.5. Polymers Exhibiting Unusual Phase Behavior

Main-chain liquid crystalline polymers having the structure showed in Scheme 8 formed low-temperature smectic mesophases, identified as S_{CA} [38]. This is a layered mesophase in which mesogens in adjacent layers tilt in an alternating sense with respect to the layer normals and its schlieren texture exhibits singularities having both two and four brushes.

Poly[methyl-11-*p*-byphenyloxyundecanoyl]siloxane and its copolymer exhibited very interesting phase behavior [39]. A highly ordered crystalline smectic phase, SmE was observed for the homopolymer in the second heating as an exothermal transition, which is similar to the crystallization of amorphous polymers on heating. Annealing of the polymers resulted in the growth of the additional peak at the first heating scan, corresponding to the partial crystallization in homopolymer and to the highly ordered crystalline smectic phase SmH. The annealing temperature influences the sequence of mesophases, the phase transition temperatures and enthalpies.

The transition between LC phases in these polymers was investigated by the Avrami analysis of DSC data. The crossover of the kinetic mode for the formation of smectic phase in a side-chain copolymer was observed. The phase transition was considered to combine the diffuse-controlled dynamic growth and the liquid crystallization. The same method was used to study the kinetics of phase transitions Sm C* / Sm B in a chiral siloxane polymer [40].

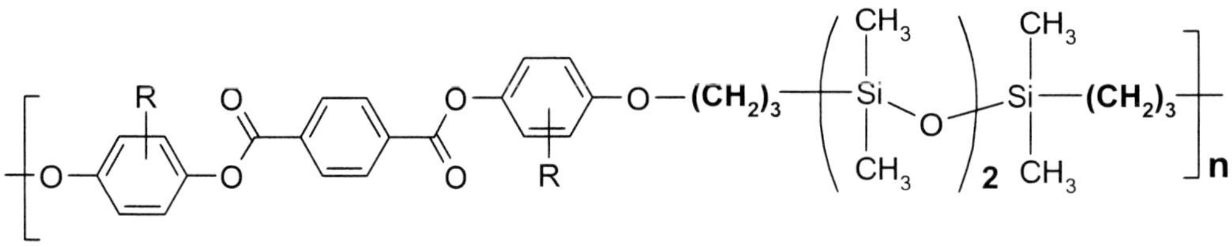

Scheme 8.

The fact that non-mesogenic compounds linked by non-chemical bonds may lead to LC behavior seems very attractive. Defined by J.M. Lehn as "chemistry beyond the molecule" [41], the supramolecular chemistry refers to assemblies of molecules obtained using a combination of secondary chemical interactions rather than covalent bonding [42]. Although hydrogen bonds between neutral organic molecules are not very strong, they play a very important part in molecular recognition, self-assembling in bio-macromolecules, increasing miscibility in polymers blends, organization in liquid crystals. The directionality and versatility of the hydrogen bonds are the major keys for their implication in supramolecular polymers design.

The self-assembly of carboxylic acids as proton donors with pyridyl fragments as proton acceptors is most frequently used in the formation of H-bonded structures [43]. Such supramolecular polymers have been obtained and investigated, especially for liquid crystalline properties [43-48]. The reverse influence has also been observed: that hydrogen bonding in supramolecular polymers is enhanced by liquid crystallinity and phase separation [49].

Siloxane-containing benzoic acids and bipyridine form interesting supramolecular structures, with cubic mesophases [47]. The formation of this rare type of LC phase has been explained by the chemical incompatibility between the polar aromatic part and the bulky, flexible, non-polar siloxane moiety in the same mesogen, which are nanophase separated [48].

A series of supramolecular polymers have recently been prepared, based on chemical recognition between carboxylic acids and bipyridine [50]. Owing that the starting compounds do not exhibit liquid crystallinity, it was interesting to observe the role of the hydrogen bonds in obtaining ordered mesophases. The structures of the siloxane diacids were chosen as to ensure different levels of chain mobility, molecular polarizability and free volume. Different thermotropic behavior was observed, which allows the tailoring of the mesophase range. The extremely easy preparation and the fact that these compounds showed interesting sensing capacity make them possible candidates for application as gas sensors, especially for nitrogen oxides and CO_2.

1.6. Miscellaneous

Side-chain liquid crystalline polymers, which contain carbazolyl derivatives as a part of mesogen, and polymethacrylate, polyester, polymethylsiloxane, polymaleimides as a polymer backbone have been synthesized and studied [51, 52]. Carbazole and carbazolyl derivatives exhibit interesting properties, such as photoconductive, luminescent, hole-transporting properties. In this series of side-chain polymers, polysiloxane derivatives exhibited a smectic A1 phase, which had the widest range –about 180°C-.

The synthesis and LC behavior of side chain polysiloxanes containing (S)-(+)-2-methyl-1-buthanol derivatives has been described [53]. Different types of mesophase were observed; the siloxane backbone and long flexible spacer induced low glass transition temperature, high thermal stability and wide mesophase range.

LC polyorganosiloxanes containing laterally linked mesogenic units on the main chain have been prepared by a multi-step procedure: introduction of hydroquinone units on the main chain by hydrosilylation polymerization after previous protection of phenols with tetrahydropyranyl groups, followed by a grafting reaction with 4-n-octyloxybenzoyl chloride [54]. High molecular weight and smectic mesophases were reported.

Siloxane-containing copolyesters with arylidene mesogenic moieties have been obtained by copolycondensation reactions between bis(hydroxybutyl)siloxanes, 2,6-bis(4-hydroxybenzylidene)cyclohexanone, and diacid chlorides (sebacoyl or terephthaloyl) [55]. The LC behavior was influenced by the nature and length of the spacer. Nematic mesophases were observed and the transition temperatures were lower than those reported for a polyester having the same mesogen, but poly(ethyleneoxide) spacer.

Liquid crystalline siloxane-containing arylidene polyethers have been synthesized [56], having lower transition temperatures than similar polymers with polymethylene spacers and reasonable thermal stability. The photobehavior of the siloxane–arylidene polyethers was influenced by the siloxane spacer length. A higher photostability was obtained in the case of siloxane–arylidene compounds, compared with that of poly(ethylene oxide) homologues reported in the literature.

Side-chain azo- polyesters with different siloxane-organic backbones have been synthesized in solution by direct polycondensation reactions involving Disperse Red 19, using an activating system, at room temperature [57]. The thermotropic LC behavior depends on the content of mesogenic groups, the length of siloxane segment and the chemical structure of the organic parts of the backbone. With increasing siloxane chain length, a decrease in transition temperatures was noticed within the polysiloxane series. The glass transition and melting temperatures decreased, while the isotropization temperature increased with main chain flexibility.

Hemiphasmidic monomers 4-[(3,4,5-triethoxy)benzoyl]-4`-[(p-allyloxy)benzoyloxy]-biphenyl, 4-[(3,5-diethoxy)benzoyloxy]-4`-[(p-allyloxy)-benzoyloxy]biphenyl, and the rod-like homologue 4-[(p-ethoxy)-benzoyloxy]-4`-[(p-allyloxy)benzoyloxy]biphenyl have been used to obtain the corresponding side-chain liquid-crystalline polysiloxanes. The resulted LC polymers possessed broad range enantiotropic nematic phases and high thermal stability [58].

Side-chain liquid crystal polysiloxanes containing mesoionic sydnonyl moieties have been synthesized [59]. Due to the unique structure, sydnone possesses both the conjugated and polar character, which makes it sensitive to both electric and magnetic fields. The polymers displayed a typical smectic schlieren texture between 120 and 200 °C.

A study of the rheological properties in the smectic phase of a side-chain LC polymer with polysiloxane backbone [60] showed that the activation energies of the shear viscosity and the moduli were smaller than that of similar polymers with polymethacrylate backbone.

2. Elastomers and Networks

LC elastomers (LCE) are weakly crosslinked polymers containing mesogenic units incorporated into main- or side chains; they have both elasticity –like the rubbers- and LC anisotropic behavior between the glass transition temperature and the LC-isotropic state transition temperature. The first LCE was synthesized in 1981 [61] and since then these polymers atracted much interest for their technological potential but also for the scientific challenge. LCEs are materials with unique optical, mechanical and piezoelectric properties. Their best characteristic is the mechanical orientability [62].

Usually a polysiloxane-based LCE consists of a polysiloxane backbone with mesogenic side chain groups and a controlled amount of crosslinkable side groups. The crosslinking may be achieved thermally, by photocurring, or by addition (hydrosilylation).

Depending on the chemical structure of the mesogenic groups, composition, crosslinking density, nature of the crosslinking agent and other factors, there have been identified different types of mesophases: nematic, cholesteric, smectic (SmA, SmC), chiral (N*, SmC*).

In the recent years, this subject is continouosly developed, given its tremendous applicative potential (for example the groups of Finkelmann, Mauzac, Terentjev, Zhang). From the synthesis point of view, there was no spectacular change: the same classical hydrosilylation reaction is used in almost 100% of the cases. There are however some structural variations, due to newly employed mesogens, but the most important trends in this area are the improved technique of preparation in order to obtain monodomain LCs, the physico-chemical investigations and the applications in nanotechnology and medicine.

The method for obtaining monodomain LC elastomers, which is largely used in various studies was established by prof. Finkelmann et al. [63] and can be briefly described as follows: „samples under preparation are first weakly cross-linked, then stretched unidirectionally to align the mesogens and establish nematic order, and then finally fully cross-linked to form a transparent birefringent monodomain".

2.1. Nematic Elastomers

There are two key equilibrium effects in LC elastomers, especially in monodomain nematic networks (see [64] and references therein). The first effect is the anomalous thermal expansion, which means the spontaneous elongation of the material along the director axis on cooling from the isotropic phase. The other effect is the so-called ''soft elasticity'', which determines the equilibrium mechanical and optical responses of monodomain nematic elastomers, as well as the properties of polydomain-monodomain transition. The reason for these phenomena is in the direct coupling between the average polymer chain anisotropy and the nematic order of the mesogens.

The influence of the cross-linker geometry on equilibrium thermal and mechanical properties of monodomain nematic elastomers was studied [64]. Three types of crosslinks were analysed for elastomers containing the same siloxane backbone, mesogenic group and concentration of crosslinking group: short siloxane linkage, short aliphatic chains miscible with the mesogenic side chains, and long segments of main-chain nematic polymer. Very different results were found for spontaneous thermal expansion and anisotropic stress-strain response along and perpendicular to the uniform nematic director. The experimental data were compared quantitatively with theoretical models of nematic elastomers. The average effective anisotropy and hence the magnitude of spontaneous strain was greatly influenced by the increase of main chain LC crosslinker amount. Networks crosslinked by siloxane linkers have very low mechanical anisotropy and practically no soft stress regime on stretching perpendicular to the uniform nematic director n. Very short relaxation times were found for elastomers crosslinked with flexible units, while in the case of LC crosslinker a pronounced slowing down of all response processes was observed. The nematic elastomers described in this paper have a great potential for use in actuator design, in the region of low stress and extremely high deformations – over 300% [64].

The so-called „soft (semi-soft) elasticity" was studied [65, 66], with the specific question whether the dynamics of the mesogens and that of the network is synchronous with respect to a mechanical excitation. For this purpose experimental methods such as XRD, NMR, DMA, FT-IR have been employed. For example, the segmental orientation and mobility of LCEs with a monodomain structure as a response to external mechanical fields parallel and perpendicular to the initial director, was studied by FT-IR spectroscopy, based on the specificity of the absorption bands corresponding to the main chain network, the spacer and the mesogenic groups [67]. It was found that stretching the elastomer in a direction parallel to the initial mesogen orientation causes nearly no angular reorientation and only slight increases of the molecular order parameters. On the other hand, stretching the elastomer perpendicular to the initial mesogen orientation results in a threshold-like dependence: for elongation ratios below the critical value no reorientation and only negligible changes of the order parameters were observed, while for elongation ratios higher than the critical value the system became unstable.

Nematic LCEs have been studied [68-77], mainly due to their very important form change with temperature as a result of order-disorder transition of the mesogens. The most spectacular applications of this feature are in fabrication of artificial muscles and mechanical actuators [78, 79].

The nematic order parameter can be manipulated by temperature, by applying an external stress or electric field, but also by photo-irradiation. When photo-isomerizable groups, like N=N bonds are introduced into a nematic elastomer, there is a large photo-mechanical response. This effect was reported for the first time in a monodomain nematic elastomer by Finkelmann et al. [71]. They demonstrate a remarkable photo-elastic response for side chain LSCEs containing Disperse Orange 1 dye covalently linked.

Depending on the proportion and position of the photosensitive groups, different types and magnitudes of response were found [80]. The photoisomerization destabilizes the nematic phase, and the effect is enhanced near the critical temperature, being closely related to the reduction of the nematic order parameter. Consequently, a large uniaxial contraction of the sample takes place.

A nematic elastomer containing azobenzene groups in the side chain (thus avoiding the direct mechanical coupling to the network) was investigated. The photo-mechanical response was found to depend on temperature and was generally very long [81]. It was proven that the actuation speed is dominated by the photoisomerization dynamics, which is very slow in a dense polymer matrix.

A much larger effect was obtained when the dye was simply dissolved into the elastomeric sample [82]. Monodomain LCE samples which consist of siloxane backbone and mesogenic benzoic acid phenyl ester as side-chain units were prepared and crosslinked using a tri-functional cross-linker. Disperse Orange I molecules were disolved in the elastomer, knowing that they are able to align with the nematic director in liquid crystals. The authors demonstrate that in this way, the mechanical deformation as a consequence of visible light was large (bending of more than 60°) and very fast (tens of milliseconds), which is more than two orders of magnitude faster than in the case of dye-functionalized elastomer [71] or [81].

Even more, an astonishing observation was made, that when a laser spot irradiated from above a dye-doped LCE sample floating on water, the LCE "swims" away from the light, and thus the LCE sample and the fluid below it constitute an unusual motor.

The magnitude of changes in the dimensions of nematic liquid-single-crystal elastomers (LSCEs) with temperature is determined by the coupling between the nematic state of order and the conformation of the polymer main chains [79]. In nematic side chain polymers the backbone conformation is very little influenced, while in nematic main chain polymers, large conformational changes should occur.

When a nematic main chain polymer was used as a crosslinker in a monodomain nematic siloxane elastomer, its proportion influenced the uniaxial equilibrium thermal expansion on cooling from the isotropic state. At higher concentration, its behavior dominates the elastomer properties, while at lower concentration, the main chain and the side chain components have their own effect. The nematic order parameter is given by an average of the two components and the thermal expansion has a two-transition nature [83].

Nematic „co-elastomers" were synthesized, consisting of side-chain polysiloxane and nematic main-chain polyethers, which act as a macromolecular crosslinking component [79].

A mechanical field was applied during crosslinking, so that the director of the nematic phase becomes macroscopically uniformly aligned. These networks strongly modified their shape (by a maximum factor of 3.8) within a narrow temperature range, in the vicinity of the clearing point. This result was much better compared to the „classical" nematic LCEs and was progressivelly improved by the increase of the crosslinker amount. The very large spontaneous thermally induced reversible deformation is promising for applications as artificial muscles or mechanical actuators.

A significant electro-mechanical effect in a LCE system filed with a very low concentration of aligned carbon nanotubes has been reported [74]. The presumed mechanism of the response was the very large anisotropy of polarizability of the carbon nanotubes. When an electric field is applied perpendicular to the equilibrium orientation, the result is a high local torque, and the rotation of the nanotubes is transmitted to the nematic elstomer network. Thus, a uniaxial mechanical response is registered, which indicate potential applications of such composites as electrically driven actuators.

2.2. Cholesteric LCEs

Cholesteric liquid crystalline elastomers combine the characteristics of elastomers with the properties of cholesteric liquid crystals. Due to their helical superstructure, these elastomers exhibit interesting optical properties, such as selective reflection and angular dependence of the reflected wavelength, which obeys the Bragg law. Networks containing chiral side groups have been shown to exhibit piezoelectric and non-linear optical activity in the cholesteric and smectic C* phases.

Cholesteric LCEs may show memory effects, so that the reflection wavelength may be adjusted by thermal treatment and electrical field [84]. Such materials have applications as switchable smart windows for the control of solar light spectrum and white–or-black polarizer-free reflective displays.

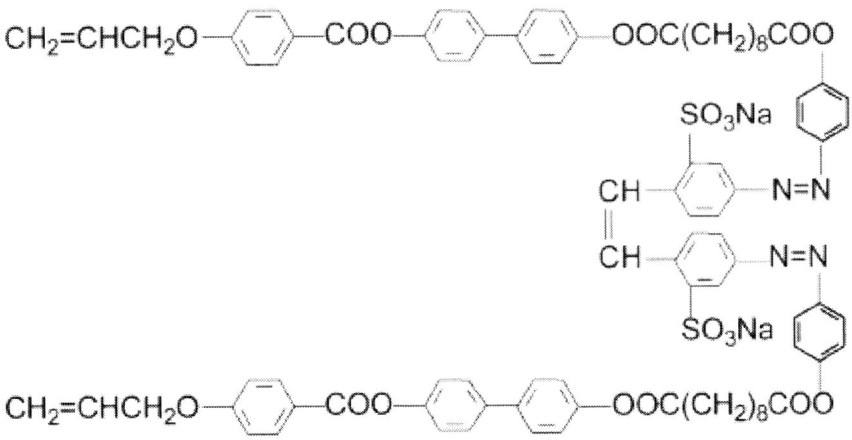

Scheme 9.

Side chain cholesteric LCEs containing flexible ethanediol-diundecylenate as crosslinking units and cholesteryl 4-allyloxybenzoate as mesomorphic units were synthesized and characterized [85]. The transition temperatures (Tg and Ti) decreased with the crosslinking density (here that means with the amount of flexible crosslinking agent). Cholesteric Grandjean texture was observed irrespective of the crosslinking density, and the crosslinking agent helped the adjustment of the cholesteric LC pitch. Reversible selective reflection was observed for highly crosslinked polymers.

Ionic LCEs combine the mechanical properties of ionic polymers with the mechanical orientability of conventional LCEs. Siloxane LCEs were obtained using chemical crosslinking agents with sulfonic acid groups [86]. Smectic or cholesteric mesophases were observed, but also two melting temperatures, corresponding to siloxane matrix and LC segments respectivelly, which denotes a microphase separation induced by ionic aggregates.

Siloxane-based LCEs containing biphenyl benzoate mesogenic units and ionic Brilliant Yellow moieties were prepared [62], using the bifunctional monomer shown in Scheme 9. Smectic mesophase was observed. The glass transition temperatures and melting points of the polymers increased slightly with the content of ionic and mesogenic groups, while the liquid crystal mesophase range of stability decreased slightly.

2.3. Smectic Elastomers

Smectic elastomers, due to their layered structure, exhibit distinct anisotropic mechanical properties and mechanical deformation processes parallel or perpendicular to the smectic layer normal, causing different responses of the networks [87].

The influence of molecular parameters (amount of mesogen, crosslinking density, crosslinker length) on the elastic and viscoelastic properties of polydomain smectic A side chain LCEs have been investigated [88]. It was found that the $\omega^{0.5}$ scaling law characterizing the isotropic phase and the $\omega^{0.3}$ scaling law characterizing the smectic phase are independent of the crosslinking density. The $\omega^{0.5}$ scaling law is also independent of the crosslinker length and the amount of mesogen (except for very small amounts), but the frecvency at which this law starts is dependent on the crosslinking density and crosslinker length. The purely elastic response of the elastomers is only visible in the isotropic phase.

In order to avoid perturbation of the liquid crystal orientation, a bifunctional unsaturated mesogen which was at the same time the crosslinking agent was used to obtain network LCPs with smectic mesophase [89]. For low contents of the mesogenic crosslinker, elastomeric behavior was observed, while elasticity was lost when higher amounts of crosslinker were used and thus thermosetting polymers were obtained.

Smectic elastomers were prepared by nonlinear polymerization of terephthalic acid, bis(4-allyloxyphenyl) ester mesogens with 1,1,3,3,5,5-hexamethyltrisiloxane and tetrakis(dimethylsiloxy)silane as a tetrafunctional cross-linker [38].

The dynamic mechanical response of unoriented, polydomain elastomers showed their possible use as vibration damping materials, in vibration isolation or impact absorption. Minor modifications in chemical structure or crosslinking density may tune the dynamic mechanical response.

In a chiral material, a static electric field applied in the plane of the smectic layers induces a tilt angle in the preferential orientation of the mesogens with respect to the layer normal. This electroclinic effect reduces the smectic layer thickness. The crosslinking of ferroelectric smectic polymers prevents the macroscopic flow and allows preparation of "true piezoelectric soft matter".

The electro-mechanical response of siloxane smectic ferroelectric elastomers has been investigated and rather different results have been obtained by interferometry and reflection microscopy [90, 91]. It seems that the film preparation and thickness are of great importance in such studies.

A series of LCEs has been prepared using a nematic mesogen and a non-mesogenic, chiral crosslinking agent [92]. Different LC phase behavior was observed, depending on composition. The effect of the crosslinking unit content on glass transition and mesophase range was investigated. Elastomers containing less than 15 mol% crosslinking agent displayed elasticity, reversible phase transitions with wide mesophase temperature ranges, and high thermal stability. The Tg values were influenced by the two anthagonic trends: plasticization and cross-linking, while the isotropization temperature was influenced by the non-mesogenic character of the crosslinking agent, which dilutes the LC phase and thus decreases the clearing point. The chirality of the crosslinking agent had a strong effect on the polymer phase, and favoured formation of the chiral smectic C phase.

A novel crosslinked smectic C main-chain liquid-crystalline elastomer has been reported very recently [87]. It has been obtained by hydrosilylation polyaddition using tetramethyldisiloxane as chain extender, pentamethylcyclopentasiloxane as crosslinking agent and vinyloxy-terminated mesogens. The use of the functional vinyloxy group eliminates the side reactions occuring in the case of vinyl groups, thus allowing the synthesis of well-defined networks with low soluble content and good mechanical properties. The SmC elastomer is a monodomain with respect to the director orientation and has a conical orientation distribution of the layer normal with respect to the director. This is the first main-chain smectic LCE analyzed with respect to the anisotropic mechanical properties, since this reliable source states that up to date only side-chain liquid-crystalline elastomers with SmA and SmC phases have been studied. The authors conclude that the basic difference between main- and side-chain elastomers is that in main-chain elastomers the mechanical field always couples to the tilt angle of the SmC phase structure. This is particularily important for chiral SmC* elastomers, where the tilt angle determines the optical and ferro-electrical properties.

2.4. Imprinted Networks

In molecular imprinting technique a polymer network is formed arround a template which is chemically linked with the polymer; after removal of the template, functional cavities capable of molecular recognition remain inside the network. In order to reduce the amount of crosslinkining agent, polymer networks that are at the same time liquid crystalline and molecularly imprinted have been synthesized using side-chain LC polysiloxanes and covalent or non-covalent linkages with the template molecules [93, 94]. The presence of the mesogenic groups allowed the decrease of the crosslinking agent amount from 70-80% to 5-10%, due to the specific interactions. The memory effect was also improved and better specificity towards the template as well as higher capacity were observed than in the non-imprinted polymers.

Chiral networks were obtained long time ago [95] by a template method using a chiral solvent, but they remain actual due to their „memory function". Interesting applications are related to stereo-selective separation. A more recent approach for the preparation of chiral networks is molecular imprinting, which uses as template a molecule that is associated with the polymer by chemical or physical interactions.

The introduction of mesogenic units reinforces the memory effect and allows that small amounts of crosslinking agent to be used. A chiral template was used to induce the helical structure inside a nematic LC network that has been macroscopically oriented and frozen by crosslinking [96]. Thus a molecular chirality and a phase chirality have been simultaneously imprinted. The imprinted polymer had a pronounced selectivity towards the template enantiomer. In this case, hydrogen bonds between carboxylic groups of the elastomer network and phenylalanine groups of the chiral template were used in the imprinting process.

In the same way, when theophyline was used as a template, high capacity molecularly imprinted polymers were obtained and they proved potential applications as antibody analogues [97]. The molecular separation is possible due to both conformational polymer micropores that are formed around templates and to the interactions between functional groups inside the cavities.

2.5. Miscellaneous

The subject of metal nanoparticles has been developed in the recent years, due to potential applications in microelectronics, catalysis, magnetical and optical devices. There are different approaches for their synthesis, one of them involving LC polymers or networks, either thermotropic or lyotropic. A polysiloxane-based LCE, in its polydomain and monodomain forms was used as a template for the preparation of gold nanoparticles [98] and the mutual effect of the mesomorphic matrix and the particles have been investigated. On the one hand, the presence of gold particles slightly decreased the mesophase stability of the LC state and, on the other hand, the order state of the matrix strongly affected the size and repartition density of the particles. Much smaller particles were obtained in the smectic phase than in the isotropic state. Polydomain elastomer gave larger particles with broader distribution.

3. Hyperbranched and Dendritic Structures

Dendrimers and structurally related multipodes represent a class of macromolecules with a perfectly branched uniform structure in contrast to the hyperbranched polymers in which the branching occurs in a random fashion [99].

Dendritic liquid crystals and multipodes are „large but discrete molecular systems created when many mesogenic feet are tied covalently to a central core (scaffold), with precisely defined molecular nature and topology" [100]. They combine the properties of low molecular weight LCs and of conventional polymeric structures, being at the same time monodisperse, which is a great advantage over the polymers due to the reproducibility of the properties and their independence on molecular weight. In such systems containing siloxane units, both sterical hindrance and nanosegregation determine the structure of the self-assembly phase.

A classification of the dendritic and hyperbranched LCPs was made [100]: - hyperbranched polymers and LC dendrimers having mesogenic groups in each branching unit; - dendrimers without mesogenic groups, in which the mesophases are formed by self-assembling aided by microphase segregation; - LC dendrimers in which the mesogenic units are located on the surface of the dendritic core.

LC functional dendrimers have attracted much interest in the recent years, due to their supramolecular chemistry and self-assembling properties and to their applications in nanoscience and biology. As industrial standards for optoelectronic devices are being continuously pushed forward, new LC materials with precisely tunable properties are continuously required. The precise control of the molecular architecture and the in-depth understanding of structure-properties relationships are critical for high technological applications [101].

The order parameters in LC systems are very important for the determination of the relationship between the molecular properties and the macroscopic properties. The polarized IR spectroscopy was used to determine the order parameters of an organo-siloxane tetrapode liquid crystal of zero dendritic order in its nematic and smectic phases [101].

Different types of really beautiful LC dendrimers and multipodes have been designed and investigated by Saez and Goodby and are summarized in ref. [100]. Structural variations of mesogens and spacers were studied in order to observe the influence on the LC behavior. Some interesting results have been obtained with siloxane-containing compounds. In symmetrical tetramers, the flexible but bulky disiloxane moiety disrupts the interaction between the aromatic cores, thus decreasing the stability of the mesophase. On the other hand, the replacement of one mesogen in a symetric polypede suppressed or diminished the LC order.

Octasilsesquioxane was used as a core with a rigid framework, with 8 primary radial branches for derivatization. In general, the nature of the mesogenic units, the topology of the attachment of the mesogens to the core, generation number and microphase segregation strongly influence the LC behavior. A dendritic structure (first generation) having cyanobiphenyl mesogens showed lower transition temperatures compared to the homologous zeroth generation dendrimer [102].

Octamers containing side-on chiral mesogens showed wide temperature range enantiotropic chiral nematic phases and glass transitions near room temperature [103]. The explanation for the glassy state and low melting points seems to be the reduced flexibility of the octasilsesquioxane central core, which prevents the close packing of the mesogens, and thus disrupts the order of the system.

Very impressive first generation dendrimeric structure containing 16 side-on attached mesogenic units has been obtained [104] and the following sequence of LC phases has been observed: chiral nematic, hexagonal disordered columnar and rectangular disordered columnar phases. It was shown that the silsesquioxane core and the relatively short methylene spacer help the formation of the hexagonal and rectangular disordered structures, probably through phase segregation of the siloxane cores and mesogenic units, but without fully decoupling of the mesogenic motion.

The so-called „mesogen-jacketed liquid crystalline polymers" (MJLCPs) have similar chemical structures with conventional side-chain LCPs, but the mesogenic groups are laterally attached to the backbone through their center of gravity with short or no spacers [105 and references therein]. The fact that the rigid side groups are densely linked to the flexible main chain makes the whole molecule to act as a semirigid entity that assembles into columnar phases.

A series of eight-arm star poly{2,5-bis[(4-methoxyphenyl)oxycarbonyl]styrene} were synthesized by atom transfer radical polymerization (ATRP) with octakis(2-bromo-2-methylpropionoxypropyldimethylsiloxy)-octasilsesquioxane as initiator [105] –Scheme 10.

The starlike architecture of the macromolecules was found to affect the mesophase structure. The phase behavior was a function of molecular weight. A hexatic columnar nematic phase (Φ_{HN}) was found for star polymers with higher Mn, while amorphous state was observed for lower Mn.

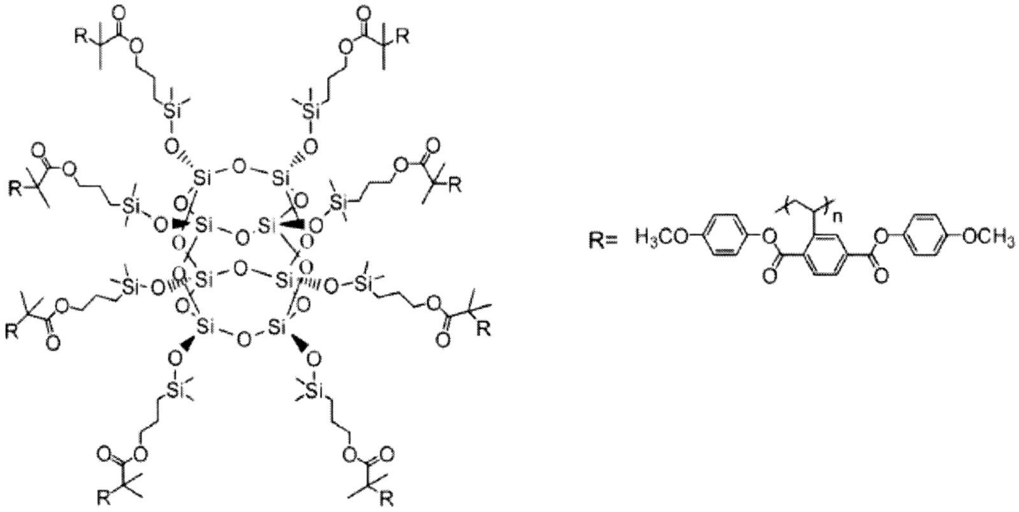

Scheme 10.

Janus dendrimers containing disiloxane units in the spacer were studied [106]. They are monodisperse materials, with enantiotropic chiral SmC* and chiral nematic or purely chiral nematic phases, depending on the topology.

The living anionic polymerization is an effective pathway for the synthesis of well-defined polymers with very low polydispersity. Among other functionalities, vinyl substituted polysiloxanes of almost any type of main chain topology—linear, comb-like, star shaped and dendritic systems have been obtained [107, 108]. Such compounds have been used in so-called *reverse hydrosilylation* of vinyl substituted polysiloxanes with $Si(Me)_2H$ terminated mesogens in order to obtain LC polysiloxanes with different shape and controlled structure [101].

The fact that these dendrimeric systems contain functional vinyl groups not only in the outer sphere but also inside the core, made it possible to achieve higher content of mesogenic moieties in a macromolecule, with respect to the main polymer skeleton, although important sterical hindrance precluded sometimes the complete addition of the mesogens.

It was found that star shape and dendritic LC structures exhibited various calamitic mesophases (SmA, N* and SmC*) depending on the type of mesogenic groups. On the other hand, in comblike structures, hexagonal phase was observed, even though they contain typical rod-like mesogenic moieties. The dendritic type topology of polymers led to narrower temperature range of mesophase and lower molecular order in comparison with analogous brush-like systems.

The incorporation of the siloxane group in multipodes based on disc-shaped mesogens resulted in strong decrease of transition temperatures, which led to room temperature nematic phase or suppression of the mesophase. In this last case, the formation of supramolecular structures –charge transfer complexes with 2,4,7-trinitro-9-fluorenone- induced the LC properties, with best results obtained for 1:1 donor-acceptor complexes [109]. This is an example of how non-covalent bonds can induce the liquid crystallinity, by increasing the system's order.

4. Polymer-Dispersed Liquid Crystals PDLCs

Polymer-dispersed liquid crystals (PDLCs) consist of micron-sized nematic droplets of low molecular weight LCs dispersed in an optically isotropic polymer matrix.

The most important applications of PDLCs are in electro-optical devices, like large area flat panel displays, privacy windows, electro-optic shutters [110]. LCs dissolved in a solvent or dispersed in a carrier polymer are also used in coatings, dyes or analytical chromatography stationary phases.

PDLCs are prepared by premixing a LC with a monomer, which is capable of curing. The two-phase morphology that results in the preparation process is crucial from the technological point of view, since it controls the electro-optical properties. This is the reason why in the last years great attention has been given to the miscibility or interface morphology in PDLC systems.

Polysiloxanes as the polymer matrix in these systems have some advantages over their counterparts –polyacrylate and polymethacrylate-, namely the greater flexibility and low glass transition temperature.

The interactions in LCs dispersed in a polysiloxane fluid were studied and quantified in a few papers [111-113]. Inverse gas chromatographic (IGC) measurements were done using a LC (4-(n-hexyloxy)-4'-cyanobiphenyl, HCB) dispersed in PDMS and the results were compared with the retention properties of a side-chain LC polysiloxane bearing the same mesogen. The retention of the LC polymer was significantly different from a mixture with the same composition. DSC and IGC values of the transition temperatures indicate that PDMS does not blend with HCB and does not affect the melting temperature [111].

Phase diagrams of blends of poly(methylphenylsiloxane) and two low molecular weight liquid crystals (namely 4-cyano-4'-*n*-pentyl-biphenyl and an eutectic mixture of paraphenylenes, E7) were obtained by polarized optical microscopy [112]. It was found a good agreement between the theory (Flory–Huggins theory of isotropic mixtures and the Maier–Saupe theory of nematic order) and experiments and a pronounced enhancement of miscibility with respect to similar systems involving poly(dimethylsiloxane).

A remarkable increase of the nematic to isotropic transition temperature T_{NI} was found for systems containing the eutectic mixture E7, which was interpreted as a tendency toward preferential solvation. Evidence that support this hypothesis was brought by HPLC studies

[113]. LC composition of nematic droplets was analyzed and it was observed that this composition differs clearly from that of bulk LC.

Polymer dispersed ferroelectric LCs proved to be more efficient for display applications than the nematic based systems due to their faster response and superior contrast ratio and viewing angle.

Polysiloxane- dispersed ferroelectric liquid crystals composites were prepared by solvent-induced and polymer-induced phase separation techniques [114]. The droplets size increased with increasing viscosity and the most uniform size and distribution was obtained for similar densities of the LC and polysiloxane. The film morphology and electro-optical properties were also influenced by the polymer viscosity. The spontaneous polarization decreased and the response time increased with increasing PDMS viscosity. The switching time of these composite systems was in the range of microseconds.

Conclusion

Although they have a history of over 100 years, liquid crystals find more and more their way towards high technology applications. This is partly due to the developments made in their structure, and especially to different kinds of polymeric architectures, that bring improved properties. Siloxane-containing LCs are an important part of this challenging research trend, being present in practically all the great discoveries of last generation technique. Either we discuss about low molecular weight LCs, or main chain, side chain, hyperbranched or crosslinked structures, the siloxane moieties have something interesting to offer and this adventure continues...

Acknowledgement

The financial support from Romanian Ministry of Education and Research, under grant no. 5/28.09.2007 („Idei" Program, ID_233) is greatefully ackowledged.

References

[1] Teyssier, D.; Boileau, S. in *Silicon-Containing Polymers,* Jones, R.G.; Ando, W.; Chojnowski, J. (eds.), Kluver Academic Publishers, Dordrecht, Boston, London, 2000, Chapter 22, 593-613.

[2] Shashidhar, R.; Geer, R. E. *International Journal of Engineering Science* 2000, *38*, 1049-1056.

[3] Rössle, M.; Braun, L.; Schollmeyer, D.; Zentel, R.; Lagerwall, J. P. F.; Giesselmann, F.; Stannarius, R. *Liq. Cryst.* 2005, *32*, 533-538.

[4] Rössle, M.; Zentel, R.; Lagerwall, J. P. F.; Giesselmann, F. *Liq. Cryst.* 2004, *31*, 883-887.

[5] Diele, S.; Oelsner, S.; Kuschel, F.; Hisgen, B.; Ringsdorf, H.; Zentel, R. *Makromol. Chem.* 1987, *188,* 1993-2000.

[6] Gardiner, D. J.; Coles, H. J. *J. Appl. Phys.* 2006, *100,* 124903-124911.
[7] Gardiner, D. J.; Coles, H. J. *J. Phys. D: Appl. Phys.* 2006, *39,* 4948-4955.
[8] Moment, A.; Hammond. P. T. *Polymer* 2001, *42,* 6945-6959.
[9] Shepperson, K. J.; Meyer, T.; Mehl, G. H. *Mol. Cryst. Liq. Cryst.* 2004, *411,* 1227-1233.
[10] Elsaesser, R.; Mehl, G H.; Goodby, J. W.; Photinos, D. J. *Chem. Commun.* 2000, 169, 851-852.
[11] Bracon, F.; Guittard, F.; Taffin de Givenchy, E.; Géribaldi, S. *Polymer* 2000, *41,* 7905-7913.
[12] Mu, J.; Okamoto, H.; Yanai, T.; Takenaka, S.; Feng, X. *Colloids and Surfaces A: Physicochemical and Engineering Aspects* 2001, *181,* 303-313.
[13] Hsiue, G. H.; Sha, Y. A.; Jeng, R. J. *Macromol. Chem. Phys.* 2001, *202,* 287-296.
[14] Robinson, W. K.; Kloess, P. S.; Carboni, C.; Coles, H. J. *Liq. Cryst.* 1997, *23,* 309-312.
[15] Robinson W. K.; Carboni , C.; Kloess, P.; Perkins, S. P.; Coles , H. J. *Liq. Cryst.* 1998, *25,* 301-307 .
[16] Noot, C.; Perkins, S. P.; Coles, H. J. *Ferroelectics* 2000, *244,* 331-338.
[17] Olsson, N.; Andersson, G.; Helgee, B.; Komitov, L. *Liq. Cryst.* 2005, *32,* 1125–1138.
[18] Olsson, N.; Dahl, I.; Helgee, B.; Komitov, L. *Liq. Cryst.* 2004, *31,* 1555–1568.
[19] Pivnenko, M. N.; Lehmann, P.; Komitov, L.; Coles, H. J. *Liq. Cryst.* 2005, *32,* 173–181.
[20] Guillon, D.; Osipov, M.A.; Méry, S.; Siffert, M.; Nicoud, J.-F.; Bourgogne, C.; Sebastião, P. *J. Mater. Chem.*, 2001, *11,* 2700–2708.
[21] Hsiue, G. H.; Sha, Y. A.; Hsieh, S. J.; Jeng, R. J.; Kuo, W. J. *Liq. Cryst.* 2001, *28,* 365-374.
[22] Zhang, B. Y.; Meng, F.B.; He, X. Z.; Lin, D. *Liq. Cryst.* 2005, *32,* 1161-1167.
[23] Hu, J. S.; Zhang, B.Y.; Pan, W.; Zhou, A. J. *Liq. Cryst.* 2005, *32,* 441-447.
[24] Wang, Y.; Zhang, B. Y.; He, X. Z.; Wang, J. W. *Colloid Polym. Sci.* 2007, *285,* 1077-1084.
[25] Lin, C.H.; Hsu, C. S. *Polym. Bull.* 2000, *45,* 53–59.
[26] Kimura, K.; Zhuang, J.H.; Shirabe, K.; Yamashita, Y. *Polymer* 2003, *44,* 4761–4764.
[27] Matsumoto, T.; Matsuoka, T.; Suzuki, Y.; Miyazawa, K.; Kurosaki, T.; Mizukami, T. *Macromol. Symp.* 2003, *199,* 83–96.
[28] Morgan, P.W.; Kwolek, S.L.; Pletcher, T.C. *Macromolecules* 1987, *20,* 729–739.
[29] Wojtkowski, P.W. *Macromolecules* 1987, *20,* 740–748.
[30] Iwan, A.; Sek, D. *Prog. Polym. Sci.* 2007, *doi: 10.1016/j.progpolymsci.2007.09.005*
[31] Racles, C.; Cozan, V. *High Perform. Polym.* 2002, *14,* 169-181.
[32] Racles, C.; Cozan, V.; Cazacu, M.; Foldes, E.; Sajo, I. *High Perform. Polym.* 2002, *14,* 397-413.
[33] Vasiliu, M.; Cazacu, M.; Marcu, M.; Ioan, S. *Polym. Plast. Techn. Eng.* 2005, *44,* 993-1002.
[34] Racles, C.; Cazacu, M.; Vasiliu, M.; Cozan, V. *Polym. Plast. Techn. Eng.* 2005, *44,* 1049-1058.
[35] Cazacu, M.; Marcu, M.; Vlad, A.; Toth, A.; Racles, C. *J. Polym. Sci. Part A –Polym. Chem.*, 2003, *41,* 3169-3179.
[36] Racles, C.; Cozan, V.; Sajo, I. *High Perform. Polym.*, 2007, *19,* 541-552.
[37] Bronnikov, S.; Racles, C.; Nasonov, A.; Cazacu, M. *Liq. Cryst.* 2006, *33,* 1015–1019.

[38] Patil, H.P.; Liao, J. Hedden, R.C. *Macromolecules* 2007, *40,* 6206-6216.
[39] Kozlovsky, M. V. *J. Polym. Sci. B: Polym. Phys.* 2001, *39,* 1055-1060.
[40] Kozlovsky, M. V.; Skarp, K. *J. Polym. Sci. B: Polym. Phys.* 2005, 43, 1779-1784.
[41] Lehn, J. M. *Supramolecular Chemistry: Concepts and Perspectives,* VCH, Weinheim, 1995.
[42] Armstrong, G.; Buggy, M. *J. Mater. Sci.* 2005, *40,* 547– 559.
[43] Parra, M.; Hidalgo, P.; Alderete, J. *Liq. Cryst.* 2005, *32*, 449-455.
[44] Lee, M.; Cho, B.K.; Kang, Y.S.; Zin, W.C. *Macromolecules* 1999, *2*, 8531-8537.
[45] Han, H.; Molla, A.H.; Bhowmik, P.K. *Polym. Prepr.* 1995, *36*, 332-333.
[46] Bhowmik, P.K.; Wang, X.; Han, H. *J. Polym. Sci., Part A: Polym. Chem.* 2003, *41*, 1282-1295.
[47] Nishikawa, E.; Samulski, E. T. *Liq. Cryst.* 2000, *27*, 1457-1462.
[48] Nishikawa, E.; Samulski, E. T. *Liq. Cryst.* 2000, 27, 1463-1471.
[49] Brunsveld, L.; Folmer, B. J. B.; Meijer, E. W.; Sijbesma, R. P. *Chem. Rev.* 2001, *101*, 4071-4097.
[50] Racles, C. Cazacu, M. *J. Appl. Polym. Sci.*, 2008, *in press*.
[51] Matsumoto, K.; Arai, K.; Mihara, T.; Koide, N. *Mol. Cryst. Liq. Cryst.* 2004, *411*, 1481–1492.
[52] Matsumoto, K.; Koide, N. *Mol. Cryst. Liq. Cryst.*, 2001, *365*, 81–89.
[53] Zhang, B.-Y.; Hu, J.-S.; Yang, L.-Q.; He, X.-Z.; Liu, C. *Eur. Polym. J.* 2007, *43*, 2017-2027.
[54] Guo, G.; Ba, C.; Li, H.; Zhou, X.; Xie, P.; Zhang, R. *Liq. Cryst.* 2002, *29,* 1247-1250.
[55] Racles, C.; Cozan, V.; Cazacu, M. *High Performance Polymers* 2003, *15*, 231-242.
[56] Racles, C.; Airinei, A.; Cozan, V.; Cazacu, M.; Sajo, I.E. *J. Appl. Polym. Sci.* 2003, *90*, 3093–3099.
[57] Racles, C.; Filip, D.; Cazacu, M.; Cozan, V.; Toth, A.; Ioanid, G. *Macromol. Chem. Phys.* 2006, *207*, 1599–1609.
[58] Yao, D. S.; Zhang, B. Y.; Sun, Q. J.; Zhang, L. F. *J. Appl. Polym. Sci.* 2005, *95*, 946-952.
[59] Chan, W. L.; Yana, H.; Szeto, Y. S. *Mater. Lett.* 2004, *58*, 882-884.
[60] Yang, I.-K.; Chang, S. H. *J. Polym. Res.* 2002, *9,* 163-168.
[61] Finkelmann, H., Kock, H.J. *Macromol. Chem. Rapid Commun.* 1981, *2*, 317-322.
[62] Meng, F.B., Zhang B.Y., Jia, Y.G., Yao, D.S., *Liq. Cryst.* 2005, *32*, 183-189.
[63] Kupfer, J.; Finkelmann, H. *Macromol. Chem. Phys.* 1994, *195*, 1353-1367.
[64] Clarke, S. M.; Hotta, A.; Tajbakhsh, A. R.; Terentjev, E. M. *Phys. Rev. E* 2001, *64*, 061702.
[65] Terentjev, E.M.; Warner, M. *Eur. Phys. J. E* 2004, *14*, 323-327.
[66] Lubensky, T.C.; Stenull, O. *Eur. Phys. J. E* 2004, *14*, 333-337.
[67] Li, J.; Tammer, M.; Kremer, F.; Komp, A.; Finkelmann H. *Eur. Phys. J. E* 2005, *17*, 423-428.
[68] Greve, A.; Finkelmann, H. *Macromol. Chem. Phys.* 2001, *202,* 2926-2946.
[69] Schonstein, M.; Stille, W.; Strobl, G. *Eur. Phys. J. E* 2001, *5*, 511-517.
[70] Clarke, S. M.; Tajbakhsh, A. R.; Terentjev, E. M.; Warner, M. *Phys. Rev. Lett.* 2001, *86*, 4044-4047.
[71] Finkelmann, H.; Nishikawa, E.; Pereira, G. G.; Warner, M. *Phys. Rev. Lett.* 2001, *87*, 015501.

[72] Conti, S.; Desimone, A.; Dolzmann, G. *Phys. Rev. E.* 2002, *66,* 061710.
[73] Finkelmann, H.; Greve, A.; Warner, M. *Eur. Phys. J. E* 2001, *5*, 281-293.
[74] Courty, S.; Mine, J.; Tajbakhsh, A.R.; Terentjev, E. M. *Europhys. Lett.* 2003, *64,* 654-660.
[75] Stenull, O.; Lubensky, T.C. *Phys. Rev. E.* 2004, *69*, 021807.
[76] Martinoty, P.; Stein, P.; Finkelmann, H.; Pleiner, H.; Brand, H.R. *Eur. Phys. J. E* 2004, *14*, 311-321.
[77] Simoes, M.; De Campos, A. *Phys. Rev. E.* 2004, *69*, 061704.
[78] deGennes, P.; Hebert, M.; Kant, R. *Macromol. Symp.* 1997, *113*, 39-49.
[79] Wermter, H.; Finkelmann, H. *e-Polymer,* 2001, *013* http://www.e-polymer.org
[80] Hogan, P. M.; Tajbakhsh, A. R.; Terentjev, E. M. *Phys. Rev. E* 2002, *65*, 041720-041729.
[81] Cviklinski, J.; Tajbakhsh, A. R.; Terentjev, E. M. *Eur. Phys. J. E*, 2002, *9,* 427-434.
[82] Camacho-Lopez, M.; Finkelmann, H.; Palffy-Muhoray, P.; Shelley, M. *electronic-Liquid Crystal Communications* November 26, 2003, http://www.e-lc.org/docs/2003_11_25_00_32_43
[83] Tajbakhsh, A. R.; Terentjev, E. M. *Eur. Phys. J. E –Soft Matter* 2001, *6*, 181-188.
[84] Mitov, M.; Nouvet, E.; Dessaud, N. *Eur. Phys. J. E*, 2004, *15*, 413-419.
[85] Zhang, B. Y.; Jia, Y. G.; Hu, J.S.; Meng, F. B. *Liq. Cryst.* 2004, *31,* 387-392.
[86] Meng, F. B.; Zhang, B. Y.; Liu, L. M.; Zang, B. L. *Polymer,* 2003, *44,* 3935-3943.
[87] Sanchez-Ferrer, A.; Finkelmann, H. *Macromolecules*, Published on Web 01/05/2008
[88] Zanna, J. J.; Stein, P.; Marty, J. D.; Mauzac, M.; Martinoty, P. *Macromolecules* 2002, *35,* 5459-5465.
[89] Jia, Y. G.; Zhang, B. Y.; Zhou, E.; Feng, Z. L.; Zang, B.L. *J. Appl. Polym. Sci.* 2002, *85,* 1104-1109.
[90] Köhler, R.; Stannarius, R.; Tolskdorf, C.; Zentel, R. *Appl. Phys. A* 2005, *80*, 381–388.
[91] Skupin, H.; Tolksdorf, C.; Gebhard, E.; Zentel, R.; Krüger, P.; Lösche, M.; Kremer, F. *Nature* 2001, 410, 447-450.
[92] He, X. Z.; Zhang, B. Y.; Sun, Q. J.; Lu, H. W.; Li. L. *Liq. Cryst.* 2005, *32*, 431–439.
[93] Marty, J. D.; Tizra, M.; Mauzac, M.; Rico-Lattes, I.; Lattes, A. *Macromolecules* 1999, *32,* 8674-8677.
[94] Marty, J. D.; Mauzac, M.; Fournier, C.; Rico-Lattes, I.; Lattes, A. *Liq. Cryst.* 2002, *29,* 529-536.
[95] De Gennes, P.G. *Phys. Lett. A* 1969, *28,* 765.
[96] Marty, J. D.; Gornitzka, H.; Mauzac, M. *Eur. Phys. J. E* 2005, *17,* 515-520.
[97] Marty, J. D.; Mauzac, M.; Lavabre, D. *Mol. Cryst. Liq. Cryst.* 2005, *437,* 1307-1314.
[98] Gascon, I.; Marty, J. D.; Gharsa, T.; Mingotaud, C. *Chem. Mater.* 2005, *17,* 5228-5230.
[99] Merkel, K.; Kocot, A.; Vij, J. K.; Mehl, G. H.; Meyer, T. *J. Chem. Phys.* 2004, *121*, 5012-5021.
[100] Saez, I. M.; Goodby, J. W. *Liq. Cryst. Today* 2004, *13*, 1–15.
[101] Ganicz, T.; Pakula, T.; Fortuniak, W.; Białecka-Florjanczyk, E. *Polymer* 2005, *46,* 11380–11388.
[102] Saez, I. M.; Goodby, J. W. *Liq. Cryst.* 1999, 26, 1101-1105.
[103] Saez, I. M.; Goodby, J. W. *J. Mater. Chem.* 2001, *11*, 2845-2851.
[104] Saez, I. M.; Goodby, J. W.; Richardson, R. M. *Chem. Eur. J.* 2001, *7,* 2758-2764.
[105] Pan, Q.; Gao, L.; Chen, X.; Fan, X.; Zhou, Q. *Macromolecules* 2007, *40,* 4887-4894.

[106] Saez, I. M.; Goodby, J. W. *Chem. Commun.* 2003, 1726-1727.
[107] Chojnowski, J.; Cypryk, M.; Fortuniak, W.; Rozga-Wijas, K.; Scibiorek, M. *Polymer* 2002, *43*, 1993–2001.
[108] Chojnowski, J.; Cypryk, M.; Fortuniak, W.; Scibiorek, M.; Rozga-Wijas, K. *Macromolecules* 2003, *36*, 3890–3897.
[109] Kouwer, P. H. J.; Jager, W. F.; Mijs, W. J.; Picken, S. J.; Shepperson, K. J.; Mehl, G. H. *Mol. Cryst. Liq. Cryst.* 2004, *411,* 1419-1427.
[110] Justice, R. S.; Schaefer, D. W.; Vaia, R. A.; Tomlin, D. W.; Bunning, T. J. *Polymer* 2005, *46,* 4465-4473.
[111] Shillcock, I. M.; Price, G.J. *Polymer* 2003, *44*, 1027–1034.
[112] Gogibus, N.; Benmouna, F.; Ewen, B.; Pakula, T.; Coqueret, X.; Benmouna, M.; Maschke, U. *J. Polymer Science: Part B: Polym. Phys.*, 2003, *41*, 39–43.
[113] Bedjaoui, L.; Gogibus, N.; Ewen, B.; Pakula, T.; Coqueret, X.; Benmouna, M.; Maschke, U. *Polymer* 2004, *45,* 6555–6560.
[114] Sumana, G.; Raina, K. K. *J. Appl. Polym. Sci.*, 2004, *94,* 159-166.

In: Advances in Functional Heterochain Polymers
Editor: Maria Cazacu
ISBN: 978-1-60456-598-0

Chapter 3

THERMOTROPIC LIQUID CRYSTALLINE POLYAZOMETHINES

Vasile Cozan[1] and Luminita Marin[2]
"Petru Poni" Institute of Macromolecular Chemistry,
Aleea Gr. Ghica Voda 41A, Iasi, 700487, Romania

ABSTRACT

Aromatic polyazomethines are the most available thermotropic liquid crystalline (TLC) materials, which can be used in optics and optoelectronics. Polymers containing both rigid azomethine units and amorphous polysulfone segments – called poly(azomethine-sulfone)s – are interesting because of the proper contribution of each unit to their properties: good thermal and mechanical resistance, increased level of extended conjugation, good solubility and improved processability, etc. The present chapter describes mainly the influence of structural features on the thermotropic behavior in case of some polyazomethines and poly(azomethine-sulfone)s.

ACRONYMS

AM - azomethine
DMF – N,N-dimethylformamide
DMSO – N,N-dimethylsulfoxid
DSC – differential scanning calorimetry
LC – liquid crystalline
NMP – N-methyl-2-pyrrolidone
PSF – polysulfone
PAM – polyazomethine
OPM – optical polarizing light microscopy

[1] vcozan@icmpp.ro.
[2] lmarin@icmpp.ro.

PI – polydispersity index (PI = M_w/M_n)
TGA – thermogravimetrical analysis
THF - tetrahydrofurane
TLC – thermotropic liquid crystalline
WAXD – wide angle x-ray diffraction

1. Introduction

Polyazomethines, polyimines, or *poly(Schiff bases)* are well-known polymers having high thermal stability and chemical structures built by alternation of azomethine (-CH=N-) groups and phenylene rings in their backbone. The main characteristic of these materials is that they have electronic delocalization along the backbone, which confers interesting physical and chemical properties, such as: good thermal stability and mechanical strength, good environmental resistance and stability in hot acids, liquid crystalline and fiber-forming properties, interesting non-linear optical semiconductor properties, etc. During time, all these attractive features led to various applied researches like battery anodes or cathodes, semiconductors, advanced technology materials, energy storage, conversion devices and integrated electro-optics for switching and display.

Initially this review was thought to be dedicated to the reported data about thermotropic polyazomethines, but the last hour online publication of a nice review about polyazomethines and polyketanils by Iwan and Sek [1] obliged us to change somewhat the idea of the review, avoiding thus the inherent uninspired repetitions. So, this review will focus especially on our own contributions in thermotropic poly(azomethine-ether)s and poly(azomethine-sulfone)s, which were not discussed by Iwan and Sek. We are glad to recommend the review of Iwan and Sek for all the other aspects (historic of synthesis, properties and applications etc.) of polyazomethines.

Two main categories of semirigid thermotropic polyazomethines have been discussed in the present work: poly(azomethine-ether)s and poly/copoly(azomethine-ether-sulfone)s. Poly(azomethine-ether)s containing both calamitic azomethine mesogens and non-mesogenic bisphenol A units proved to be an interesting example of how molecular weight of the polymer could influence its LC behavior [5]. By introducing aromatic sulfone moieties, besides azomethine mesogens, a variety of poly(azomethine-ether-sulfone)s have been prepared and investigated [3, 19, 37, 44, 55]. In spite of unfavourable effects given by the kinked diphenylsulfone units, distinct LC properties were recorded for each item and appropriate comments related to their structure-LC property relationships have been made.

2. Poly(Azomethine-Ether)s – Influence of the Molecular Weight on the Lc Behavior

The study of thermotropic polymers which contain disruptive kinking groups demonstrated the favorable effect of such groups on the decreasing of the transition temperatures and solubility, without affecting the formation of the liquid crystalline (LC) state [2, 3]. On the other hand, a factor which affects dramatically the transition temperatures of the LC state is the molecular weight of the polymer [4]. The analysis of the literature data gave us the conclusion that molecular weight of the polymer influences selectively its mesomorphism, depending on the intrinsic chemical structure of the polymeric chain. Starting from this assumption we synthesized, fractioned and thermally characterized a poly(azomethine-ether) by Williamson polycondensation of the bisphenol A with a thermotropic monomer oxo-bis(4-chlorobenzylideneiminophenylene [5] (see Schemes 1 and 2). The polymer PAM 0 was fractionated into four polymeric fractions PAM 1–PAM 4 by extraction with methanol (Scheme 3). In view to confirm the synthetic pathway and to understand better the physico-chemical properties of the polymeric fractions PAM 1–PAM 4, a model compound (M 2) was synthesized by the reaction of the monomer M 1 with 4-*tert*-butylphenol (Scheme 2).

The UV-vis spectra of samples M 1, M 2 and PAM 1 – PAM 4 recorded in DMF solutions, showed an intense absorption band in the range 334 nm and 340 nm due to the electronic transitions π- π^* of the conjugated system Ph-CH=N-Ph (Table 1). Similar absorption maxima have been reported for polyazomethines containing benzylideneketone moieties [6]. By measuring the inherent viscosity of DMF solutions of the polymeric fractions PAM 1 – PAM 4, gave values comprised between 0.13 dL/g and 0.95 dL/g. The glass transition temperatures determined from the DSC curves showed values ranging from 35 °C (PAM 1) up to 100 °C (PAM 4) (Table 3). These data, associated with the different solubility of the polymeric fractions into methanol, suggested the following series ordered by the molecular weight: PAM 1 < PAM 2 < PAM 3 < PAM 4. Later measurements by gel permeation chromatography (GPC) in chloroform gave M_n values comprised between 1900 (PAM 1) and 3700 (PAM 4), which confirmed the proposed series (Table 1).

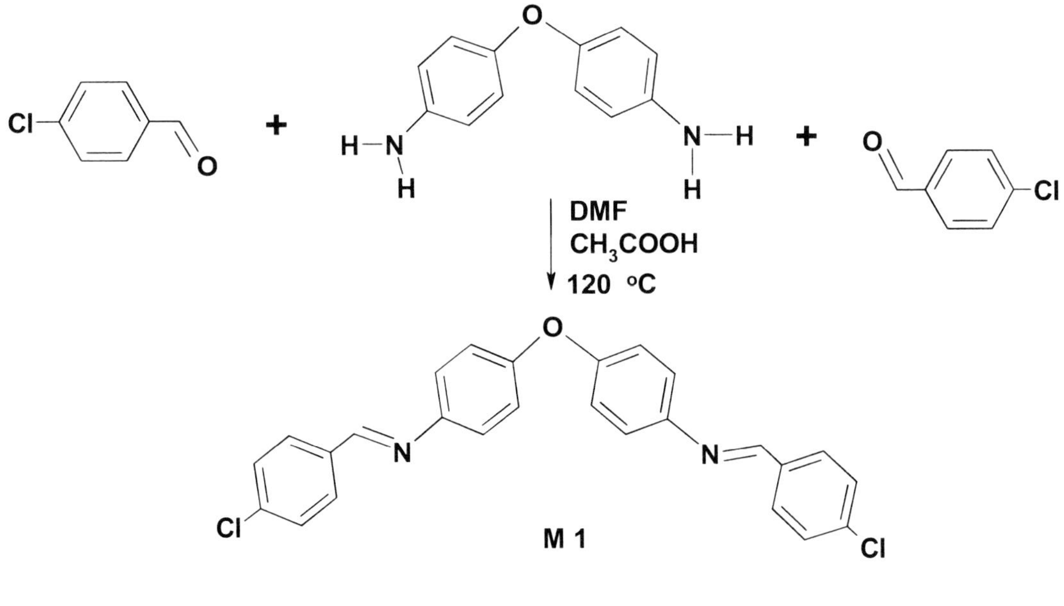

Scheme 1. Synthesis of the azomethine monomer M 1.

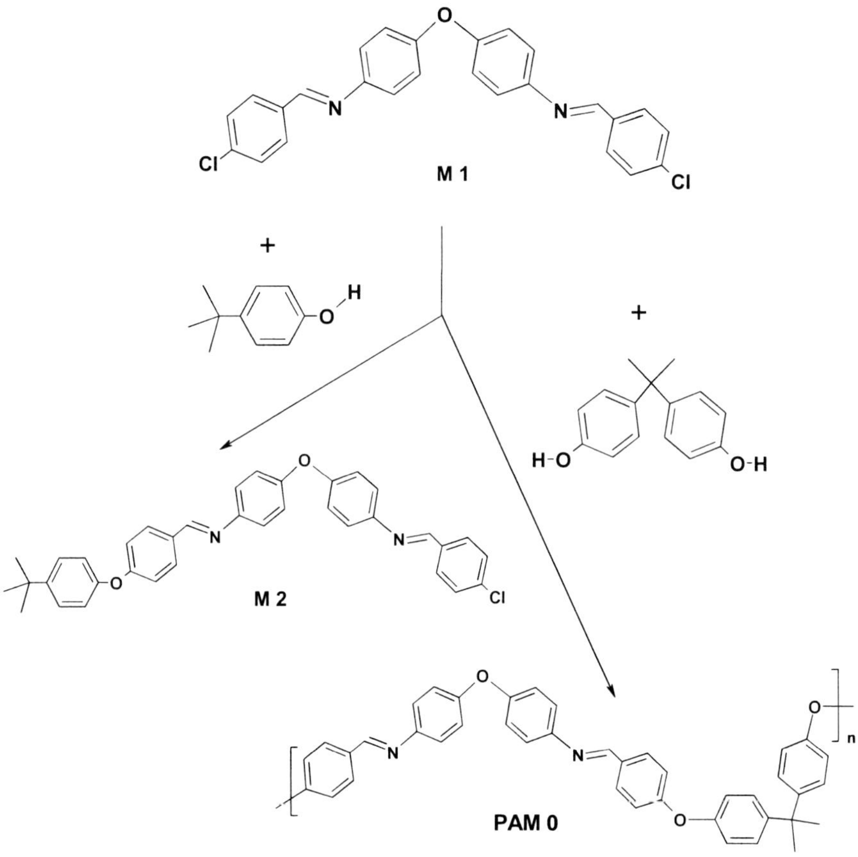

Scheme 2. Synthesis of the model compound M 2 and polymer PAM 0.

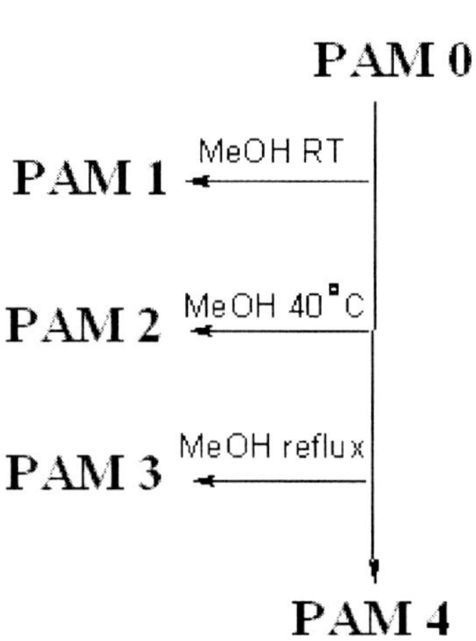

Scheme 3. The fractionation of the polymer PAM 0 into polymeric fractions PAM 1 – PAM 4 by extraction with methanol..

Table 1. Physico-chemical characteristics of the model compound M 2 and the polymeric fractions PAM 1 – PAM 4

Code	λ^{a}_{max} (nm)	η^{a}_{inh} (dLg^{-1})	$M_n^{\ b}$	PI^c	$T_0^{\ d}$ (°C)	$T_{10}^{\ d}$ (°C)
M 2	334	-	-	-	190	350
PAM 1	340	0.13	1900	1.65	123	231
PAM 2	338	0.42	2500	1.85	131	273
PAM 3	337	0.59	2900	1.68	169	312
PAM 4	334	0.95	3700	1.44	193	351

[a] Measured in DMF; [b] From GPC measurements in chloroform; [c] PI: polydispersity index (PI = M_w / M_n); [d] From TGA measurements in air: T_0 = onset temperature of the decomposition; T_{10} = temperature of the 10 % wt. loss.

Table 2. Solubilities of the monomer M 1, model compound M 2 and the polymeric fractions PAM 1 – PAM 4

Code	MeOH	DMSO	DMF	NMP	$CHCl_3$	Acetone	CCl_4
M 1	-[a]	+(h)	+(h)	+(h)	-	-	-
M 2	-	+	+	+	+	+	-
PAM 1	+	+	+	+	+	+	-
PAM 2	+(h)	+	+	+	+	±	-
PAM 3	+(h)	+	+	+	+	±	-
PAM 4	-	+	+	+	+	-	-

[a] +: soluble at the room temperature; ±: partially soluble at the room temperature; -: insoluble at the room temperature; +(h): soluble at slow heating.

The solubility of the monomer M 1, model compound M 2 and polymeric fractions PAM 1 – PAM 4 was tested in various common organic solvents and the data are presented in table 2. All the polymers showed excellent solubility in aprotic dipolar solvents like DMF, DMSO, or NMP. In other solvents (methanol, acetone, carbon tetrachloride) the polymers presented poor solubility, except for PAM 1. The monomer M 1 was soluble only in DMSO, DMF and NMP at heating. The model compound M 2 was soluble in all the solvents, with the exception of methanol. The better solubility of the polymers and of the model compound in comparison with that of the monomer could be explained by the presence of kinked or voluminous units in the chemical structure: bisphenol A for polymers, and *tert*-butyl in case of the model M 2. Additionally, the presence of a great number of ether-aromatic linkages in the polymeric chain and in the model compound, versus monomer, could affect favourably the solubility [7].

The TGA analysis showed good thermal stability of the polymeric fractions due to the presence of the azomethine groups in their backbone. At the isotropization temperature level it was observed only 0.1 % - 0.2 % wt. loss which practically didn't affect the mesophase formation.

Table 3. Thermotropic behavior of the monomer M 1, model compound M 2 and the polymeric fractions PAM 1 – PAM 4

Code	Transition temperatures OPM (° C)	DSC		
		Rate heating/cooling (°C/min)	Transition Temperatures (°C)	Corresponding enthalpies[b] (J / g)
M 1	Cr[a] 243 I I 226 LC[c]	1H[a] / 5	Cr 243 I	ΔH_m : 98.6
		1C / 5	I 231.7 LC[c]	$\Delta H_{i\text{-}lc}$: 136.8
		2H / 10	Cr 242.7 I	-
		2C / 10	I 230.6 LC[c]	-
		3H / 15	Cr 245.4 I	ΔH_m : 97.0
		3C / 15	I 227.2 LC[c]	$\Delta H_{i\text{-}lc}$: 114.0
M 2	Cr 135 LC 195 I I 186 LC[c] ΔT^d : 60	1H[a] / 1	Cr 133 LC 190 I	$\Delta H_m + \Delta H_{lc\text{-}i}$: 26.0
		1C / 1	I 181 LC[c]	$\Delta H_{i\text{-}lc}$: 20.1
		2H / 5	Cr 133 LC 191 I	$\Delta H_m + \Delta H_{lc\text{-}i}$: 34.2
		2C / 5	I 176 LC[c]	$\Delta H_{i\text{-}lc}$: 38.4
		3H / 15	Cr 133 LC 193 I	.$\Delta H_m + \Delta H_{lc\text{-}i}$: 39.8
		3C / 15	I 172 LC[c]	$\Delta H_{i\text{-}lc}$: 36.6
PAM 1	Cr 85 LC 113 I ΔT^d : 20	2H / 10	G 35 Cr 95 LC 115 I	ΔH_m : 1.9
PAM 2	Cr 105 LC 121 I ΔT^d : 23	1H / 10 2H / 10	Cr 105 LC 128 I G 62 Cr	ΔH_m : 6.6
PAM 3	Cr 126 LC 165 I ΔT^d : 32	2H / 10	G 78 Cr 135 LC 167 I	ΔH_m : 11.0
PAM 4	Cr 165 LC 207 I ΔT^d : 45	2H / 10	G 100 Cr 162 LC 207 I	ΔH_m : 24.9

[a] G = glassy state; Cr = crystalline state; LC = liquid crystalline state; I = isotropic state; 1H = first heating cycle; 1C = first cooling cycle; [b] Enthalpy subscript: m = melting; lc-i = liquid crystalline-to-isotropic transition; i-lc = isotropic -to-liquid crystalline transition; [c] Sample frozen in the LC state at room temperature; [d] Mesophase stability range = $T_{lc\text{-}i}$ - T_m.

The transition temperatures of the polymeric fractions measured by DSC method (figures 1 and 2) were compared with those obtained by observations under polarizing light optical microscopy (OPM) and a good concordance between these two methods could be seen in table 3. The differences could be explained by variations of the heating/cooling rates, amounts of sample used and presence of an inert gas (nitrogen) in case of the DSC measurement, vs. air in case of OPM one.

The data from table 3 showed that all the polymeric fractions exhibited enantiotropic LC behavior. The monomer M 1 showed in DSC thermogram a single peak at the heating and cooling. However, in OPM it was observed under cooling the formation of a fluid texture which proved the existence of a mesophase (figure 3a). From this reason the sample M 1 was exposed to successive heating and cooling cycles with various rates (°C/min): 5, 10 and 15.

The transition temperatures recorded at the cooling 231.7 °C; 230.6 °C and 227.2 °C, respectively showed a shift of the exothermal peak of only 4.5 °C fact that support the OPM observations and proved the existence of a mesophase for the monomer M 1.

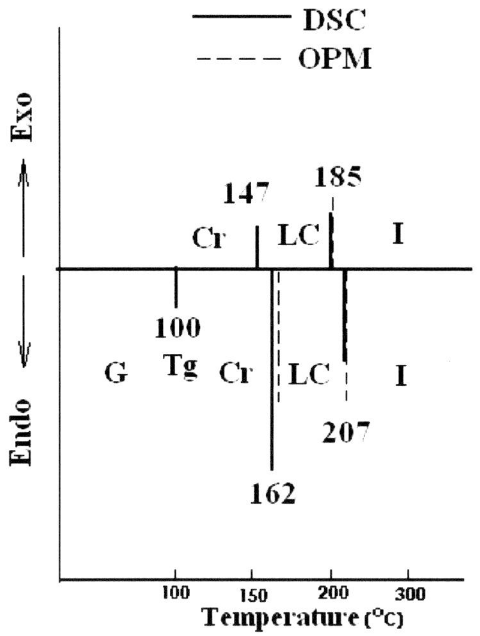

Figure 1. Composite DSC/OPM diagrams of the sample PAM 4 : first heating cycle, 10 °C/min; first cooling cycle, 10 °C/min.

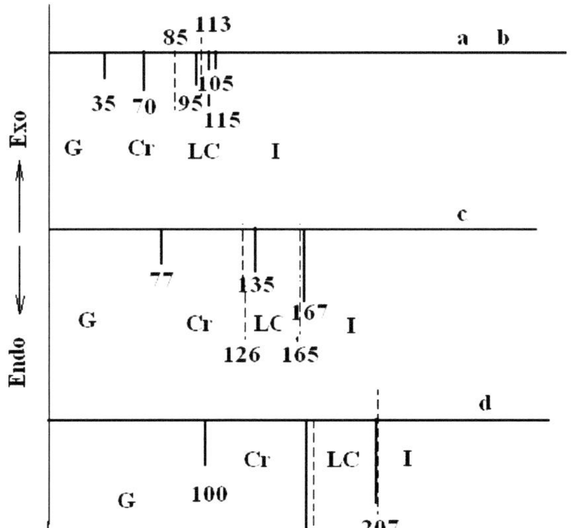

Figure 2. Composite DSC/OPM diagrams of the polymeric fractions: a) PAM 1, second heating cycle, 10 °C/min; b) first heating cycle, 10 °C/min; c) PAM 3 second heating cycle, 10 °C/min; d) PAM 4 second heating cycle, 10 °C/min.

The OPM observations for sample M 2 evidenced a birefringent texture at 135°C with a viscous fluid feature under shear (figure 3b). The birefringence disappeared completely at 195°C and a mosaic texture occurred instantaneously under cooling at 186 °C (figure 3c). The DSC thermograms showed in the heating cycle two endothermal peaks with different enthalpies: a small and broad one at 133 °C and another bigger one at 191 °C. Under cooling only an exothermal peak appeared at 176 °C. After three successive heating/cooling cycles with various rates (°C/min): 1; 5; 15 the exothermal peak at the cooling ranged between 172 °C and 181 °C. The peak from 133 °C was assigned to a crystalline-LC transition. Because the enthalpy difference between peaks from 191 °C and 176 °C is small, and based on OPM observations, we assigned these peaks to the isotropization and isotropic-LC transitions, respectively. The polymeric fractions PAM 1 and PAM 2, presented in OPM a fine grainy texture (figure 3d) similar to the textures reported for thermotropic polymers having rigid chains, without flexible spacers [8]. The polymer PAM 3 presented two textures at the cooling: a fine grainy one and a mosaic one (figure 3e) [9]. The sample PAM 4 was investigated by OPM in two different ways: a) during the cooling with 5°C/min three successive textures difficult to assign appeared (figure 3f). After annealing the sample at 200 °C for 1 h interesting droplets appeared (figure 3g) which coalescenced into a fine texture (figure 3h). After cooling at 194.5 °C and annealing for 75 min then cooled at 180 °C and annealed for 3 h a clear mosaic texture appeared (figure 3i). On basis of all these observations it is obviously that the mesomorphic evolution of this polymer was strongly affected by its thermal history.

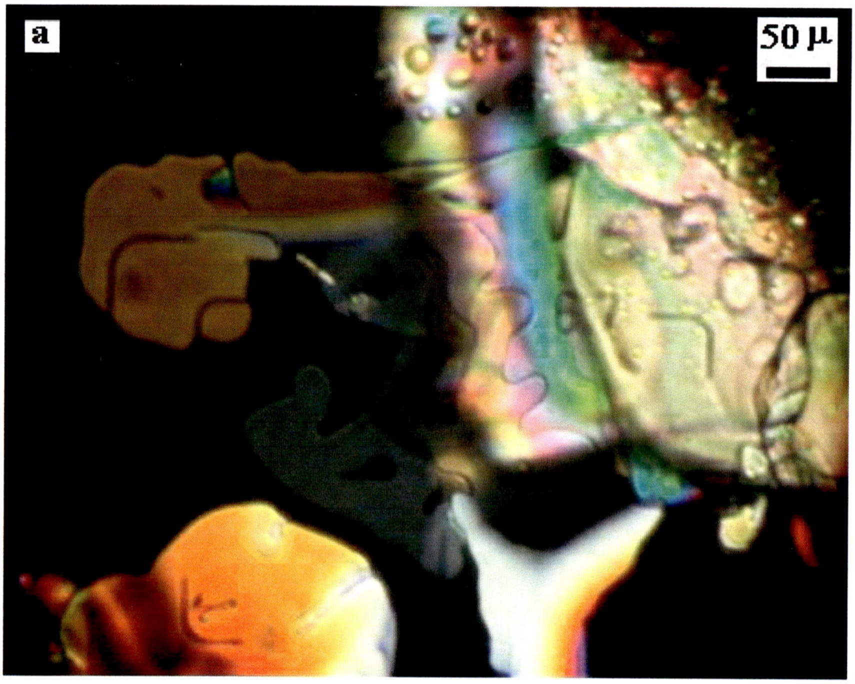

a) M 1 first heating 223 °C.

Figure 3. (Continued on next page.)

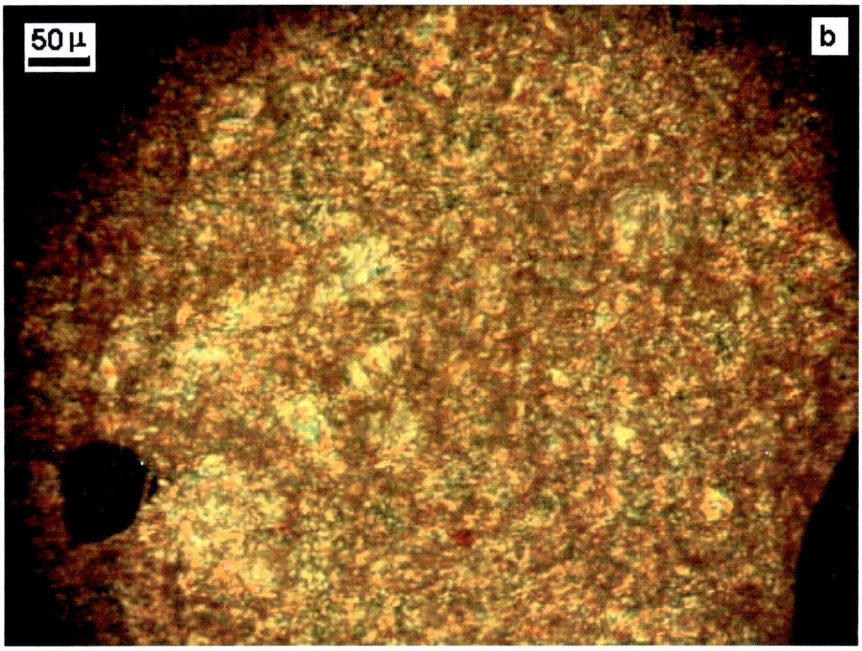

b) M 2 first heating 137 oC;

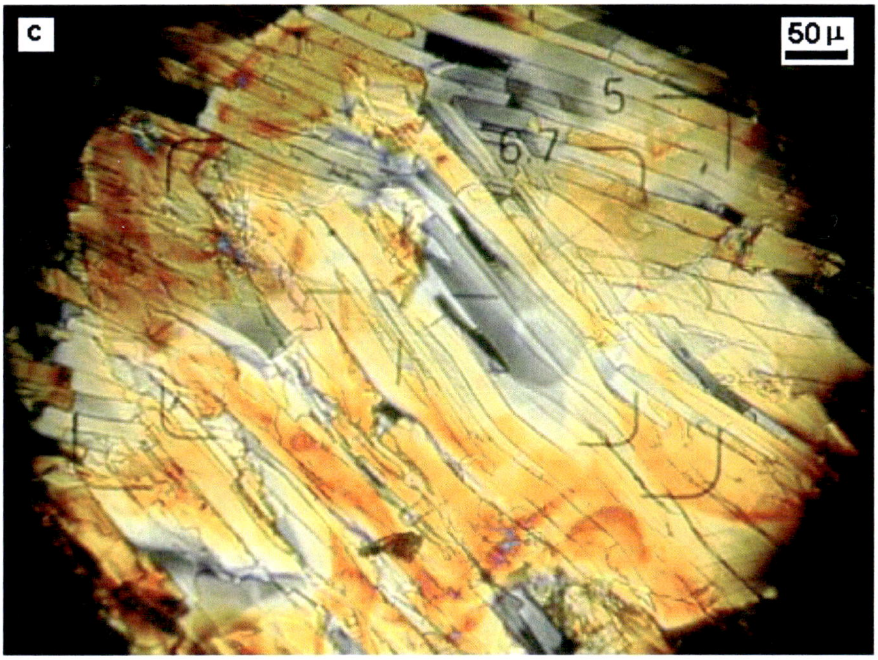

c) M 2 first cooling 185 oC

Figure 3. (Continued on next page.)

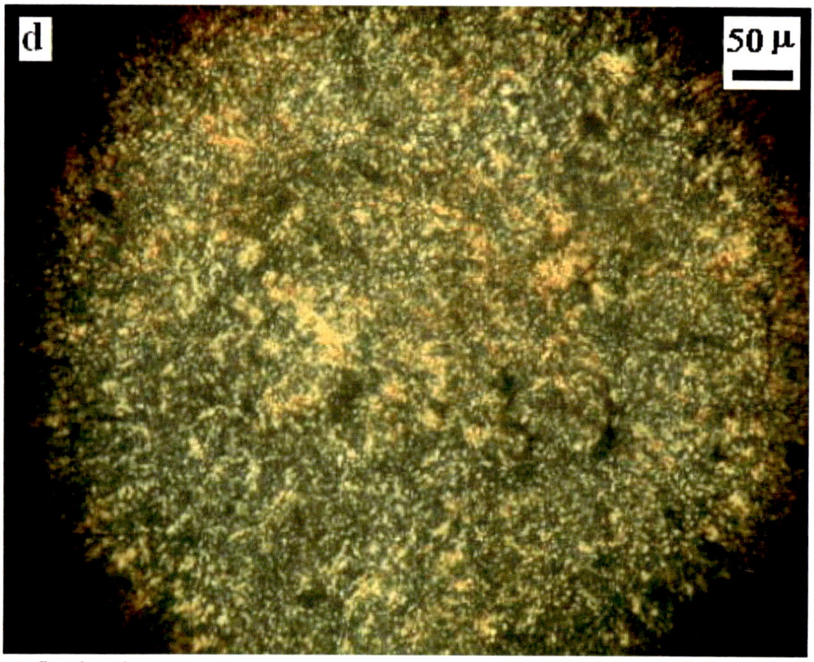

d) PAM 1 first heating 103 oC.

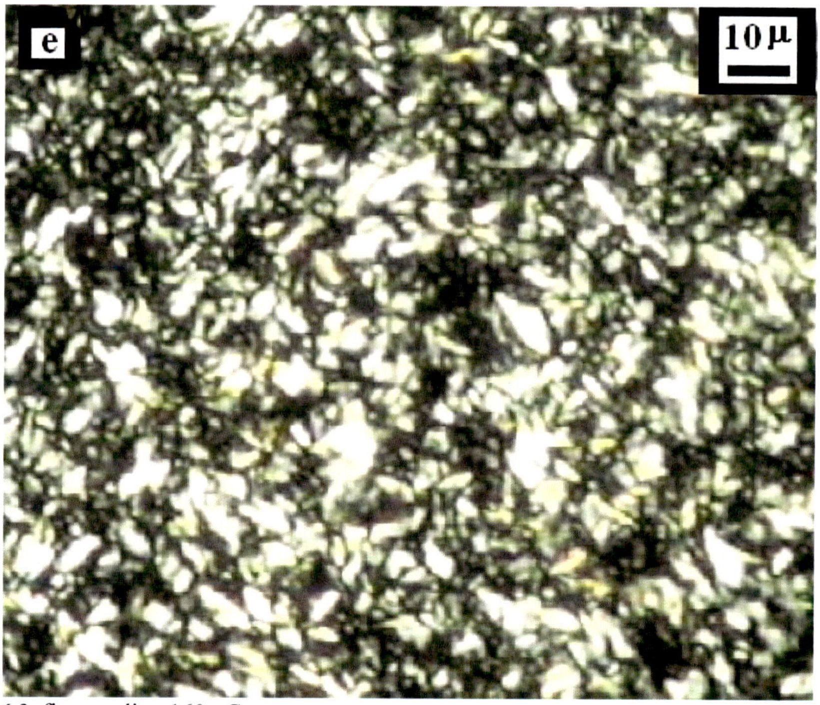

e) PAM 3, first cooling 160 oC

Figure 3. (Continued on next page.)

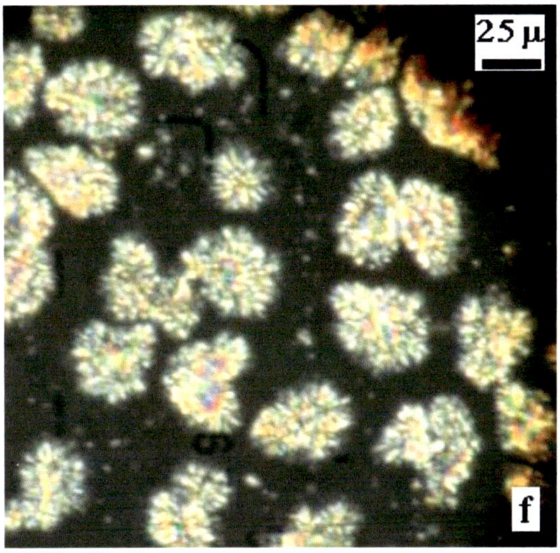

f) PAM 4, first cooling 187 oC

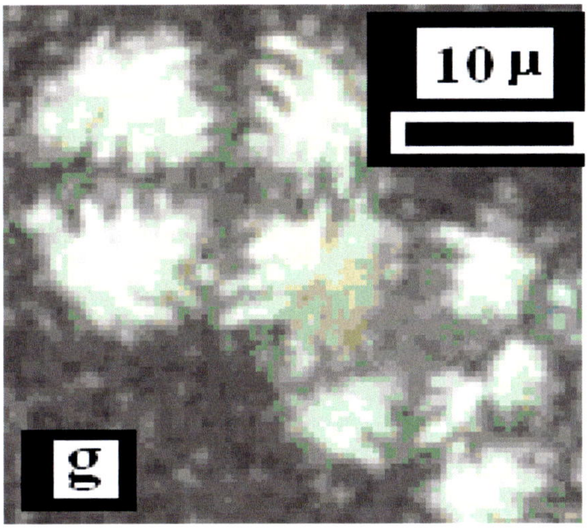

g) PAM 4 first cooling 200 oC, after 60 min annealing

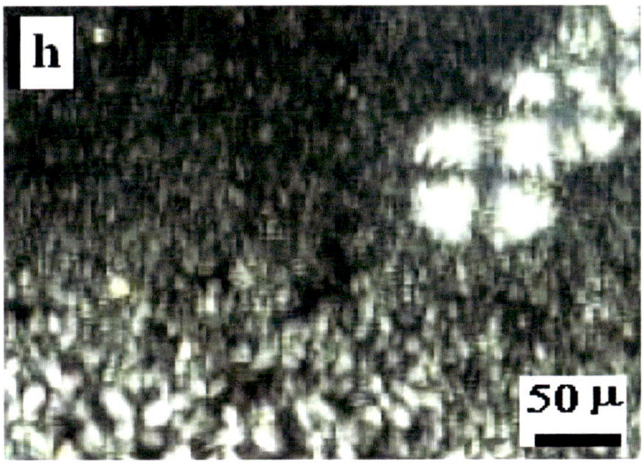

h) PAM 4 first cooling 194.5 oC, after 75 min annealing; cooling at 180 oC, after 3 h of annealing.

Figure 3. (Continued on next page.)

i) PAM 4 first cooling 180 oC.

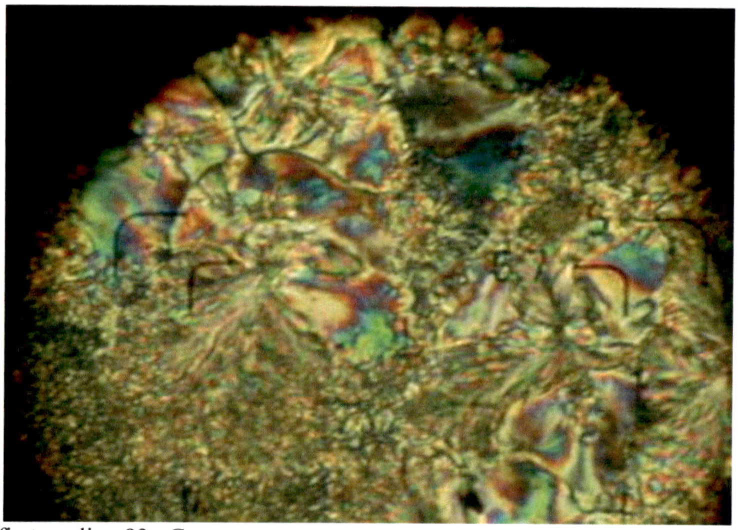

j) PAM 4, first cooling 83 oC

Figure 3. Cross polarizers OPM images of samples: a) M 1; b,c) M 2; d) PAM 1; e) PAM 3; f-j) PAM 4

At the further cooling from the LC state the samples PAM 1 – PAM 4 had different behaviors: PAM 3 and PAM 4 crystallized (figure 3j), PAM 1 froze into the mesophase while PAM 2 froze into the isotropic state.

The powder X-ray diffraction spectrum of the sample PAM 4 (figure 4) at room temperature, showed a semicrystalline pattern. Two intense peaks are present at angle (2θ ^{o}) of 20 and 23 and a more diffuse one at 29. Similar data were reported for thermotropic poly(azomethine-ester)s [10].

The influence of the molecular weight on the thermotropic behavior is depicted in figure 5. It was observed an increasing number of distinct textures with the increasing of the molecular weight. Figure 5 shows a decreasing of the transition temperatures of the polymeric fractions PAM 1 – PAM 4 in comparison with the monomer M 1. This could be due to the presence of bisphenol A moieties which induce a flexibilization of the macromolecular chain. With the increasing of the M_n it was noted an increase of the transition temperatures while the mesophase range was also enlarged.

The geometrical characteristics of the discussed polymers were investigated by molecular modeling method (Mm+) [11]. Figure 6 presents the possible conformation illustrated on a hypothetical model compound. One can observe that the disturbing influence of the kinking units of bisphenol A on the linearity of the chain is diminished by the rotation around ether bridges. A tendency to more linear structures appeared thus, favoring the packing into supramolecular structures specific of the LC state.

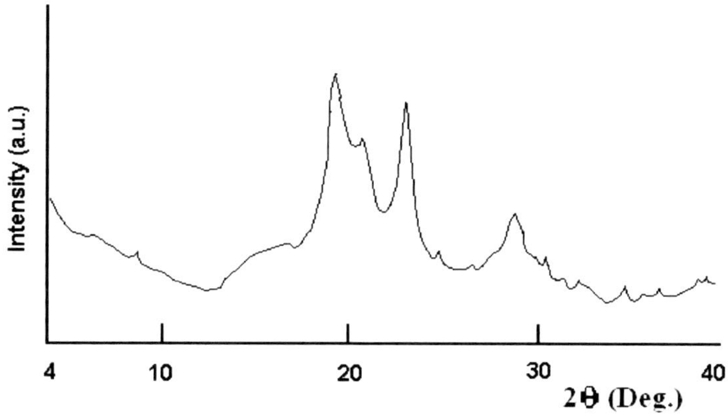

Figure 4. X ray diffractogram of the sample **PAM 4** at room temperature.

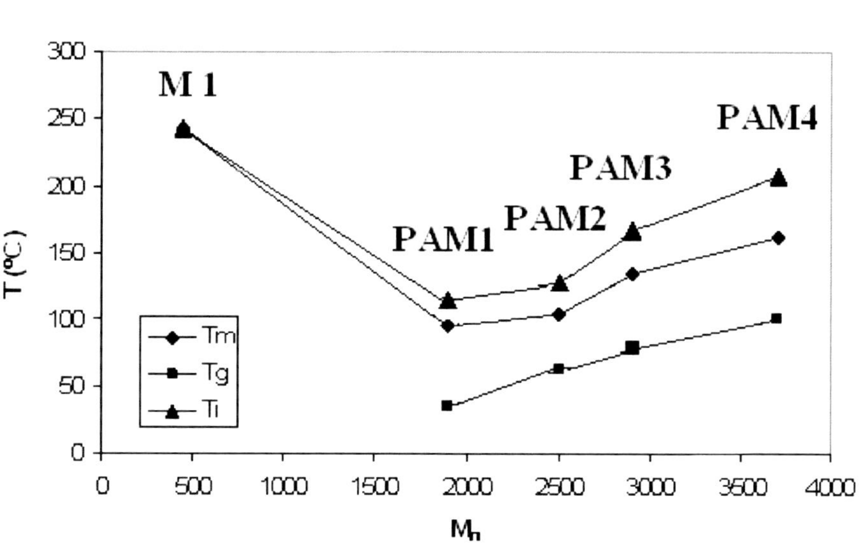

Figure 5. Dependence of the transition temperatures (T_g, T_m, $T_{lc\text{-}I}$) on the molecular weight of the monomer M 1 and polymers PAM 1 – PAM 4.

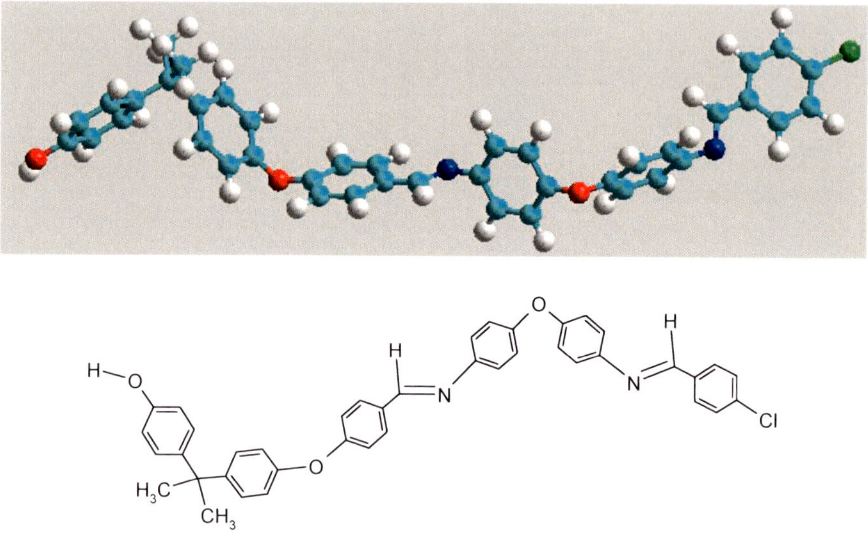

Figure 6. Model molecule used for simulation of geometry conformation of poly(azomethine-ether)s.

As textures specific to smectic phases (like mosaic) were obtained for the higher molecular weight polymeric fractions, an increased of the linearity with the increase of the chain length was confirmed.

Some conclusions could be stated as follows: a poly(azomethine-ether) derived from an azomethine monomer 4,4'-oxy-bis(4-chlorobenzylideneiminophenylene) and bisphenol A was prepared in order to obtain more information about properties of polyazomethines containing bisphenol A, and to investigate the relationship between molecular structure and mesomorphic behaviour in polyazomethines having different molecular weight. Although the bisphenol A is not favourable for liquid crystalline behaviour, due to its structural and geometric features, thermotropic properties occurred when it was incorporated into the polymeric backbone. The melting and isotropization temperatures strongly decreased when the relatively rigid geometry of mesogenic monomer **M 1** is disturbed by the introduction of kinked units of bisphenol A, allowing to obtain mesomorphic behaviour within rather large temperature range. So, the incorporation of bisphenol A by ether linkages into the main chain of a polymer is a suitable way for decreasing the transition temperatures. When comparing four fractions of the poly(azomethine-ether)s, the effect of the length of the macromolecular chain on the thermotropic behaviour was proved to be significant. In addition, the polymers present good solubilities in common organic solvents, due to the presence of the bulky bisphenol A units.

3. Poly(Azomethine-Ether-Sulfone)s Obtained by In-Situ Polycondensation. Influence of the Nature of the Internal Linking Groups on the TLC Properties

As is well known, PSFs are a class of engineering thermoplastics, generally amorphous, and have an isotropic behavior under polarizing light in the melt [12], so there are some difficulties in obtaining TLC properties for the PSF structure, mainly due to the bulkiness of the sulfonyl (-SO_2-) group but also due to the deviation from the linearity of the macromolecular chains caused by the existence of etheric and methylenic links on the backbone [16-18].

There are a few studies concerning poly(azomethine-ether-sulfone)s [12, 13]. Smith et al. Showed that the use of polysulfone (PSF) segments along the polyazomethine (PAM) backbone would result in an isotropic-anisotropic system with property contribution from each segment. Also, the use of azomethine units along the PSF backbone should result in a product which shows improved solvent resistance [12].

Three reaction paths were chosen to obtain the desired structures: a) in-situ reaction of 4-hydroxybenzaldehyde (I), sulfonyl-bis(4-chlorophenyl) (II), and diamines 4,4'-oxy-bis(aniline) (IIIa), and 4,4'-methylene-bis(aniline) (IIIb) (scheme 4), according to the literature [14, 15], led to polymers PAMS 1; b) two-step reactions: b1: synthesis of dialdehyde sulfone monomer M 4; b2: reaction of monomer M 4 with IIIa (or IIIb) resulting polymers PAMS 2, PAMS 4 (scheme 5); c) classical SN_2 polycondensation reaction between monomer II and the model compound M 7 led to polymer PAMS 6 (scheme 6).

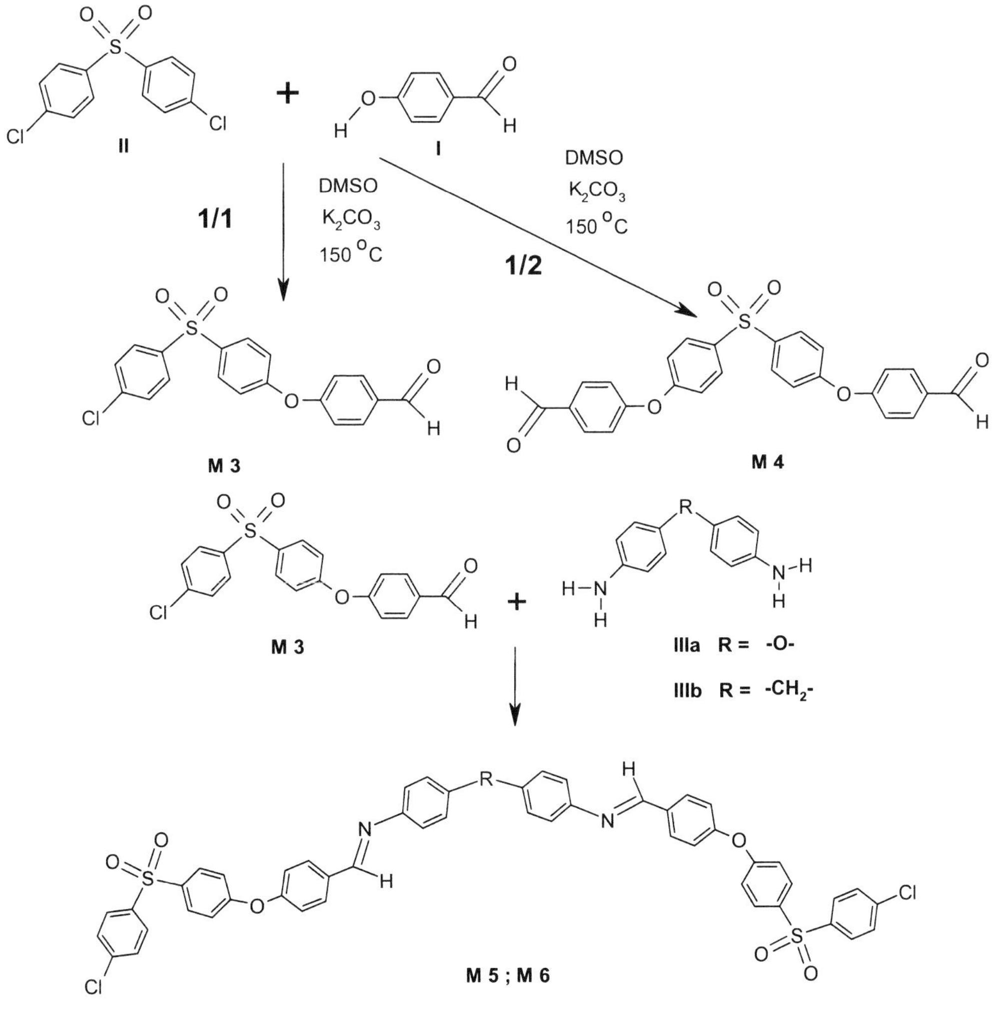

Scheme 4. (Continued on next page.)

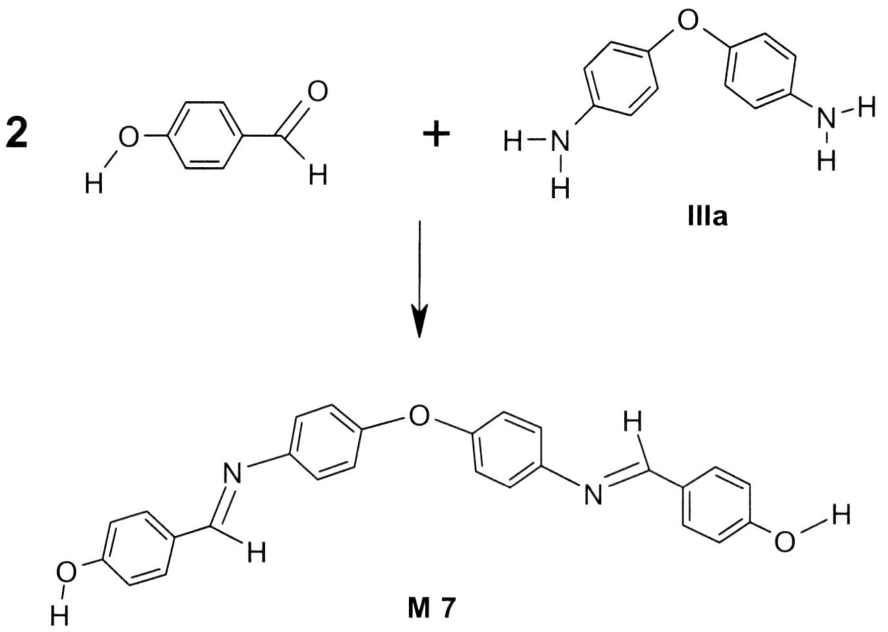

Scheme 4. Synthesis of aldehyde sulfones M 3, M 4 and azomethine model compounds M 5, M 6 and M 7.

Before any thermotropic investigations, the thermal stability of the polymers was checked by TGA in air [19]. It was found that the onset temperature of thermal decomposition ranged between 165 °C and 255 °C. After comparison with the thermal transition temperatures obtained latter by OPM, DSC and TOA methods, it was concluded that thermal degradation didn't affect the mesophase stability, with the exception of sample PAMS 2a1 for which the isotropization temperature of 240 °C (OPM) exceeded by 15 °C the onset decomposition temperature of 225 °C.

The OPM observations (table 4) showed nematic textures for both model compounds M 5, M 6 and polymers PAMS 1, PAMS 2, but no birefringence for those polymers obtained by using the diamine IIIb (PAMS 3, PAMS 4). With the exception of samples PAMS 2a1, PAMS 3, PAMS 4 and PAMS 6, all other samples showed isotropization, as reflected by OPM and DSC.

Surprisingly, the ‚b' series of polymers (PAMS 3, PAMS 4) didn't exhibit TLC properties. The main difference in the chemical structure is that two calamitic mesogenic units are bridged by a $-CH_2-$ link in the 'b' series while in the 'a' series they are bridged by a –O- linkage. Therefore, we supposed that the ether linked sequences are more favorable for inducing TLC behavior than the methylene ones. In view to verify this assumption, a new polymer (PAMS 5) was synthesized by using a polycondensation reaction between a non-TLC polymer (PAMS 4) which has chlorine end functional groups and the model compound M 7 which contains two calamitic mesogenic units linked by an ether bridge (scheme 6).

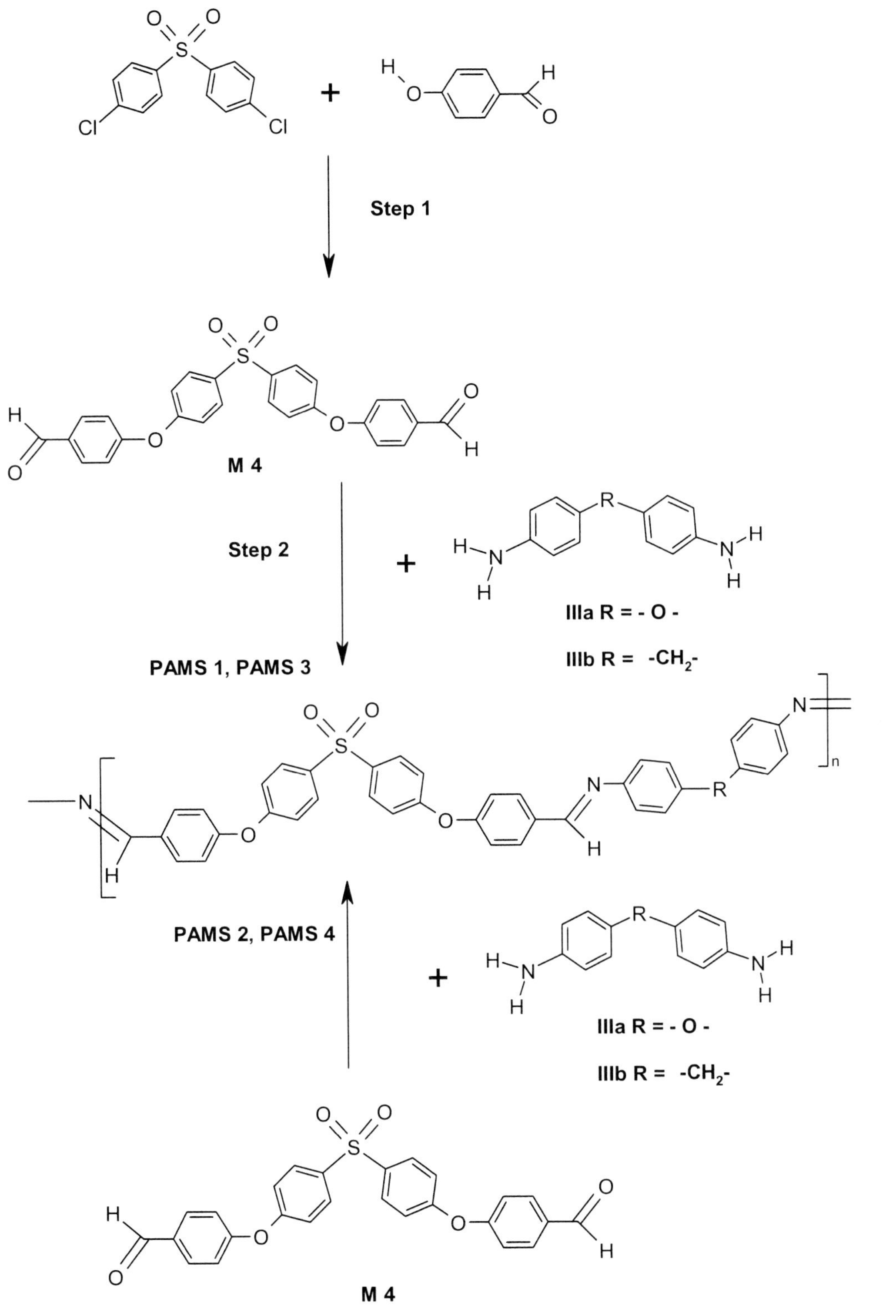

Scheme 5. Synthesis of poly(azomethine-sulfone)s: two steps (PAMS 1, PAMS 3) and one step (PAMS 2, PAMS 4).

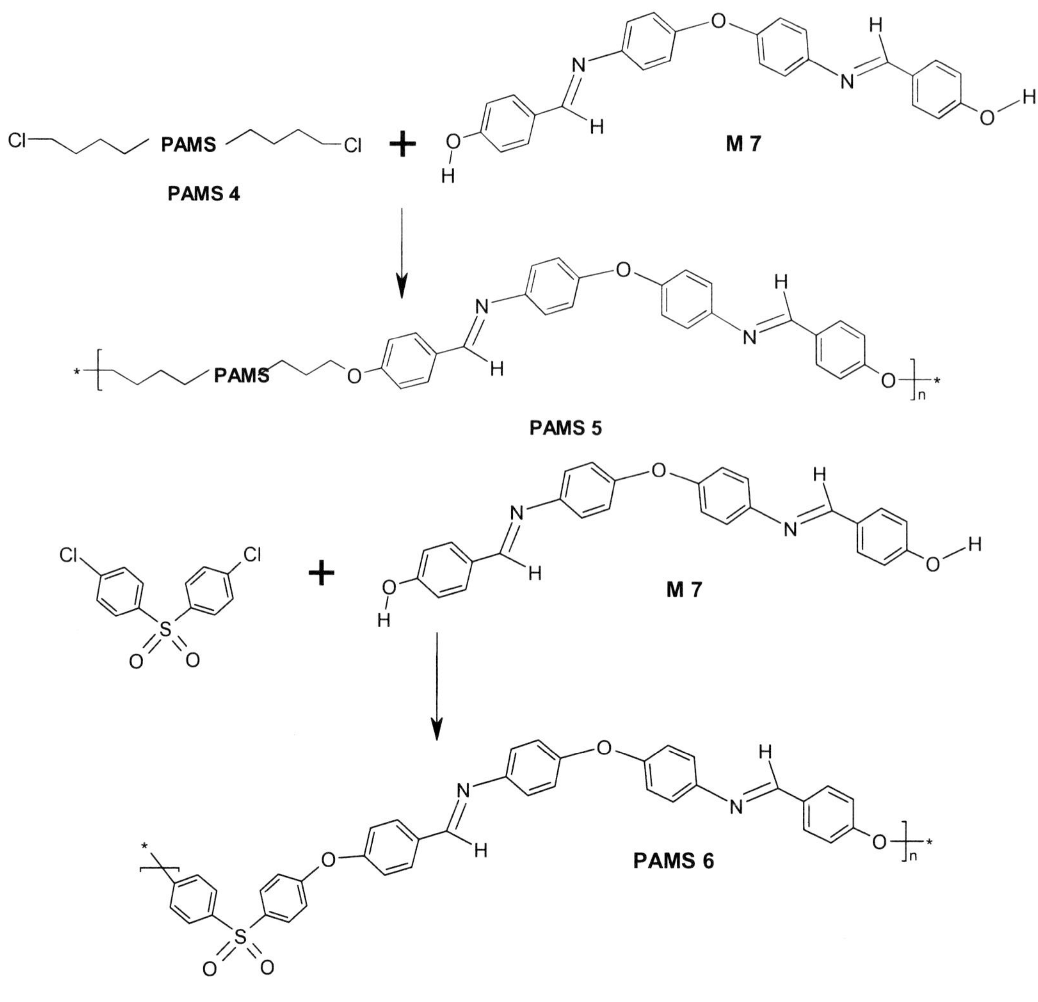

Scheme 6. Synthesis of poly(azomethine-sulfone)s PAMS 5 and PAMS 6.

Table 4. Thermotropic behavior of model compounds M 5, M 6, M 7 and polymers PAMS 1- PAMS 6

Sample code	Color	Yield, %	Inherent[a] viscosity (dL/g)	LC phase[b] (°C)	Birefringence[c]	Transition temperatures, °C, DSC and (TOA)		
						T_g	T_m	T_i
M 5	white	66	-	138-230	s	108 (110)	148 (140)	240 (220)
M 6	yellow	55	-	128-200	w	106 (-)	146 (130)	186 (-)
M 7	yellow	71	-	244-245[d]	-	-	-	-
PAMS 1	white	38	0.03	180-220	vs	85 164 (-) (160)	198 (185)	228 (220)
PAMS 2a1	yellow	38	0.07	150-[f]	vs	112 (-)	150 (138)	- (-)
PAMS 2a2	yellow	75	0.11	140-220	vs	128 (-)	145 (135)	200 (-)
PAMS 3	yellow	80	0.07	140-195[e]	-	-	153	-

Sample code	Color	Yield, %	Inherent[a] viscosity (dL/g)	LC phase[b] (°C)	Biref-rin-gence[c]	Transition temperatures, °C, DSC and (TOA)		
						T_g	T_m	T_i
						(-)	(158)	(-)
PAMS 4	yellow	67	0.11	145-200[e]	-	114 (-)	131 (130)	- (-)
PAMS 5	yellow	70	0.26	125-160	s	97 (108)	143 (138)	160 (168)
PAMS 6	yellow	87	0.34	140-200	s	- (-)	147 (140)	- (-)

[a] Measured at a concentration of 0.2 g/dL in NMP at 25 °C; [b] As observed under the OPM; [c] Birefringence's intensity as observed under the OPM: vs = very strong, s = strong, w = weak, - = none; [d] Melting point; [e] Softening range; [f] No isotropization temperature was observed.

a)

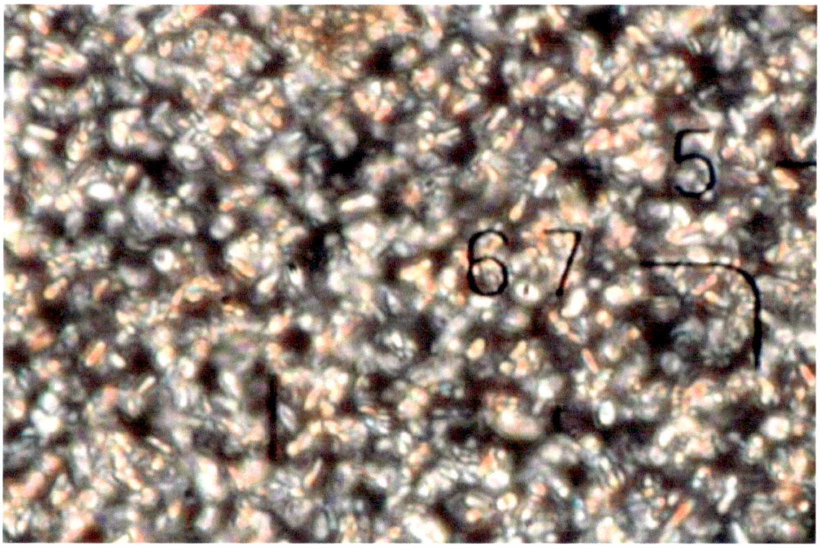

b)

Figure 7. (Continued on next page.)

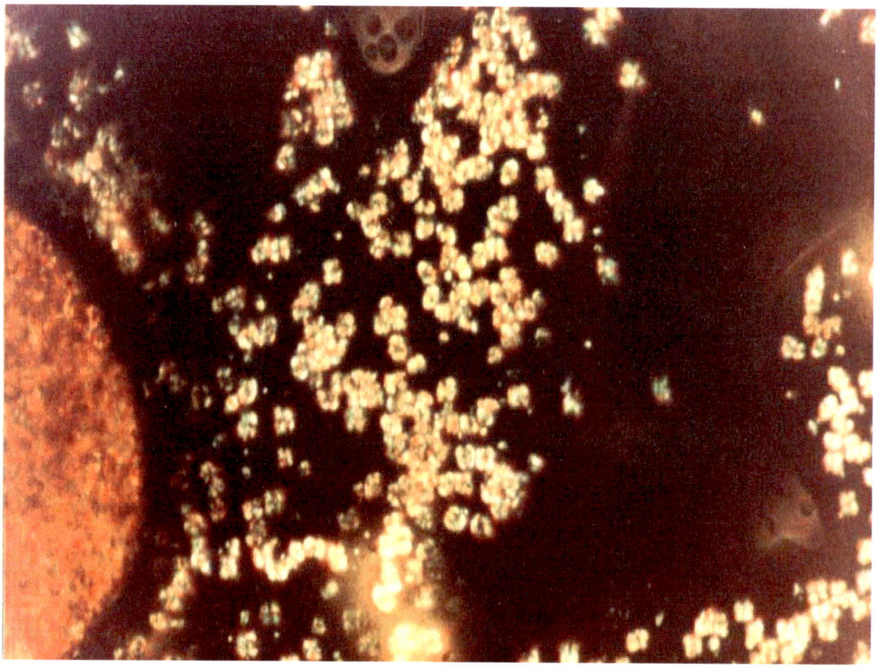

c)

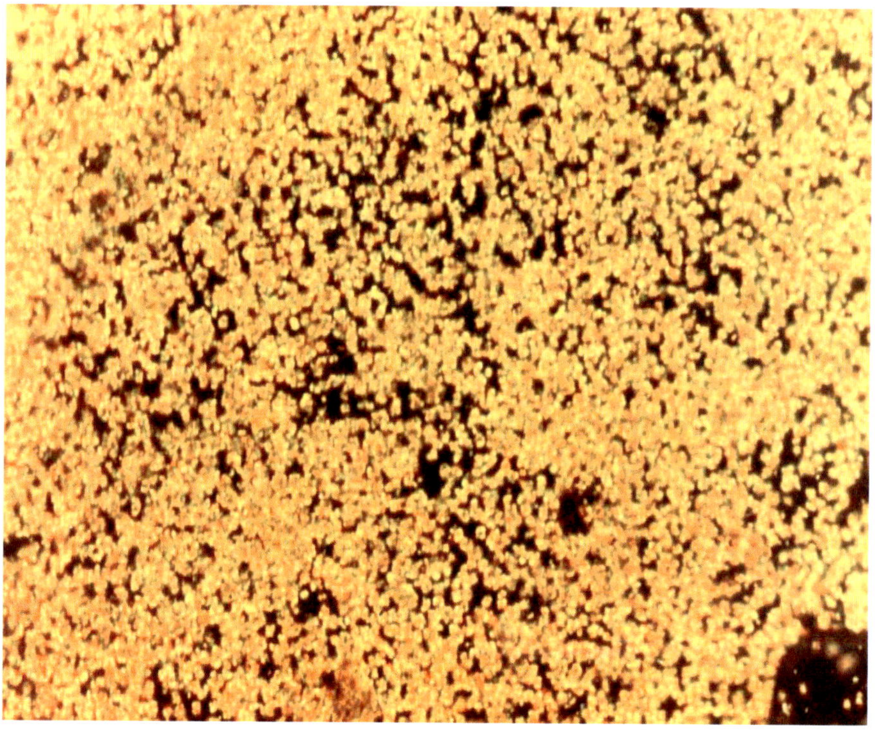

d)

Figure 7. OPM texture images of samples: a) M 5 heating 200 °C, 200x; b) M 5 cooling 215 °C, >400x; c) PAMS 5 heating 150°C, 400x; d) PAMS 6 heating 160 °C, 200x

Using OPM observations for polymer PAMS 5, a strong birefringence with grainy droplets was found in the melt between 125 °C and 160 °C (table 4, figure 7c). Also, by DSC measurements, a glass transition temperature at 97 °C, a crystalline-nematic transition temperature at 143 °C, and a nematic-isotropic transition temperature at 160 °C were recorded. The thermo-optical (TOA) measurements for this sample agree well with the DSC ones (table 4). Additionally, a direct polycondensation was made by using the reaction of sulfone II with the model compound M 7, obtaining the polymer PAMS 6. The OPM observations showed similar grainy texture in the range 140 °C to 200 °C (table 4, figure 7d). It was confirmed thus the above assumption and outlined the importance of the nature for the internal linkages which connect the mesogens. Also, it was demonstrated that such synthetic procedures could be successfully used to prepare thermally stable semirigid polymers which, instead of lack of flexible spacers, exhibit TLC behavior.

4. Side Chain Poly(Azomethine-Sulfone)s, without Flexible Spacers Prepared by Chemical Modification of Commercial Polysulfone

There are many references on aromatic LC polysulfones, but especially as polyethers or segmented copolyethers [20,21], segmented copolyesters [22–25] and block copolymers with LC segments [26–34]. In our knowledge a few reports were noted about LC homopolyether sulfones [35a] and some about side chain LC aliphatic polysulfones [35b-e]. We consider the LC systems in which are involved both isotropic aromatic ether sulfone units and anisotropic azomethine rigid rod-like moieties to be of great interest from two points of view. Firstly, the balance between isotropic/anisotropic units can be manipulated to combine the high performance properties of aromatic polysulfones [36] with highly ordered LC materials and providing new materials more economically. Secondly, it would be interesting to study what are the factors limiting the LC behaviour in such systems from polarity, chemical environment, geometry and other structural points of view. Additionally, the influence of the chemical structures on the optical properties of the polymers outlines the aim of the present work. Our previous papers [19,37] regarding linear azomethine sulfone materials pointed out that LC behaviour for such systems was possible to exhibit only at a low molecular weight scale. Some recent results reported both on chemically modification reactions of polysulfones [38-42] and synthesis and LC behaviour of new AM dimesogens [43] suggested an interesting idea to built up a new type of side chain LC polymer. A chemical modification reaction of chloromethylated Udel polysulfone was performed with a rigid AM dimesogen end-capped with polar nitro group [44].

In this way would be very interesting to check if LC behaviour could be achieved by directly coupling the rigid mesogenic moiety with the semirigid polysulfone backbone, thus avoiding the usual way requiring a flexible spacer between the mesogen and polymer backbone [45].

The synthetic route for chloromethylated polysulfone (CMPSF) is shown in Scheme 7.

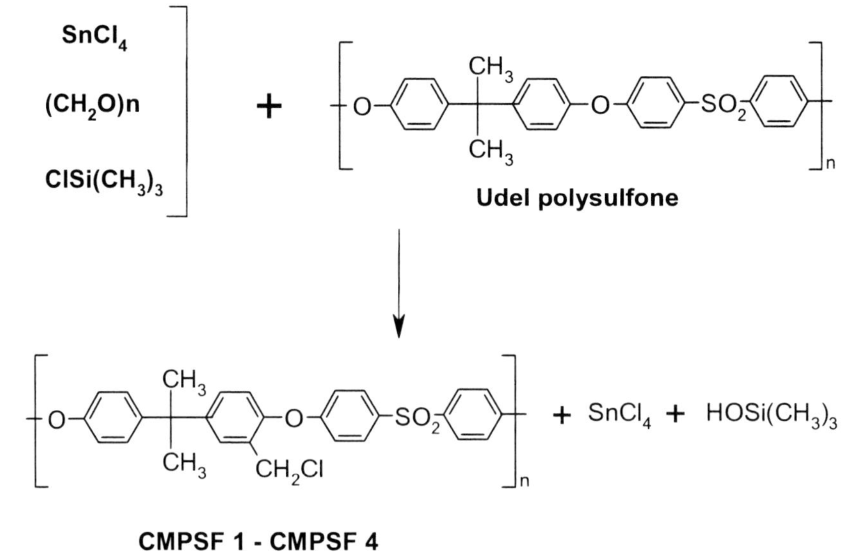

Scheme 7. Synthesis of chloromethylated polysulfones (CMPSF 1 – CMPSF 4).

^1H NMR spectra showed the characteristic peak due to CH_2Cl protons at 4.53 ppm. A more detailed discussion about the synthesis, purification and characterization of CMPSF is reported elsewhere [38]. Using the corresponding integrals of protons CH_2Cl and $(CH_3)_2C<$ the degree of substitution (SD) for starting CMPSF were calculated with the formula: $SD = 3 \cdot A_2/A_1$ where A_1 is the integrated area of the peak corresponding to $>C(CH_3)_2$ protons and A_2 is the integrated area of the peak corresponding to $-CH_2Cl$ protons (Table 5). The azomethine dimesogen (M 9) (Scheme 8) was built up after a carefully investigation of the numerous reported AM mesogens [4, 46] in view to combine both AM dimesogen structure as an electron-donor and polar nitro group as an electron-acceptor, having in the same time a hydroxyl phenolic group for coupling reaction with the macromolecular backbone. In the first step were obtained the 4-(4-hydroxybenzylideneiminophenylene)aniline (M 8a) and the 4-nitrophenoxy-benzaldehyde (M 8b) monomers. The synthesis of monomer M 8b was facilitated by the decrease of the nucleophilic character of the second amine group of benzidine after the reaction of the first amine group. In the second step the monomers M 8a and M 8b were reacted via the classical condensation reaction [43] obtaining the dimesogen M 9 (Scheme 8). For the coupling of the AM phenol dimesogen (M 9) with the CMPSF a nucleophilic substitution reaction was choosen (Scheme 9).

From IR spectra was possible to calculate the degree of substitution (z) of PAMSs, considering the ratio $A_{CH=N}/A_{C\text{-}O\text{-}C}$ as a measure of the chemical transformation. The absorption band at 1250 cm^{-1} (considered as the reference) is assigned to the stretching vibrations of the C–O–C group both in polysulfone and in dimesogenic moieties. The absorbances were calculated according to Ref. [47], taking a base line from 900 to 1800 cm^{-1}, free of any absorption. The values obtained are presented in Table 5. The degree of substitution (z) values ranged between 0.32 and 0.56. Chloromethylated polysulfones with almost one group CH_2Cl per structural unit (CMPSF 1–3) led to the highest substitution degree (z) values, taking into account the chlorine consumption, while for the other one

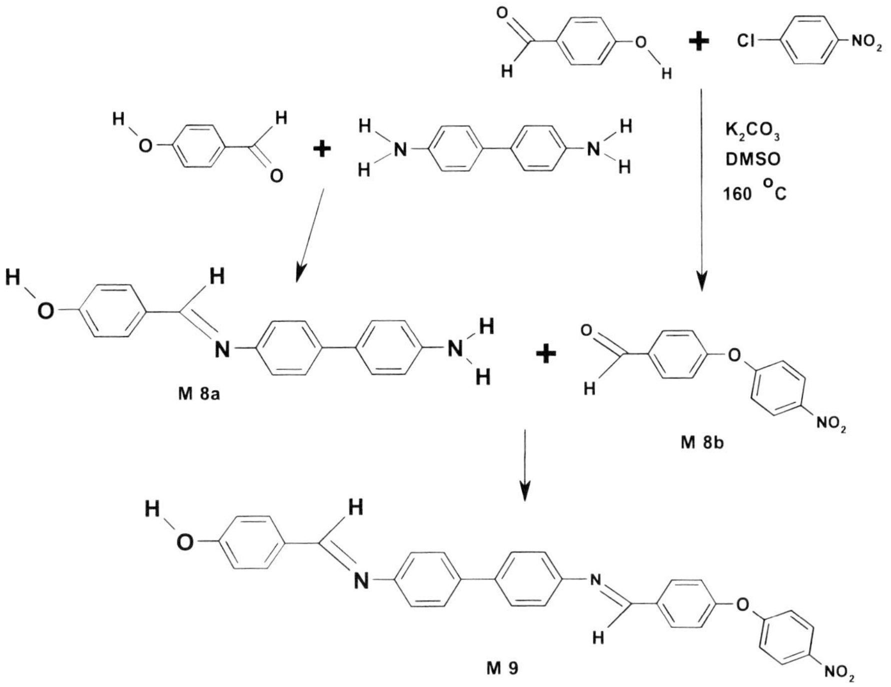

Scheme 8. Synthesis of azomethine phenolic mesogen M 9.

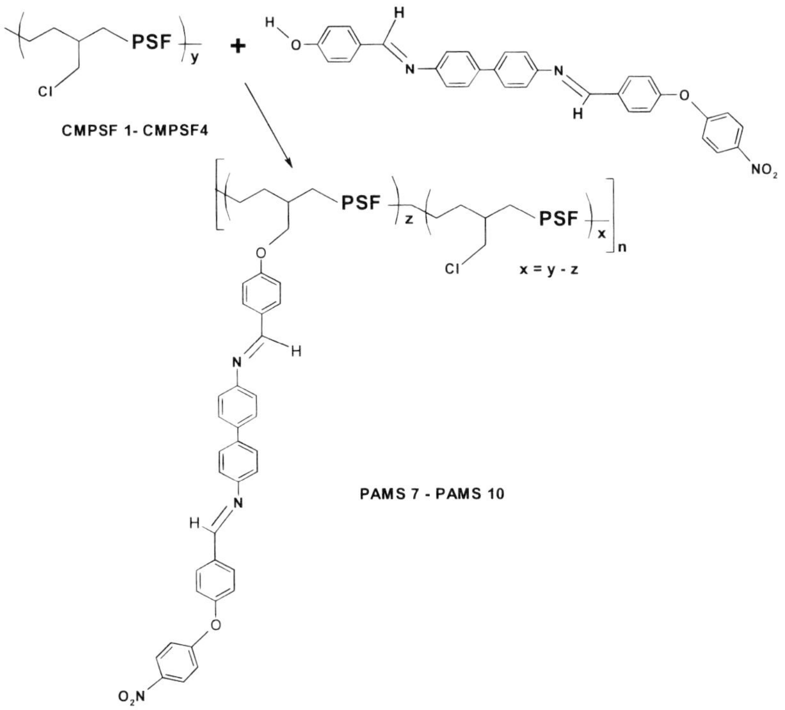

Scheme 9. Synthesis of LC polysulfones by chemical modification of the chloromethylated polysulfones.

Table 5. Characterization of CMPSF 1–4 and PAMS 7 – PAMS 10

Starting CMPSF	y^a	Final PAMSs	$A^b_{CH=N}$ 1665 cm^{-1}	$A_{C\text{-}O\text{-}C}$ 1250 cm^{-1}	z^c	$y - z^d$	λ^e_{max} nm
CMPSF 1	0.37	PAMS 7	-	-	0.32	0.05	358
CMPSF 2	0.62	PAMS 8	0.20	0.37	0.56	0.06	355
CMPSF 3	1.03	PAMS 9	0.22	0.73	0.30	0.73	360
CMPSF 4	1.49	PAMS 10	0.25	0.70	0.36	1.12	357

[a] Degree of substitution of CMPSFs, calculated from ^1H NMR spectra with the formula : $SD = 3 \cdot A_2/A_1$ [38].
[b] IR Absorbance of CH=N band at 1665 cm^{-1} calculated with the formula: $A = \log_{10} (AC/AB)$ [47]
[c] Degree of substitution of PAMSs calculated from IR measurements ($z = A_{CH=N} / A_{C\text{-}O\text{-}C}$).
[d] Content of the unreacted CH_2Cl groups in the PAMSs.
[e] Absorption maxima from UV measurements (DMF).

Table 6. Liquid crystalline behaviour of thermotropic PAMS 7 – PAMS 10

Polymer	Transition temperatures in °C [a]									Number of cycle[b]
	OPM				DSC					
	T_m	T_f	T_i	ΔT	T_g	T_m	T_f	T_i	ΔT	
Starting polymers CMPSF 1 – CMPSF 4										
CMPSF 1	180 – 200[c]	-	-	-	185	-	-	-	-	1 H
CMPSF 2	172 – 185[c]	-	-	-	175	-	-	-	-	1 H
CMPSF 3	160 – 180[c]	-	-	-	170	-	-	-	-	1 H
CMPSF 4	149 – 170[c]	-	-	-	159	-	-	-	-	1 H
Modified polymers PAMS 7 – PAMS 10										
PAMS 7	88	-	210	122	-	83	-	194	111	1 H
	-[d]	-	208	-	-	-	-	-	-	1 C
PAMS 8	120	-	136	16	-	87	-	140	53	1 H
	-	-	-	-	72	-	-	-	-	2 H
PAMS 9	147	173	250	103	-	136	185	-	-	1 H
	-[d]	-	236	-	-	115[e]	-	232	117	1 C
PAMS 10	140	165	180	40	-	131	-	202	71	1 H
	-[d]	-	168		-	139[e]	-	-	-	1 C

[a] T_g: Glass transition temperature; T_m: melting temperature; T_f: flowing temperature; T_i: isotropization temperature; ΔT: mesophase range ($T_i - T_m$);
[b] H: heating cycle; C: cooling cycle.
[c] Softening range.
[d] No crystallization but strong phase separation observed in PLM.
[e] Transitions possible to assign to a crystallization process.

(CMPSF 4), having more than one CH_2Cl group per s.u., lower z values were obtained (higher unreacted chlorine content, column (y–z) in Table 5). This behaviour probably could be due to the more pronounced steric hindrance in the case of disubstituted CMPSFs.

The electronic absorption spectra of PAMS 7 – PAMS 10 in DMF solutions are characterized by an absorption band ranging between 355 and 360 nm. The slight shift to smaller wavelengths, which occurred with the increasing of 'z' values, could be due to the greater density of chromophore groups. The thermal behaviour of CMPSFs and PAMSs was studied by means of DSC and optical OPM (Table 6). The amorphous CMPSF 1–4 exhibited glass transition temperatures in the range of 159–185 °C. A decreasing tendency of T_g values could be noted with the increasing of 'y' values (Table 6). All PAMSs exhibited melting endothermal peaks in DSC thermograms at the first heating (see for example figure 8a and 8c).

These values were compared with those obtained from OPM and proved that they really are melting transitions from solid to mesophase. All samples exhibited thermotropic LC behaviour. The textures observed were similar and seemed to be schlieren ones (figure 9a – 9c).

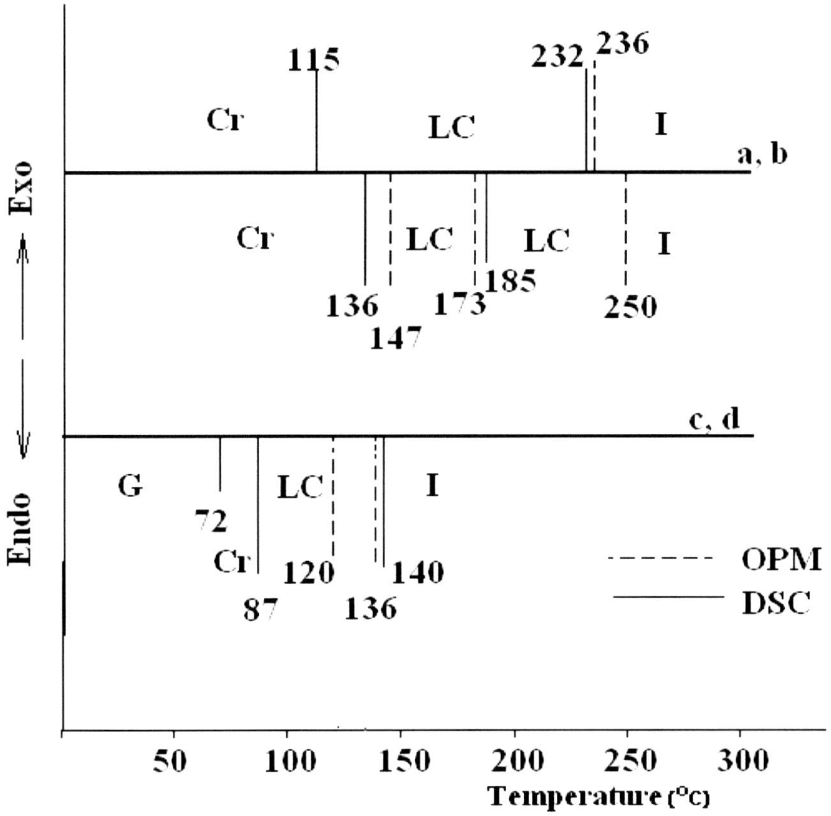

Figure 8. Composite DSC/OPM diagram of PAMSs. PAMS 9: (a) first heating cycle (top); (b) first cooling cycle (bottom). PAMS 8: (c) first heating cycle; (d) second heating cycle.

Liquid crystalline–isotropic transitions were registered both by OPM and some DSC measurements (figure 8c, Table 6). OPM observations noted for sample PAMS 8 a temperature (T_f) at which the melt flowed (Table 6). In the DSC data for the sample PAMS 9 was registered a broad endothermal peak at 185 °C which corresponds to such behaviour. In OPM (figure 9d) can be seen a strong movement of many droplets of various sizes which joined together tending to form greater domains.

The mesophases extent ranged between 16 up to 122 °C (by OPM) and 53 up to 117 °C (by DSC). The scarce experimental data available provides just a speculative correlation. However, one can see (Table 5) that samples having 'z' values about 0.3 (PAMS 7, PAMS 9,

PAMS 10) had greater mesophase ranges (Table 6) than PAMS 8 which had a greater z value. One possible but facile explanation at the moment would be that the sterical and conformational factors in the case of sample PAMS 8 overcame the ordering interactions existing in the other samples. When slowly cooling the isotropic melts (5 ^{o}C/min), the occurrence of LC droplets was observed by OPM for all samples, suggesting an enantiotropic LC behaviour. An example is presented in figure 9d for PAMS 9. In the corresponding DSC curve (Figure 8b) only a slow change of slope at 232 ^{o}C could be associated with the isotropic–LC transition.

From Tables 5 and 6 one can see that lower T_m values are obtained for samples having more dimesogenic units introduced. It seemed to be reasonable to explain such behaviour by a greater mobility of the corresponding CMPSF chains and also by a lower steric hindrance. In addition, the effect of lowering of melting transition temperatures due to the nitro groups was also reported for LC polysulfones [21]. One of the most exciting aspects of our results is the LC behaviour. The introduction of rigid dimesogenic moieties as pendant side groups on aromatic polysulfone chains proved to be very efficient in overcoming all structural barriers known in this respect [16-18]; no flexible spacer between the rigid mesogen and the backbone was used. We have no precise explanation at the moment for this phenomenon; however, probably here is involved a balance between electron-acceptor (NO_2, SO_2) groups and electron-donor (Ph-CH=N-Ph) groups content which, together with the relative semiflexibility of Ph–CH_2–O– link, may create favorable conditions for achievement of LC order.

Side chain azomethine polysulfones were obtained by chemical modification reaction of chloromethylated polysulfones, having various contents of CH_2Cl groups, with an azomethine dimesogenic phenol end-capped with polar nitro group. Enantiotropic nematic LC behaviour was observed for all samples. Lowering of the melting temperatures was influenced by the substitution degree of the starting chloromethylated polysulfones.

a

Figure 9. (Continued on next page.)

b

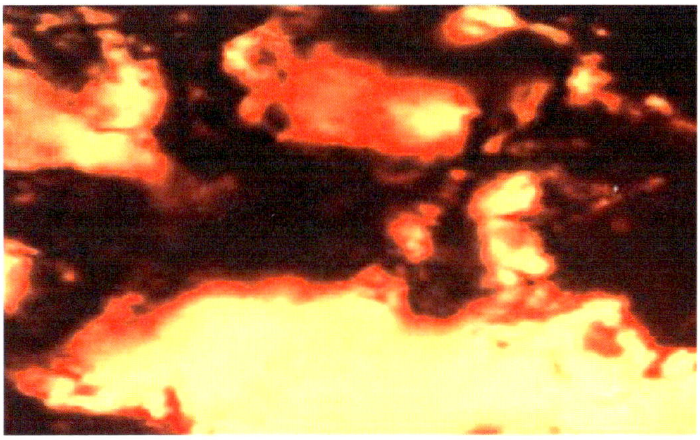

c

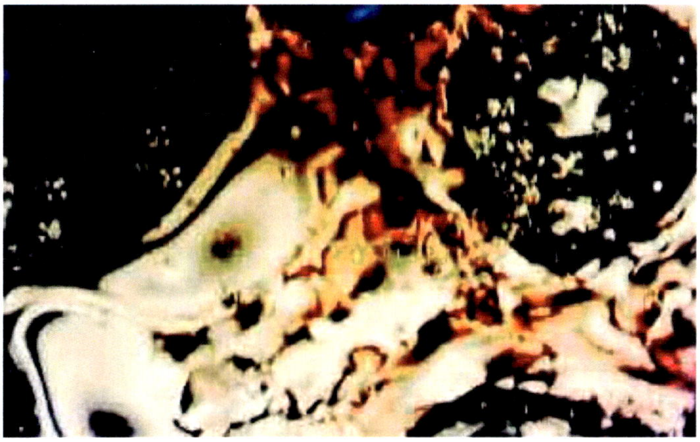

d

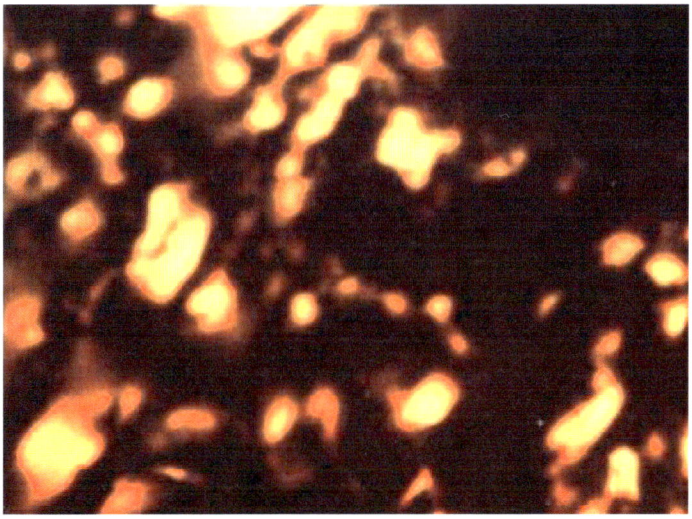

Figure 9. Microphotographs of: (a) PAMS 8 first heating cycle at 128 °C, (200x); (b) PAMS 10 first heating cycle at 173 °C (200x); (c) PAMS 9 first heating cycle, at 238 °C . (200x); (d) PAMS 9 first cooling cycle: coalescence of droplets at 236 °C (400x).

5. THERMOTROPIC COPOLY(AZOMETINE-ETHER-SULFONE)S – EFFECT OF MESOGENIC MONOMERS LENGTH AND COPOLYMERIZATION

Previous studies showed that poly(azomethine-ether-sulfone)s combine the valuable properties of both polyazomethines and poly(ether-sulfone)s, due to the contribution of the anisotropic azomethine rigid cores and amorphous isotropic arylethersulfone moieties [2, 19, 37, 44]. These polymers exhibited good solubility and accessible melting temperatures. Three new series of random and alternated copoly(azomethine)s were synthesized in view to investigate the inflence of mesogenic length and sequential distribution on the TLC properties [3]: a) poly(azomethine-ether)s, by the reaction of a dichloro-compound containing azomethine groups with a bisphenol such as isopropylidenebisphenol (bisphenol A) or fluorobisphenol (bisphenol AF); b) poly(azomethine-ether-sulfone)s, by the reaction of 4,4'-sulfonyl-bis-(*p*-chlorophenyl) with a bisphenol containing azomethine group; c) copoly(azomethine-ether-sulfone)s by the reaction of bis(*p*-chlorophenyl)sulfone with a mixture of bisphenol A and bisphenol containing azomethine groups.

The preparation of the azomethine monomers (M 7 and M 10 – M 13) was performed by acid-catalyzed condensation, according to previous reported methods [48-51]. The structures of these monomers and their characteristics are shown in Table 7. The synthetic pathway to alternated or random polymers was the classical Williamson etherification method, in which an aromatic dichloride reacted with a bisphenol as nucleophile. The poly(azomethine-ether)s PAM 5 and PAM 6 are obtained by the reaction of the monomer M 10 with bisphenol A or bisphenol AF, respectively (Scheme 10).

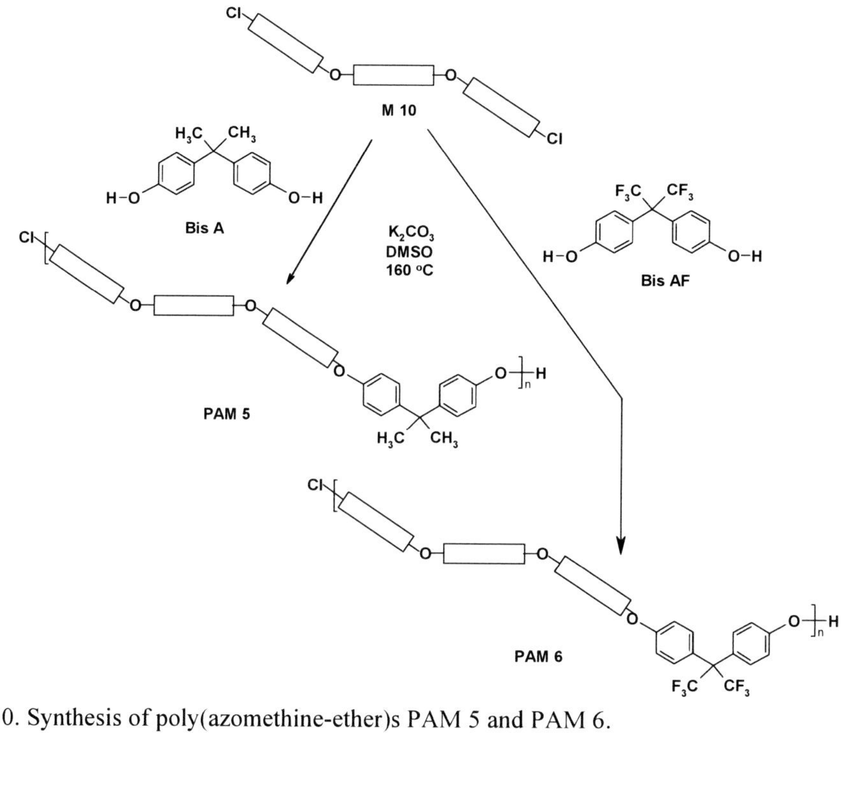

Scheme 10. Synthesis of poly(azomethine-ether)s PAM 5 and PAM 6.

The poly(azomethine-ether-sulfone)s PAMS 11 – PAMS 14 were obtained by the reaction of 4,4'-sulfonyl-bis-(*p*-chlorophenyl) with the coresponding monomers M 7, M 11 – M 13, respectively (Scheme 11). The copoly(azomethine-ether-sulfone)s PAMS 15 - PAMS 17 were obtained by the reaction of 4,4'-sulfonyl-bis-(*p*-chlorophenyl) with a mixture of bisphenol A and bisphenols containing azomethine group (M 11, M 7, M 12, respectively) in 1:0.5:0.5 molar ratio (Scheme 12).

Table 7. Chemical structures of mesogenic azomethine monomers

Code	Chemical structure	Phase transitions (OPM)[a]
M 11	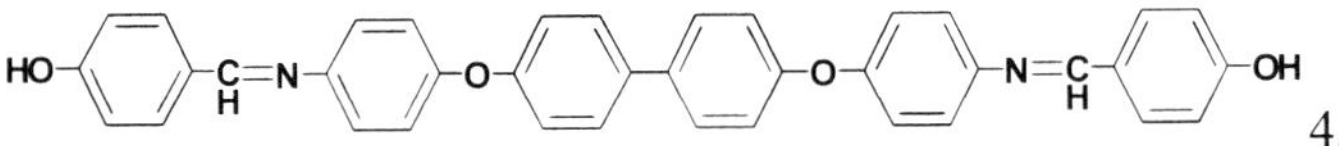 4,4'-bis(*p*-hydroxybenzylideneiminophenoxy)biphenyl	Cr 299 LC 303 I
M 7	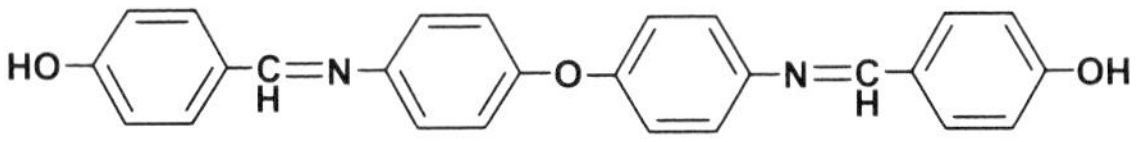 4,4'-oxo-bis(*p*-hydroxybenzylideneiminophenylene)	Cr 244 I
M 12	 4,4'-bis(*p*-hydroxyiminomethylidene)phenylene	Cr 284 I
M 13	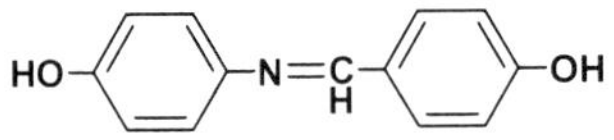 4-(*p*-hydroxybenzylideneimino)phenol	Cr 218 I
M 10	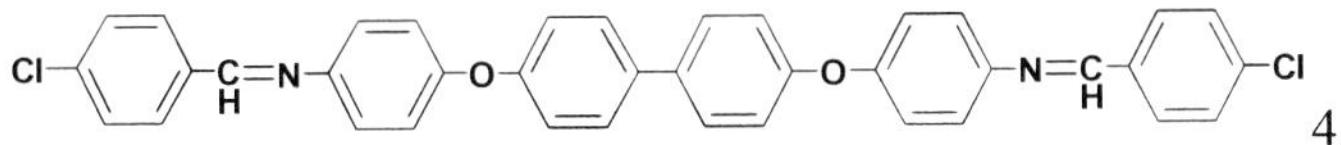 4,4'-bis(*p*-clorobenzylideneiminophenoxy)biphenyl	Cr 243 LC 246 I

[a] Thermotropic behavior determined by OPM; Cr = crystalline; LC = liquid crystalline; I = isotropic.

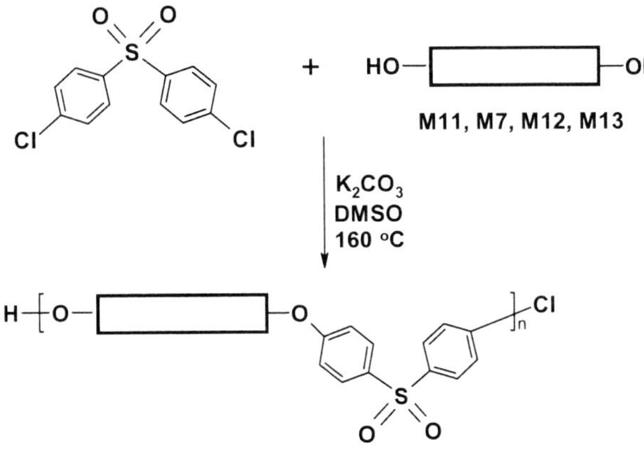

PAMS 11, PAMS 12, PAMS 13, PAMS 14

Scheme 11. Synthesis of copoly(azomethine-ether-sulfone)s PAMS 11 - PAMS 14.

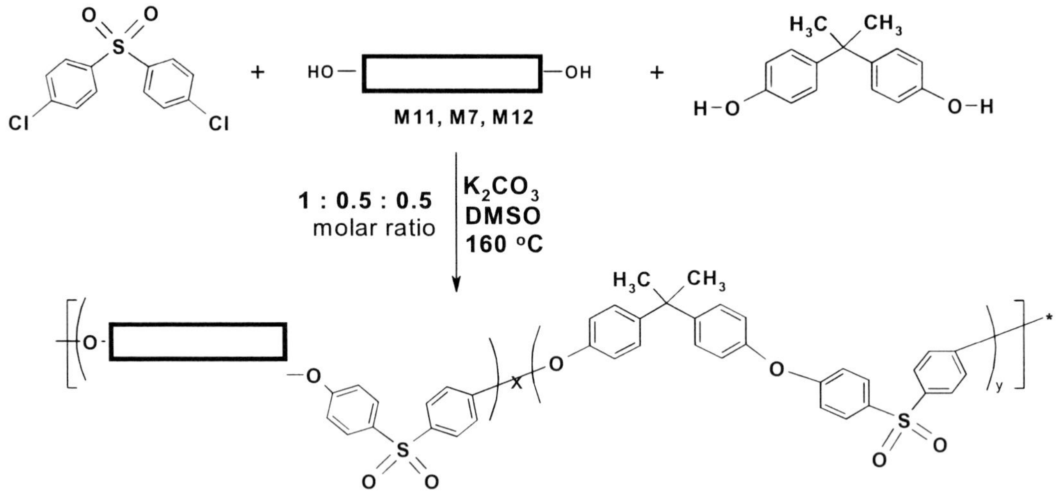

Scheme 12. Synthesis of copoly(azomethine-ether-sulfone)s PAMS 15 - PAMS 17.

Inherent viscosity measurements on DMF solutions of polymer samples gave values ranging between 0.15 and 0.44 dL/g, respectively (Table 9). GPC measurements of the same DMF samples provided numerical average molecular weight (M_n) values comprised between 5300 (PAMS 14) and 11700 (PAMS 11), which indicated fairly good molecular weight for these polymers with a rigid backbone.

The values of PI index are low and comprised between 1.08 (PAM 6) and 1.3 (PAMS 16) (except for PAMS 11), so, these polymers have a narrow molar mass distribution (Table 9). Thermal behavior of the polymers was observed by TGA, DSC and polarized light microscopy (OPM). The data are shown in Table 10. The polymers exhibit high thermal stability: slight decomposition started above 200 °C but with insignificant weight loss up to 370 °C; a 10 % weight loss was seen at temperatures above 400 °C. The highest thermal stability was recorded for the polymers PAM 5 and PAM 6, which contain bisphenol A or bisphenol AF, respectively, as a non-mesogenic unit. A birefringent fluid is the confirmation of liquid crystallinity [52] and OPM data revealed that some of the polymers exhibited TLC behavior.

The polymers PAM 5 and PAM 6 showed a strong birefringence both on heating and cooling, while the polymers PAMS 11, PAMS 15, PAMS 16 showed birefringence only at first heating, probably due to the partial decomposition of the polymer near the clearing temperature. Another explanation for this behavior could be the high rigidity of the backbone which hinders the ordering of macromolecular architectures from melting state. The thermotropic polymers (PAM 5, PAM 6, PAMS 11, PAMS 15, PAMS 16), when observed by OPM, showed a fine texture (figure 9), similar to those reported for thermotropic polymers having stiff chains without flexible spacers [8]. The polymers PAMS 12 and PAMS 14, didn't show any birefringence.

The polymers PAMS 13 and PAMS 17 didn't melted up to 350 °C, so that the thermotropic behavior could not be studied; their higher values of melting temperature (T_m) are probably due to the increase of the rigidity of the azomethinic mesogenic units, which allow a more efficient packing of the macromolecular chains.

Table 8. Spectral and elemental analysis data of the copolymers

Code	Molecular formula Formula weight	Elemental analysis N, %	S, %	IR absorption bands (cm^{-1}) and their corresponding assignment	λ^{a}_{max} / nm
PAM 5	$(C_{54}H_{43}N_2O_4)_n$ $(783.96)_n$	3.57 (4.28)	-	3060 (v=C-H aromatic), 2960, 2885 (vCH_3) 1625 (vCH=N), 1590, 1490 (vC=C, aromatic), 1230, 1250 (vC-O-C), 1085 (C-Cl para).	336
PAM 6	$(C_{55}H_{40}F_6N_2O_4)_n$ $(906.93)_n$	3.09 (3.82)	-	3075 (v=C-H aromatic), 1620 (vCH=N), 1590, 1495 (vC=C, aromatic), 1250 (vC-O-C), 1090 (C-Cl para).	328
PAMS 11	$(C_{50}H_{34}N_2O_6S)_n$ (790.90)n	3.54 (3.51)	4.05 (3.22)	3075 (v=C-H aromatic), 1630 (vCH=N), 1595, 1510 and 1500 (vC=C, aromatic), 1330 (vSO_2 asym.) and 1160 (vSO_2 sym.) 1245 (vC-O-C), 1095 (C-Cl para).	342
PAMS 12	$(C_{38}H_{26}N_2SO_3)_n$ $(622.7)_n$	4.5 (4.67)	5.15 (5.42)	3090 (v=C-H aromatic), 1625 (vCH=N), 1595, 1510 and 1500 (vC=C, aromatic), 1330 (vSO_2 asym.) and 1160 (vSO_2 sym.) 1250 (vC-O-C), 1100 (C-Cl para).	337
PAMS 13	$(C_{32}H_{22}N2O_4S)_n$ $(530.61)_n$	5.28 (3.7)	6.04 (6.49)	3080 (v=C-H aromatic), 1640 (vCH=N), 1595, 1510 and 1500 (vC=C, aromatic), 1300 (vSO_2 asym.) and 1160 (vSO_2 sym.) 1250 (vC-O-C), 1085 (C-Cl para).	342
PAMS 14	$(C_{25}H_{17}NO_4S)_n$ (427.48)n	3.28 (3.6)	7.5 (6.66)	3080 (v=C-H aromatic), 1625 (vCH=N), 1585, 1505 and 1490 (vC=C, aromatic), 1320, 1295 (vSO_2 asym.) and 1150 (vSO_2 sym.) 1245 (vC-O-C), 1100 (C-Cl para).	332
PAMS 15	$(C_{79}H_{62}N_2O_{10}S_2)_n$ $(1065)_n$	2.22 (2.87)	5.08 (6.84)	3050 (v=C-H aromatic), 2980, 2930, (vCH_3) 1630 (vCH=N), 1595, 1510 and 1500 (vC=C, aromatic), 1330 (vSO_2 asym.) and 1160 (vSO_2 sym.) 1250 (vC-O-C), 1090 (C-Cl para).	332
PAMS 16	$(C_{65}H_{48}N_2O_9S_2)_n$ $(1263.51)_n$	2.63 (3.28)	6.02 (5.84)	3080 (v=C-H aromatic), 2990, 2940, (vCH_3) 1635 (vCH=N), 1590, 1510 and 1495 (vC=C, aromatic), 1330 (vSO_2 asym.) and 1160 (vSO_2 sym.) 1250 (vC-O-C), 1070 (C-Cl para).	337
PAMS 17	$(C_{59}H_{44}N_2O_8S_2)$ $(973.15)_n$	2.88 (3.17)	6.59 (5.47)	3060 (v=C-H aromatic), 2990, 2920, (vCH_3) 1635 (vCH=N), 1595, 1515 and 1500 (vC=C, aromatic), 1305 (vSO_2 asym.) and 1160 (vSO_2 sym.) 1250 (vC-O-C), 1090 (C-Cl para).	332

[a] DMF solutions.

Table 9. Viscosity and molecular weight data of the copolymers

Code	M_n^a	M_w/M_n^b	η^a_{inh} (dLg^{-1})
PAM 5	5 403	1.13	0.32
PAM 6	5 947	1.08	0.33
PAMS 11	11 695	2.66	0.33
PAMS 12	8 362	1.19	0.15
PAMS 13	8 812	1.21	0.44
PAMS 14	5 313	1.15	0.33
PAMS 15	6 543	1.24	0.29
PAMS 16	7 100	1.30	0.31
PAMS 17	8 126	1.12	0.24

[a] GPC measurements from DMF solutions;
[b] Polydispersity Index;
[c] Inherent viscosity measured on DMF solutions at 0,2 % wt. concentration, at 20 °C.

Table 10. Thermal characterization of copolymers

Code	OPM	DSC				TGA
	Phase transitions	T_g	Phase transitions	Enthalpies[b] (J/g)	ΔT	T_o / T_{10}^a
PAM 5	Cr 262 LC 312 I (2H)	120	Cr 263 LC 295 I (1H) Cr 265 LC 285 I (2H)	ΔH_m=6.8; ΔH_i=57.6 ΔH_m=52.8; ΔH_i=1.11	32 20	248 / 411
PAM 6	Cr 278 LC 293 I	128	Cr 260 LC 303 I (1H) Cr 262 LC 280 I (2H)	ΔH_m=6.8; ΔH_i=4.6 ΔH_m=7.3; ΔH_i=3.8	43 18	282 / 438
PAMS 11	Cr 287 LC 305 I	127	Cr 284 LC 303 I	ΔH_m=3.38; ΔH_i=2.04	19	210 / 470
PAMS 12	Cr 180 I	162	Cr 190 I			205 / 475
PAMS 13	Cr, T_m > 350°C	-	-			207 / 378
PAMS 14	Cr 188 I	110	-			203 / 420
PAMS 15	Cr 225 LC 258 I	125	Cr 214 LC 261 I	ΔH_m=12.5; ΔH_i=4.6	47	200 / 450
PAMS 16	Cr 159 LC 191 I	143	Cr 153 LC 188 I	ΔH_m=9.06; ΔH_i=3.2	35	195 / 445
PAMS 17	Cr T_m > 350 °C	-	-			204 / 394

[a] Determined by TGA, in air: T_0 = onset of decomposition temperature; T_{10} = temperature at 10 % weight loss;
[b] Enthalpies subscript: m = melting; lc-i = liquid crystalline–isotropic transition.

The transition temperatures of the polymers obtained from OPM measurements were compared with those measured by DSC. The small differences observed could be explained by the variations of the heating/cooling rate, the amounts of sample used, and the presence of an inert gas (nitrogen) during the DSC measurement, versus air during OPM. The representative thermograms obtained by DSC analysis of thermotropic polymers are given in figure 10.

All the samples exhibited at least two endothermal peaks in DSC analysis, during the first heating scan (1H). The first endothermal peak was assigned to a solid–liquid crystalline transition (T_m) and the last one was assigned to a liquid crystalline–isotropic transition (T_i). During the second heating (2H) only PAM 5 and PAM 6 samples showed endothermal peaks (figure 10a), while the other polymers showed only a glass transition (T_g). The polymers PAMS 11, PAMS 15 and PAMS 16 showed two peaks in the first heating scan (figure 10b). Integrating the DSC thermogram peaks over a range of temperatures enabled the enthalpy of the samples for that transition to be obtained (table 10). For samples PAM 6, PAMS 11, PAMS 15, PAMS 16, ΔH_m and ΔH_i are low and as expected, the crystalline-liquid crystalline transition involved more energy than the liquid crystalline-isotropic transition ($\Delta H_m > \Delta H_i$).

a)

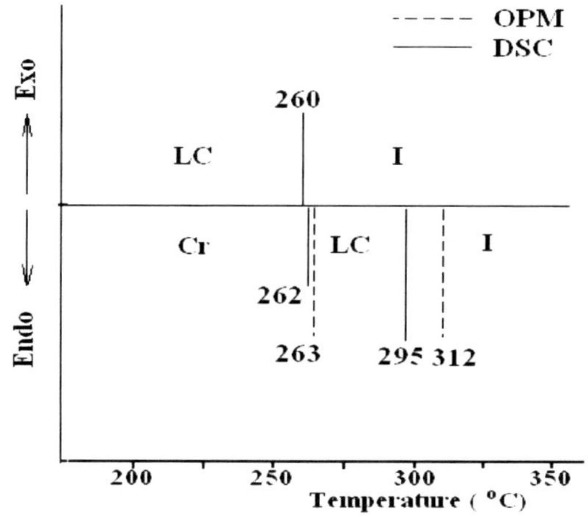

b)

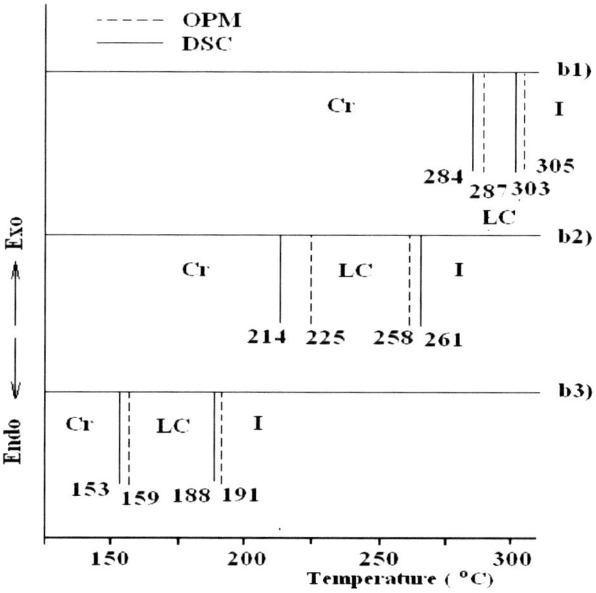

Figure 10. Composite DSC/OPM diagrams of samples: a) PAM 5, 15 °C/min first heating (1H), first cooling (1C), second heating (2H), 10 °C/min; b1) PAMS 11, b2) PAMS 15, b3) PAMS 16, first heating, 10 °C/ min.

For PAM 5 a less than usual aspect of the thermogram was observed. In the first heating scan, ΔH_m is lower than ΔH_i, while in the second scan the ΔH_m is higher than ΔH_i. The explanation for this behavior could be the thermal history of the sample and the more oriented state in a solid in the second scan. For samples PAM 6, PAMS 11, PAMS 15, PAMS 16, the transition energy is lower than for sample PAM 5. This behavior could be explained by a more oriented state for polymers containing tetracoordinated carbon atoms which are more flexible. When the samples were quenched from the melting state and heated again, T_g values were obtained. The T_g values of the random polymers (PAMS 15 - PAMS 17) were slightly lower then those of alternated polymers. By comparing the thermal data in Table 10, some interesting correlations can be observed and thus the polymers under investigation can be divided in to four series. Series 1: poly(azomethine-ether-sulfone)s PAMS 11 - PAMS 14, in which different mesogenic units are connected by the same kinking unit [4,4'-sulfonyl-bis-(*p*-chlorophenyl)]; series 2: the polymers PAM 5, PAM 6, PAMS 11, in which the same mesogenic unit is connected by the different kinking units such as bisphenol A, bisphenol AF or 4,4'-sulfonyl-bis-(*p*-chlorophenyl), respectively; series 3: the copolymers PAMS 15, PAMS 16, PAMS 17 in which 4,4'-sulfonyl-bis-(*p*-chlorophenyl) units are copolymerized with a mixture of equal amounts of bisphenol A and mesogenic bisphenols M1–M3; series 4: the polymers PAM 5, PAMS 11, PAMS 15, in which the same mesogenic unit is connected with bisphenol A (PAM 5), 4,4'-sulfonyl-bis-(*p*-chlorophenyl) (PAMS 11), or 4,4'-sulfonyl-bis-(*p*-chlorophenyl) and bisphenol A (PAMS 15). The structural features of the mesogenic groups affect the onset, nature, and stability of the liquid-crystalline mesophases through their rigidity, their polarizability and their aspect ratio [4]. In the polymer 1 series, it was observed that in the case of PAMS 13 which contains a rigid mesogenic unit (the monomer M 12 does not contain ether linkages) the melting point is higher than the decomposition temperature.

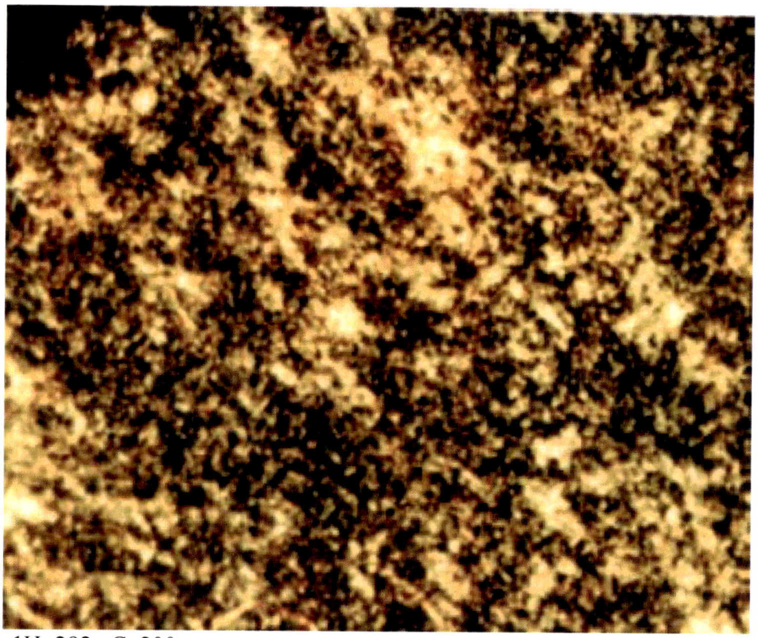

PAMS 11: 1H, 292 oC, 200x

Figure 11. (Continued on next page.)

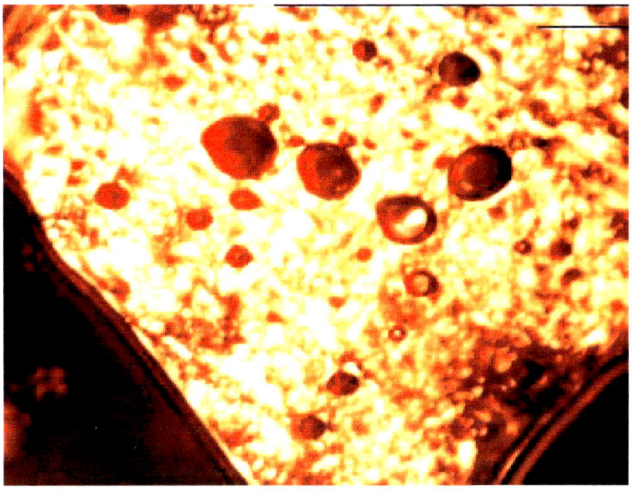

PAM 5: 1C, 187 oC, 200x

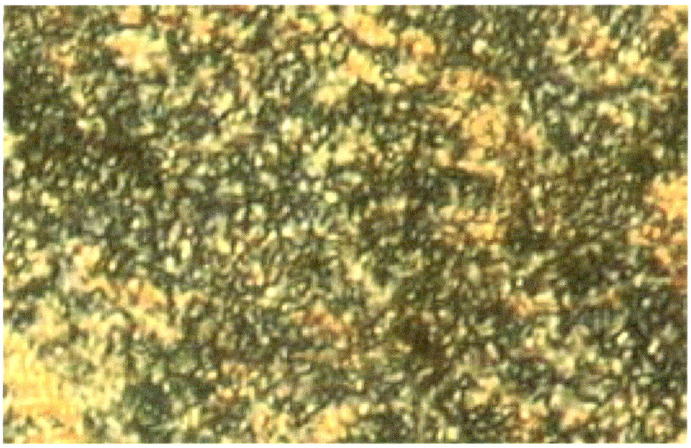

PAMS 16: 1H, 166 oC, 200x

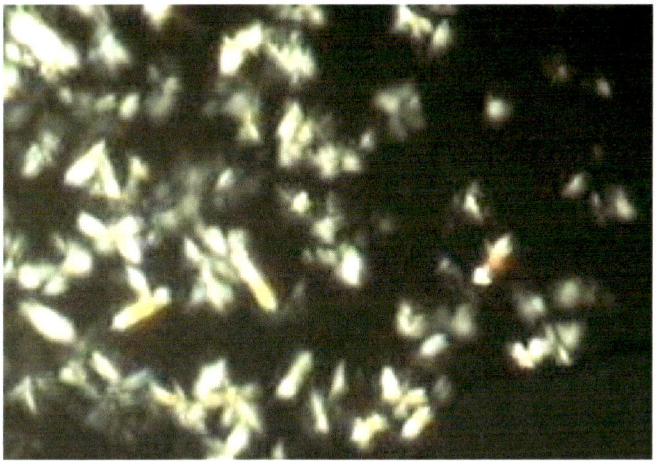

PAM 6: 1C, 276 oC, 200x

Figure 11. OPM images of representative textures of thermotropic copolymers.

For the polymer PAMS 11 which contains two ether linkages inside the mesogenic group, a stable mesophase was obtained. In contrast, no mesophase can be seen in polymers PAMS 13 and PAMS 14 which are not sufficiently rigid to be candidates for liquid crystalline behavior. In series 2, it was observed that all of the polymers have mesomorphic behavior. The melting point increases in the following order: T_m (PAM 5) < T_m (PAM 6) < T_m (PAMS 11), which is the order of the increasing rigidity of the kinking units, while the isotropization temperature is little affected: T_i (PAM 5) ~= T_i (PAM 6) ~= T_i (PAMS 11). The polymer PAM 5, which contains bisphenol A as a non-mesogenic unit, has the lowest T_m value, while the polymer PAMS 11, which contains the diphenylsulfone group as a non-mesogenic unit, has the highest T_m value. In conclusion, the stability of the mesophase domain is larger when the non-mesogenic unit is more flexible.

For series 3, a decrease in the melting temperature and an increase of the mesophase temperature range for random polymers versus alternating ones are observed. In series 4, keeping the mesogenic group constant (PAMS 11, PAMS 15, PAM 5), a schematic phase diagram shows a reduction of the melting point (figure 12). As can be seen, the incorporation of bisphenol A into the polymer backbone by polycondensation led to a decrease of melting (T_m) and isotropization temperatures (T_i), and an increase of the mesophase stability.

In order to observe the possible conformation of poly(azomethine-ether-sulfone)s, a simple model molecule was simulated by molecular modeling methods (Mm+, figure 13) [11]. One can observe that rotations around the ether bonds minimize the disruptive influence of the kinked units [bis(*p*-chlorophenyl)sulfone, bisphenol A] on the straightness of the chain. Therefore, this could be an explanation for the LC behavior of these polymers.

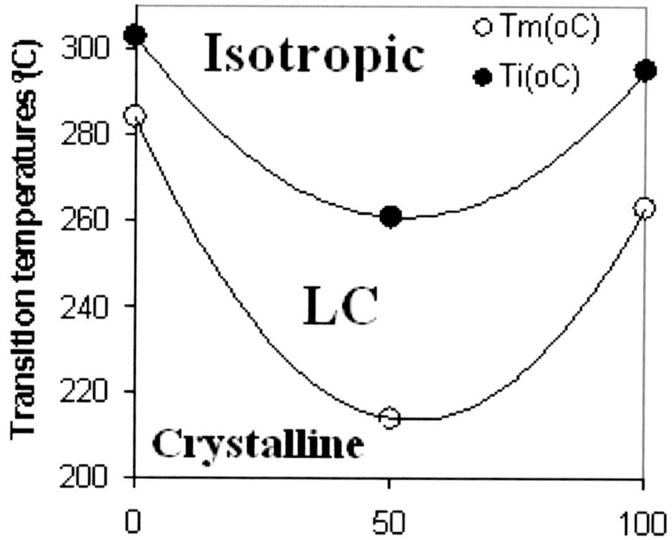

Figure 12. Dependence of transition temperatures and mesophase range on the bisphenol A content.

Figure 13. Simulation of PAMS 16 structure.

The WAXD measurements, on powder samples at room temperature, for the polymers PAMS 11, PAMS 12, PAMS 15 and PAMS 16 exhibited a semicrystalline morphology (figure 14). Some diffraction maxima are present mainly around (2θ, deg.): 20 (PAMS 11, PAMS 15), 22.2 (PAMS 16) and 26.7 (PAMS 12). This suggested that the polymers had different degrees of crystallinity in the solid state with a significant influence on the ordering ability of the mesogenic cores in the mesophase [53]. The thermotropic polymers PAMS 11, PAMS 15, PAMS 16 showed two main diffraction peaks, the first one being very intense, around 20 (2θ, deg.), whereas the isotropic polymer PAMS 12 showed one weaker broad diffraction peak, around 27 (2θ, deg.), which denoted less crystallinitty compared to the above thermotropic polymers.

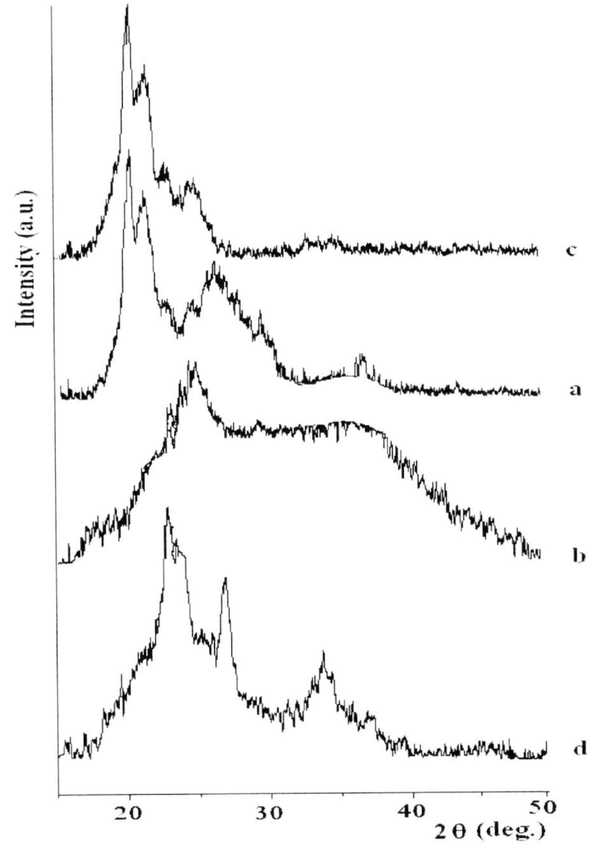

Figure 14. WAXD diffractograms of polymer samples: a) PAMS 11, b) PAMS 12, c) PAMS 15, d) PAMS 16 at room temperature.

Table 11. The temperature dependence of electrical conductivity of samples PAMS 11 and PAMS 15

Polymer	d (μm)	ΔT (K)	σ_c ($\Omega^{-1}\times cm^{-1}$)	σ_T ($\Omega^{-1}\times cm^{-1}$)	T_c (K)	ΔE (eV)
PAMS 11	1.84	300 - 475	1.1×10^{-8}	81.5×10^{-8}	340	1.5
PAMS 15	1.05	300 - 480	1.25×10^{-8}	8.1×10^{-8}	345	0.3

Note: d – film thickness; ΔT- temperature range in which the heat treatment of the samples was performed ; σ_c si σ_T – electrical conductivity at room temperature before and after heat treatment, respectively; T_c – characteristic temperature (for T>Tc the polymer shows an intrinsic conduction); ΔE – thermal activation energy of electrical conduction.

The electrical conductivity was measured for two polymers: PAMS 11 and PAMS 15. In order to see the dependence of the electrical conductivity on temperature, thin films were deposited on glass substrates from DMF solutions. To obtain polymeric films with stable structures, the samples, after preparation, were successively heated and cooled within a certain temperature range. While the polyethersulfone is considered to be an insulator, having electrical conductivity about 10^{-16} ($\Omega^{-1}\times cm^{-1}$), the studied polymers showed a high increase of electrical conductivity, with values of 10^{-8} ($\Omega^{-1}\times cm^{-1}$) (table 11) [54]. However, it was observed that polymeric films, after this thermal treatment, have a better electrical conductivity than the untreated polymeric films. It is believed that the supramolecular organization in a liquid crystalline state of these polymers increases the electrical conductivity.

New copoly(azomethine-ether)s, and copoly(azomethine-ether-sulfone)s, derived from azomethine monomers, 4,4'-sulfonyl-bis-(*p*-chlorophenyl) or/and bisphenol A and bisphenol AF were prepared in order to obtain more information about the properties of polyazomethines containing various kinking groups. WAXD showed that these polymers were semicrystalline in nature, with crystalline peaks around 20–27 (2Θ, deg.). According to TGA and DSC analysis, all the polymers have good thermal stability and insignificant weight loss until 370 °C. Furthermore, the polymers were very soluble in common organic solvents. Although the bis(*p*-chlorophenyl)sulfone, bisphenol AF and bisphenol A units are not favorable for liquid crystalline behavior, thermotropic properties could be obtained when such units were incorporated into a polyazomethine backbone. The existence of the mesophases was observed even without using a spacer between mesogenic units. The melting and isotropization temperatures strongly decreased when the relatively rigid geometry of mesogenic monomers M 7, M 11 - M 13 were disturbed by the introduction of kinking units of 4,4'-sulfonyl-bis-(*p*-chlorophenyl), bisphenol AF and especially in the case of bisphenol A, which promoted mesomorphic behavior within a large temperature range. The incorporation of bisphenol A by ether linkages into the main chain of a polymer was useful for improving the solubility and for decreasing the melting and the isotropization temperatures. An increase in electrical conductivity of 10^{-8} ($\Omega^{-1}\times cm^{-1}$) was obtained for linear copoly(azomethine-ether-sulfone)s.

6. Thermotropic Copoly(Azomethine-Eter-Sulfone)s – Influence of Molar Ratio between Mesogenic and Nonmesogenic Monomers on the TLC Behavior

The modification of the polymeric chain by random copolymerization constitutes an efficient method of decreasing the melting temperature. The destruction of the periodicity of macromolecular chain inhibits the crystallization leading to a decreasing of the melting temperature, without adiacent loss of mesogenicity or a decreasing of the isotropization temperature. In this way the statistic copolymerization proved to be an efficient method in obtaining of thermotropic polymers with large mesophase ranges and melting temperatures which afford their processability [45].

Because our copoly(azomethine-sulfone)s described previously showed lower transition temperatures than the corresponding polyazomethines counterpart [3] new statistical copoly(azomethine-sulfone)s have been prepared and the influence of the molar ratio between the mesogen/non-mesogen units and copolymerization on the TLC behavior was investigated [55]. A series of copoly(azomethine-sulfone)s (PAMS 18 – PAMS 25) have been prepared by reaction of 4,4'-sulfonyl-bis-(*p*-chlorophenyl) with bisphenols: bisphenol A as non-mesogenic bisphenol and 4,4'-oxo-bis(4-hydroxybenzylideneiminophenylene) as mesogenic monomer, in various molar ratios (scheme 13).

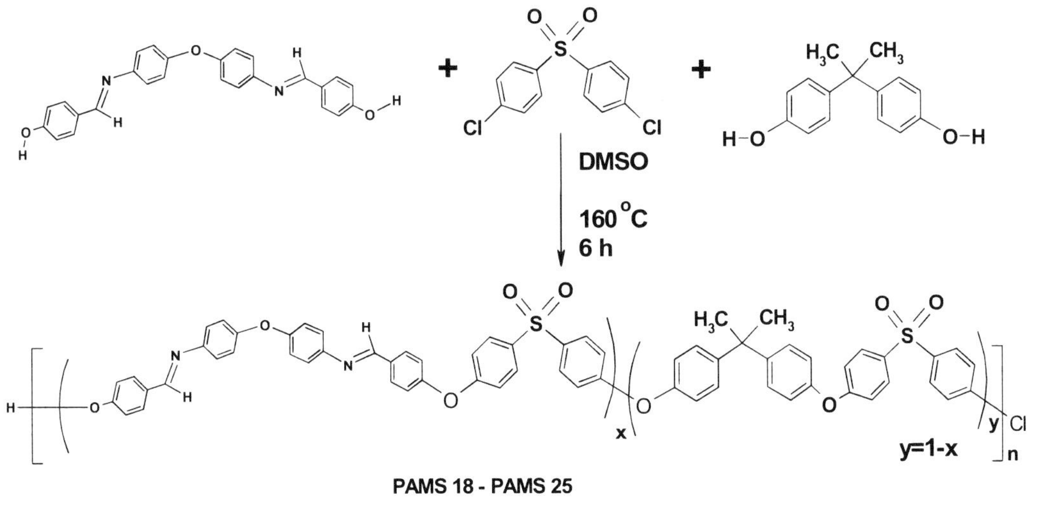

Scheme 13. Synthesis of statistic copoly(azomethine-sulfone)s PAMS 18 – PAMS 25.

The content of azomethine groups was calculated from IR spectra as a ratio between absorbances of azomethine and methyl (isopropylidene) groups $A_{CH=N}/A_{CH3}$:

$$A_{CH=N} = A_{1635}/A_{1590} \text{ and } A_{CH3} = A_{2980}/A_{3080}$$

The ratio of the two reference absorbances used (A_{1590}/A_{3080}) is generally constant and proach to the unity. Two baselines, free of any absorptions, were used: 3800 – 1800 cm^{-1} and 1800 – 915 cm^{-1}.

Each one was calculated by the formulae [47]: $A = \log_{10}(AC/AB)$, where:

- AC = distance (through the peak maximum) from the scale to the baseline;
- AB = distance from the scale to the top of the peak.

The values of the ratio $A_{CH=N}/A_{CH3}$ showed a decreasing tendency from 0.75 to 0.34, corresponding to the initial ratio of the two monomers involved (table 12). The good elemental analysis data showed a slow influence of the end functional groups on the measurements, suggesting relatively high values for the molecular weight. The M_n values, determined by GPC had values in the range 4500 – 8200 g/mol with narrow polydispersities (1.2; 1.3) (table 12).

Table 12. Characterization of statistic copoly(azomethine-sulfone)s PAMS 18 – PAMS 25

Code	Bisphenols molar ratio: azomethine / bis A	Yield (%)	M_n^a (M_w/M_n)	$A^b_{CH=N}/A_{CH3}$	Elemental analysis Calc. (Found)	
					N(%)	S(%)
PAMS 18	100 / 0	81	5400 (1.3)	-	4.50 (4.67)	5.15 (5.42)
PAMS 19	90 / 10	58	4500 (1.3)	0.75	3.64 (4.18)	5.55 (5.08)
PAMS 20	80 / 20	73	7700 (1.2)	0.86	4.03 (4.25)	5.37 (5.81)
PAMS 21	70 / 30	76	8200 (1.2)	0.49	2.59 (3.08)	6.04 (6.01)
PAMS 22	60 / 40	63	7100 (1.3)	0.41	2.22 (2.77)	6.21 (6.43)
PAMS 23	50 / 50	50	6500 (1.3)	0.52	2.72 (3.28)	5.98 (5.84)
PAMS 24	40 / 60	54	6800 (1.2)	0.41	2.22 (2.64)	6.21 (6.39)
PAMS 25	30 / 70	40	6900 (1.2)	0.34	1.89 (2.34)	6.36 (5.79)

a GPC measurements in DMF
b IR absorbance ratios calculated using the formulae: A=log10(AC/AB) [47].

These values corresponded to approximatelly ten mesogenic units on a macromolecular chain. The molecular weights of the statistic copolymers are greater than that of the liniar copolymer PAMS 18, probably duet o the increase in solubility of copolymers containing bisphenol A units. All these copolymers were completelly soluble in THF, 1,4-dioxane, NMP, DMF, DMSO and partially soluble in chloroform. Thermotropic data are presented in table 13.

Table 13. Thermotropic behavior of the copolymers

Code	Transition temperatures (°C) DSC / TOA				Transition temperatures (°C) (OPM)[b]
	T_g[a]	T_m	T_i	T_i-T_m	
PAMS 18	162	190	-[c]	-	Cr 180 LC[c]
		191	-[c]	-	
PAMS 19	133	165	187	22	Cr 169 LC[d] 189 I
		166	187	21	
PAMS 20	115	167	187	20	Cr 171 LC 185 I
		172	187	15	
PAMS 21	108	168	183	15	Cr 150 LC 184 I
		158	185	27	
PAMS 22	145	162	187	25	Cr 169 LC[d] 188 I
		158	195	37	
PAMS 23	143	153	188	35	Cr 159 LC[d]
		150	191	41	
PAMS 24	123	151	-[c]	-	Cr 152 LC[c]
		150	-[c]	-	
PAMS 25	-	143	-	-	Cr 145 I
		145	-	-	

[a] Glass transition temperature measured in the second heating scan, at a heating rate of 10°C/min; T_m = melting temperature; T_i = isotropization temperature; $T_i - T_m$ = mesophase range; [b] Cr = crystal; LC = liquid crystalline; I = isotropic; [c] Liquid crystalline only under shear stress applied; [d] Both LC and shear induced LC domains

The TGA measurements in air showed similar behavior for all the copolymers. Although the thermal decomposition started approximatelly at 200 °C, its very slow evolution gave 10 % wt. loss at 430 – 469 °C.

OPM measurements showed that some copolymers had TLC behavior. A strong birefringence was observed at heating. At cooling the birefringence diminished, probably duet o the high rigidity of the backbones which hindered the supramolecular organization.

The samples PAMS 18, PAMS 24 and PAMS 25 showed birefringence only under shear stress (table 13, figure 15a). The samples PAMS 19, PAMS 22, PAMS 23 exhibited birefringence and isotropic domains which became birefringent only under shear (figure 15b, PAMS 19). The samples PAMS 20 and PAMS 21 are totally birefringent (figure 15c). A similar behavior was reported for statistic copolyesters [16, 17].

Data for the TLC behavior of copolymers are presented in table 13. DSC, TOA and OPM data are cumulated in the composite diagrams 16. All the copolymers exhibited two endothermal peaks in DSC scans. From corelation with OPM observations, the first endothermal peak was assigned to a crystalline-LC transition, while the second one to a LC-isotropic transition. In the second heating scan, all the samples exhibited only a glass transition. The T_m values of the copolymers ranged between 143 °C and 190 °C (as measured by DSC). Under OPM the variation of T_m values ranged between 145 °C and 191 °C, which denoted a good agreement between the two methods used.

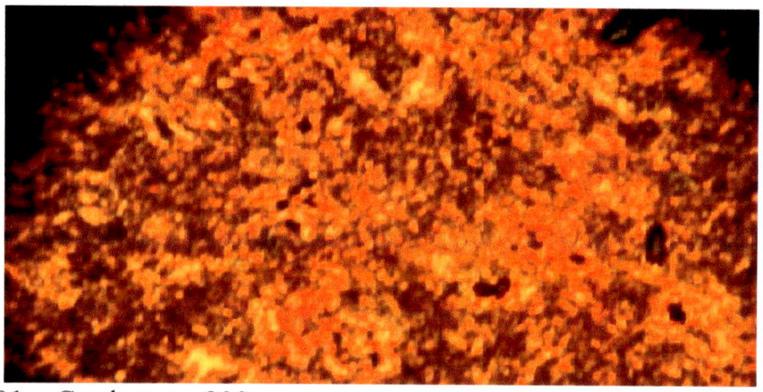

(a) PAMS 18, 201 o C sub stres, 200 x

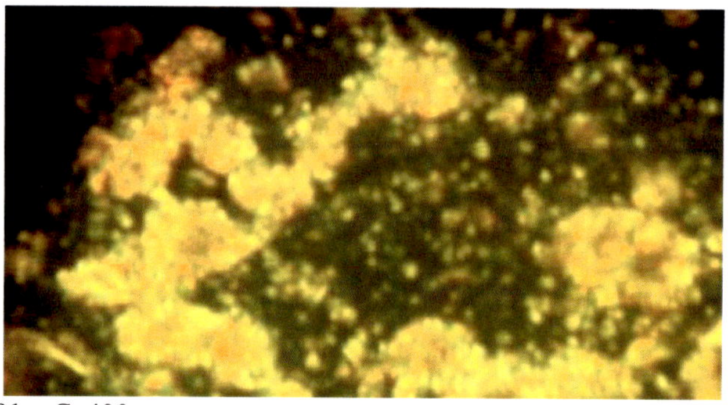

(b) PAMS 19, 181 o C, 400 x

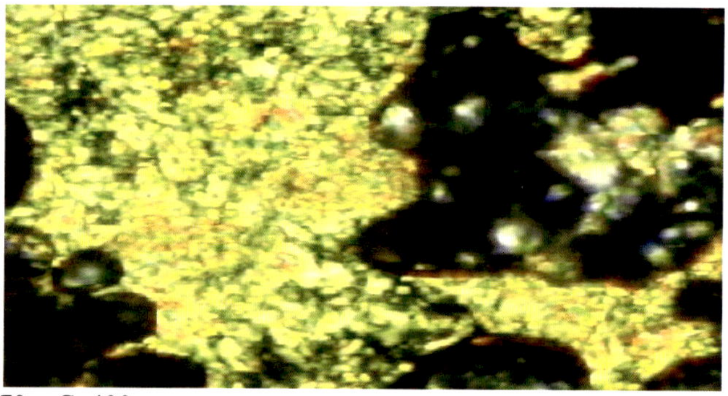

(c) PAMS 21, 170 o C, 400 x

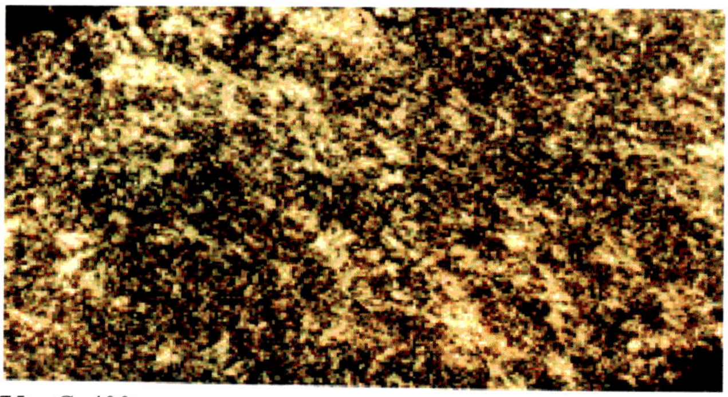

(d) PAMS 20, 175 o C, 400 x

Figure 15. OPM images of the textures of copolymers.

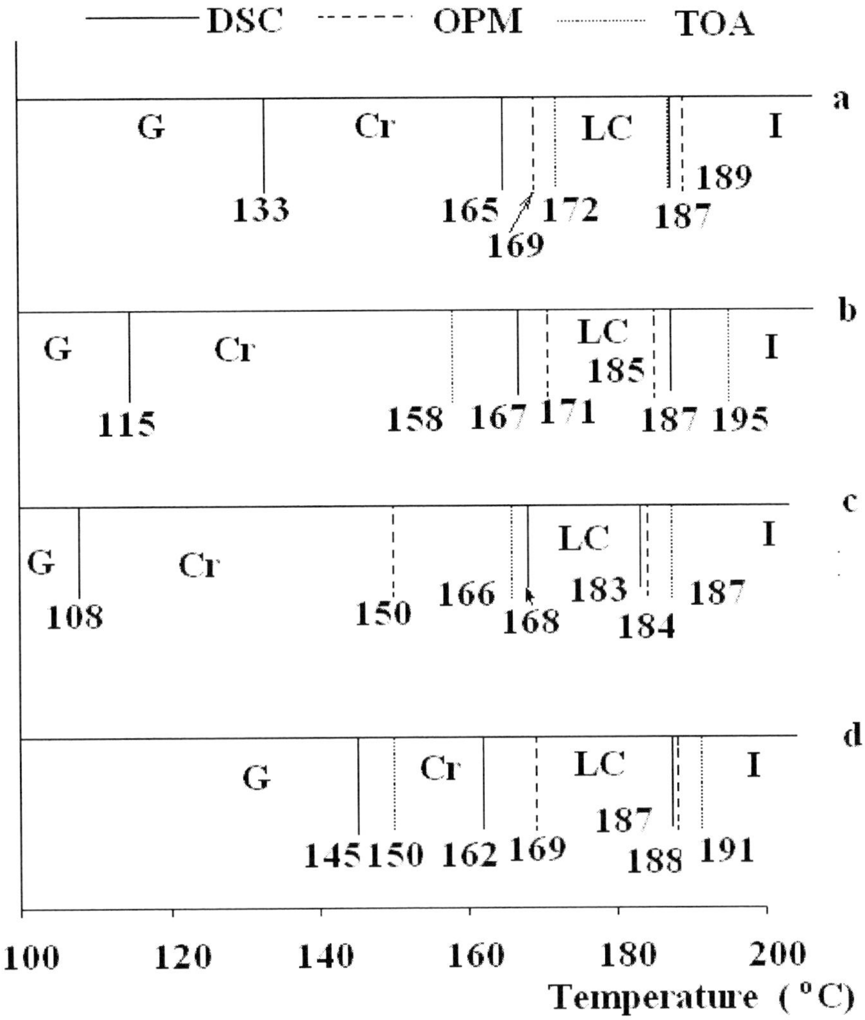

Figure 16. Composite DSC/TOA/OPM diagrams of copoly(azomethine-sulfone)s: a) PAMS 19; b) PAMS 20; c) PAMS 21; d) PAMS 22

By comparison with the aromatic polyazomethines, which had higher melting points than the onset of decomposition temperature, the efficiency of the introduction of diphenylsulfone units and/or non-mesogenic bisphenol A units and also the random copolymerization technique it is obviously.

The X-ray diffraction curves (figure 17) showed semicrystalline peaks at 2θ (deg.) angles of: 22.7; 23.6; 26.8; 33.6. These maxima are more intense in the case of samples PAMS 20 and PAMS 21, which exhibited strong birefringence in the melting state, while in the case of samples PAMS 19, PAMS 22, PAMS 23, which showed birefringent and isotropic domains under OPM, the intensity of the peaks decreased. This behavior suggested that the different degree of the crystallinity of the sample in the solid (powder) state could influences the ordering abbility of the mesogen carrying chains into the mesophase. The selected distance spacings corresponding to these peaks are listed in table 14. These distances were assigned to the lateral spacings between macromolecular chains [56, 57].

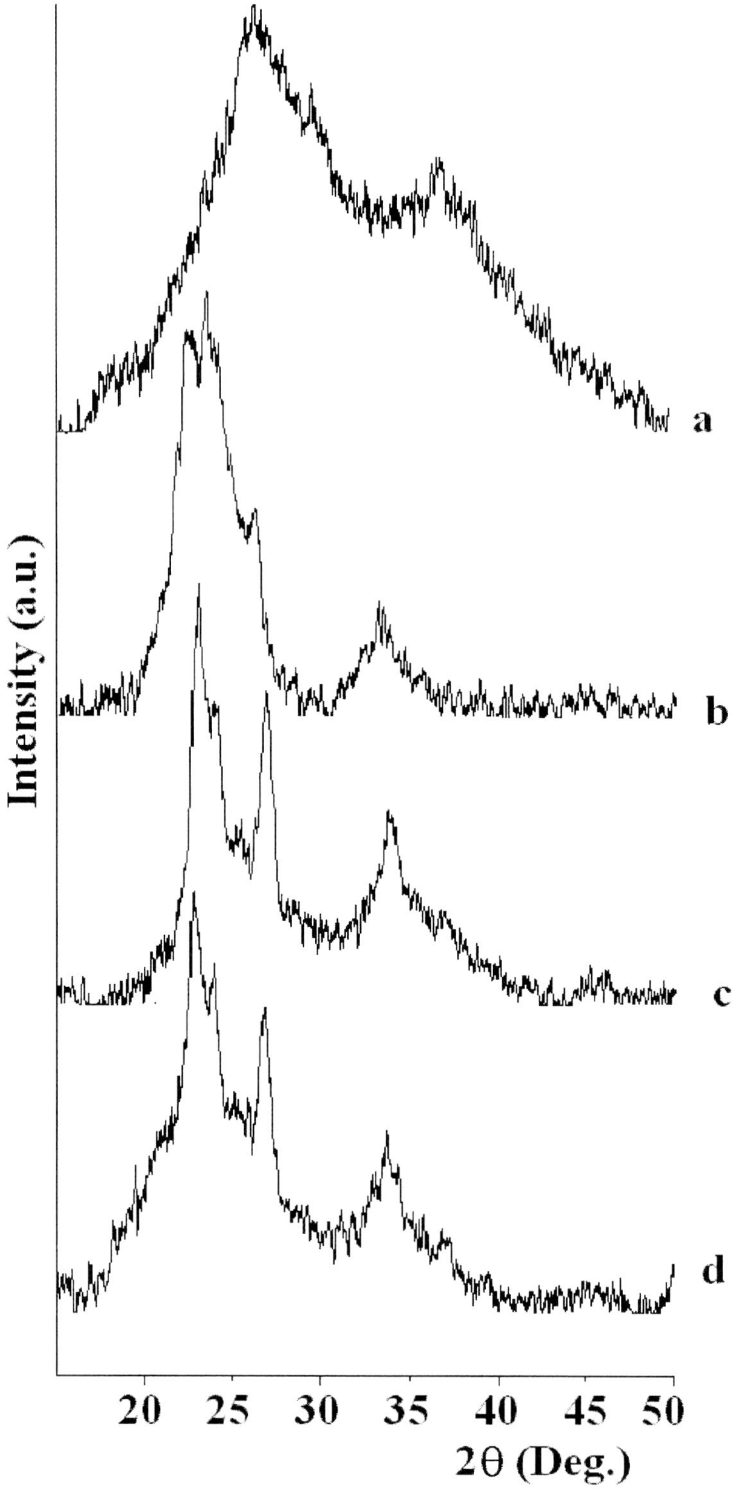

Figure 17. (Continued on next page.)

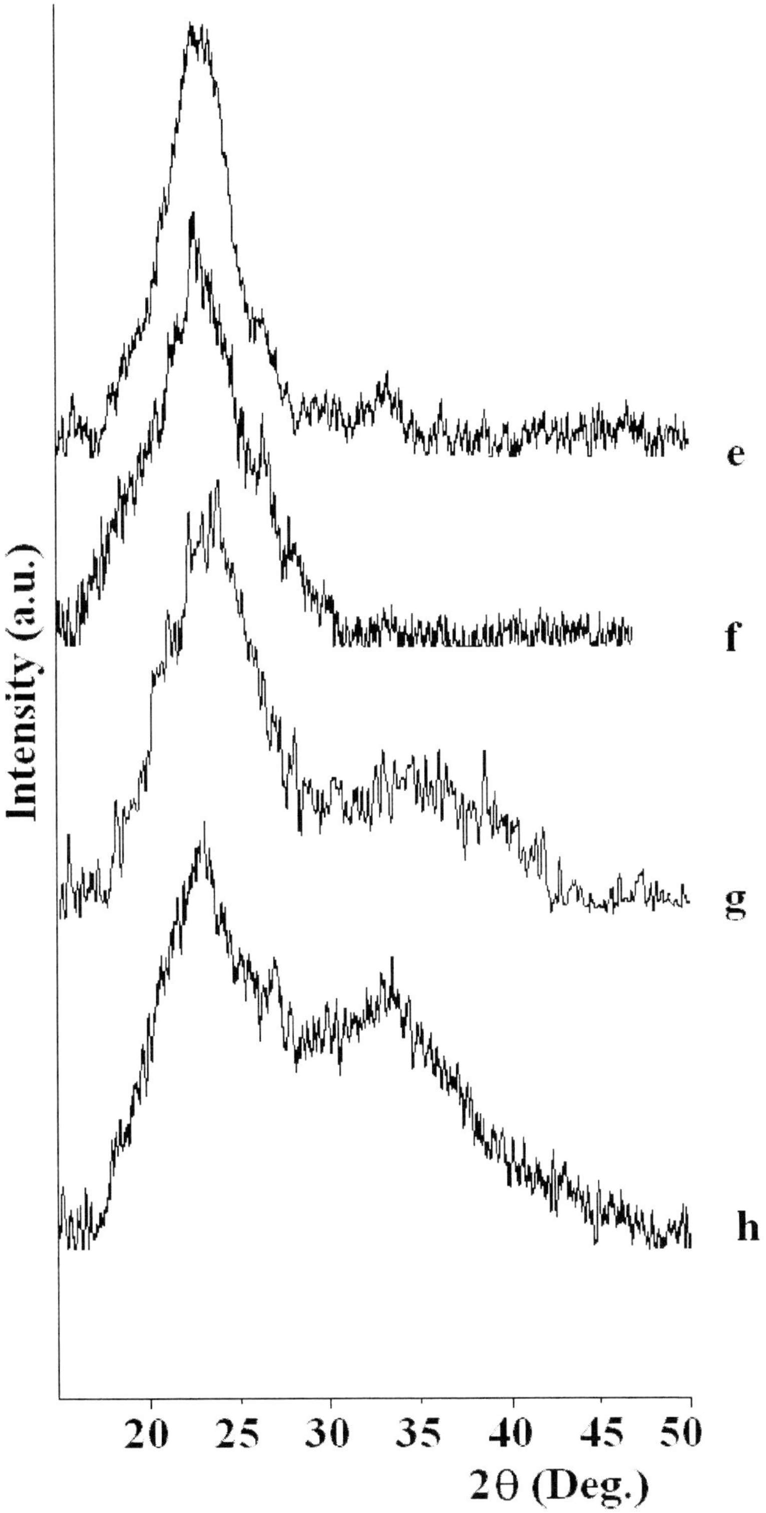

Figure 17. Room temperature WAXD patterns of copoly(azomethine-sulfone)s PAMS 18 (a) – PAMS 25 (h) .

Table 14. WAXD distance spacings of copolymers PAMS 18 – PAMS 25

Code	d_1(Å)	d_2 (Å)	d_3 (Å)	d_4 (Å)
CoP1	4,0	3,8	3,5	2,8
CoP2	4,5	4,3	3,9	3,0
CoP3	4,6	4,3	3,9	3,1
CoP4	4,5	2,0	3,8	3,0
CoP5	4,5	4,4	3,9	3,1
CoP6	4,5	4,3	3,8	3,0
CoP7	4,5	4,2	4,5	2,9
CoP8	4,5	4,2	3,8	3,0

A suggestive image of the way by which the random copolymerization influences the TLC behavior is depicted in figure 18.

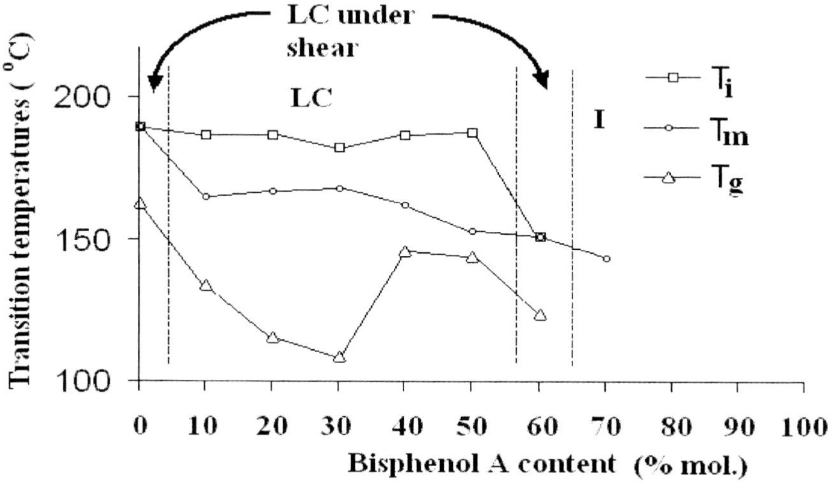

Figure 18. The variation of the transition temperatures with content of bisphenol A.

In the left part of the diagram there is the copolymer without bisphenol A (PAMS 18) which is LC only under shear. By increasing the content of bisphenol A, more flexible units are added and the T_m values decreased. At the same time, this decreasing of rigidity favours the appearance of the LC behavior. The greatest mesophase range was attained at 50 % bisphenol A. At further increasing of the non-mesogenic content, the LC was affected, too much mobility of the chains led to appearance of isotropic domains (PAMS 24). Finally, at 70 % non-mesogenic bisphenol, the ordering abbility was lost completely, only isotropic behavior being observed (PAMS 25).

7. Conclusion

Linear poly(azomethine-ether)s exhibited LC behavior, although kinked bisphenol A moiety (used as non-mesogenic counterpart) is not favourable for LC ordering. Molecular weight of the polymer strongly influenced the number of phase transitions and the mesophase range. Side chain poly(azome- thine-ether-sulfone)s were built by directly coupling the azomethine mesogen with the polysulfone backbone. LC behavior was observed and the influence of the substitution degree of the modified polymer on the LC properties was investigated. Various linear poly(azomethine-ether-sulfone)s also exhibited LC behavior while having good thermal stability. Generally, fine grainy textures were observed for the studied polymers and interesting semiconducting properties have been recorded for two copolymers.

7. AKNOWLEDGEMENTS:

authors are grateful to CNCSIS (grant 104GR/25.05.2007) and CEEX-MATNANTECH (Project 10/2005) for the financial support.

8. REFERENCES

[1] Iwan, A.; Sek, D. *Prog. Polym. Sci.* 2007, doi: 10.1016/j.progpolymsci.2007.09.005
[2] Marin, L.; Cozan, V.; Bruma, M. Proceedings of 8th International Symposium on Polymers for Advanced Technologies, Budapest, Hungary, 2005, Section A, *http://www.e-polymers.org/PAT2005ePolymers/proceedings.htm*
[3] Marin, L.; Cozan, V.; Bruma, M. *Polym. Adv. Technol.* 2006, 17, 554.
[4] Demus, D.; Goodby, J.; Gray, G. W.; Spiess, H. W.; Vill V. editors, Handbook of Liquid Crystals, WILEY – VCH, Berlin, vol 3: High Molecular Weight Liquid Crystals, 1998, p. 341.
[5] Marin, L.; Cozan, V.; Bruma, M.; Grigoras, V. C. *Eur. Polym. J.* 2006, 42, 1173.
[6] Aly, K.; Khalaf, A. A.; Alkskas, I. A. *Eur. Polym. J.* 2003, 39, 1035.
[7] Egorov, E. A.; Savitskii, A. V.; Zhizhenkov, V. V.; Gorshkova, I. *A. Physics of the Solid State* 1998, 40, 1173.
[8] Windle, A. H. in Liquid Crystalline and mesomorphic Polymers, V.P. Shibaev, L. Lam (Eds.), Springer-Verlag, New York, 1994, p. 42.
[9] Demus, D.; Richter, L. Textures of Liquid Crystals, Verlag Chemie, Weinheim, 1978.
[10] Aly, K.; Khalaf, A. A.; Alkskas, I. A. *Eur. Polym. J.* 2003, 39, 1035.
[11] Hyper Chem (TM), Hypercube, Inc., 1115 NW 4th Street, Gainesville, Florida 32601, USA.
[12] Smith, C. D.; Holzberlein, R. L.; McGrath, J. E. Synthesis and characterization of random and segmented poly(azomethine-sulfone)s, in Polymeric Materials: Science and Engineering, Proceedings of the ACS Division of Polymeric Materials, Dallas, Texas, 1989, vol. 60, p.316.
[13] Rabilloud, G.; Sillion, B. *Eur. Polym. J.*, 1990, 26 (9), 967-976.
[14] D'Alelio, G. F.; Crivello, J. V.; Schoenig, R. F.; Huemmer, T. F. *J. Macromol. Sci.-Chem.* 1967, A1 (7), 1161-1249.
[15] Wiswanathan, R.; Johnson, B. C.; McGrath, J. E. *Polymer* 1984, 25 (12), 1827.
[16] Lenz, R.W. *Faraday Disc.*, 1985, 79, 29.

[17] Lenz, R.W.; Jin, *J.I. Macromolecules*, 1981, 14 (5), 1405-1411.
[18] Mahabadi, H.K.; Alexandru, L. J. *Polym. Sci., Polym. Chem. Ed.*, 1990, 28 (2), 231-243.
[19] Cozan, V.; Butuc, E.; Stoleriu, A.; Rusa, M.; Rusu, M.; Ni, Y.; Ding, M. *J.Macromol.Sci.-Pure Appl.Chem.* 1995, A32 (7), 1243-1262.
[20] Shaffer, T.D.; Percec, *V. Am. Chem. Soc. Div. Polym. Chem. Polym. Prepr.* 1985, 26 (2), 289.
[21] Shaffer, T.D.; Percec, V. *Makromol. Chem.* 1986, 187, 1431-1439.
[22] Auman, B.C.; Percec, V. *Polymer* 1988, 29, 938-949.
[23] Cooper, K.L.; Chen, D. H.; Huang, H.-H.; Wilkes, G.L.; McGrath, J.E. Polymeric Materials: Science and Engineering. Proceedings of the American Chemical Society Division of Polymeric Materials, vol. 60. Dallas, Texas, 1989, p. 322-326.
[24] Chang, Y.-M.; Hsu, C.-H.; Liu, M.-C. *Mater. Chem. Phys.* 1996, 43, 250.
[25] Chang, Y.-M.; Hsu, C.-H.; Liu, M.-C. *Chem. Abstr.* 1996, 124: 290439t.
[26] Heitz, W.; Brenda, M.; Heyde, G.; Wallach, T. *Am. Chem. Soc. Div. Polym. Chem. Polym. Prepr.* 1992, 33 (2), 201-202.
[27] Ni, Y.; Guan, J.; Jin, X. 'Synthesis and characterization of a series of new LC copolyesters containing biphenyl-ester sulfone' in. 'Liquid Crystalline Polymers', Carfagna, C. editor New York, Pergamon Press, 1993, p.25-29.
[28] Ni, Y.; Wang, Y.H.; Li, G.; Jin, X.M.; Ding, M.X. *Chinese Chem. Lett.* 1993, 4, 125-128.
[29] Shu, H.C.; Bar, C.L. *Mater. Chem. Phys.* 1993, 34 (1), 28-34.
[30] Shu, H.C.; Bar, C.L. *Chem. Abstr.* 1993, 19: 96314e.
[31] Brenda, S.; Heitz, W.; Karbach, A.; Wehrmann, *R. Macromol. Chem. Phys.* 1994, 195, 1327-1339.
[32] Yang, H.; Wu, X.; Li, *S. Chem. Abstr.* 1996, 124:203553j.
[33] Yang, H.; Wu, X.; Li, *S. Liq. Cryst.* 1996, 20, 261.
[34] Yang, H.; Wu, X.; Li, S. *Chem. Abstr.* 1996, 124: 290428p.
[35] a) Shaffer, T.D.; Percec, V. Makromol. Chem. 1986, 187, 111-123; b) Dass, N.N.; Date, R.W.; Fawcett, A.H.; McLaughlin, J.D.; Sosanwo, O.A. 'Recognition of a New Type of Main Chain Liquid Crystalline Polymer: Poly(1-olefin sulfones)', Macromolecules 1993, 26, 4192-4195; c) Braun, D.; Arnold, N.; Liebmann, A.; Schmidtke, I. Makromol. Chem. 1993, 194, 2687-2696; d) Braun, D.; Arnold, N.; Liebmann, A.; Schmidtke, I. Makromol. Chem. Phys. 1994, 195, 1603; e) Date, R.W.; Fawcett, A.H.; Geue, T.; Haferkorn, J.; Malcolm, R.K.; Stumpe, J. 'Self-Ordering within Thin Films of Poly(olefin sulfone)s', *Macromolecules* 1998, 31, 4935-4949.
[36] El-Hibri, M.J.; Nazabal, J.; Equiazabal, J.I.; Arzak, A. in 'Handbook of Thermoplastics', Olabisi, O. Ed., New York, Dekker 1997, p. 893.
[37] Cozan, V.; Gaspar, M.; Butuc, E.; Stoleriu, A. *Eur.Polym.*J. 2001, 37, 1-8.
[38] Avram, E.; Butuc, E.; Luca, C.; Druta, I. *J. Macromol. Sci.-Pure Appl. Chem.* 1997, A34 (9), 1701-1714.
[39] Cozan, V.; Butuc, E.; Avram, E.; Airinei, A. *Appl. Organomet. Chem.* 2003, 17 (5), 282-286.
[40] Airinei, A.; Butuc, E.; Cozan, V.; Ardeleanu, R. *Appl. Organomet. Chem.* 1998; 12: 485-489.

[41] Racles, C.; Avram, E.; Marcu, M.; Cozan, V.; Cazacu, M. *Polymer* 2000, 41, 8205-8211.

[42] Airinei, A.; Butuc, E.*; Cozan, V. J. Appl. Polym. Sci. 2000, 77, 2009-2016.*

[43] *Racles, C., Cozan, V. High Perform. Polym.* 2002, 14, 169-181.

[44] Cozan, V.; Avram, E. *Eur. Polym. J.* 2003, 39, 107.

[45] Donald, A.M.; Windle, A.H. Liquid Crystalline Polymers, Cambridge University Press, 1992, p.85.

[46] Keller, H.; Hatz, R. Handbook of Liquid Crystals, Verlag Chemie, Weinheim, 1980, p.76, p.82.

[47] Rabek, J. F. Experimental methods in polymer chemistry, New York: Wiley, 1980, p.241.

[48] Sivaramakrishnan, K. N.; Blumstein, A.; Clough, S.B.; Blumstein, R.B. *Polym. Prepr.* 1978, 19, 190–192.

[49] Marin, L.; Cozan, V.; Bruma, M. *Rev. Roum. Chim.* 2005, 50, 649–653.

[50] Racles, C.; Cozan, V.; Cazacu, M.; Foldes, E.; Sajo, I. *High Perform. Polym.* 2002, 14, 396–413.

[51] Hulubei, C.; Cozan, V. *High Perform. Polym.* 2004, 16, 405–418.

[52] Baron, M. *Pure Appl. Chem.* 2001, 73, 845–895.

[53] Shukla, U.; Rao, K.V.; Rakshit, A.K. *J. Appl. Polym. Sci.* 2003, 88, 153-160.

[54] Rusu M, Airinei A, Cozan V, Marin L, Salaoru I, Caplanus I, Rusu GI. On the electrical and optical properties of some poly(azomethine-sulfone)s in thin films. 7th International Conference on Physics of Advanced Materials, June 10–12 2004, Iasi, Romania.

[55] Marin, L.; Cozan, V.; Craus, M. L.; Pavel, D. in Advanced Research in Polymer Science, Firas Awaja (ed.), Transworld Research Network, Kerala, India, 2006, 37 (2) 661.

[56] Srinivasan, R.; Radhakrishnan, G. *J. Polym. Sci. Part A: Polym. Chem.* 2001, 39, 1743.

[57] Alexander, L. E. X-ray diffraction methods in Polymer Science, Willey, New York, 1969.

In: Advances in Functional Heterochain Polymers
Editor: Maria Cazacu
ISBN: 978-1-60456-598-0

Chapter 4

HETEROATOM-CONTAINING CONJUGATED POLYMERS

Mircea Grigoras[1]
Electroactive Polymers Department, "Petru Poni" Institute of Macromolecular Chemistry, 41A Gr. Ghica Voda Alley, 700487-Iasi, Romania

ABSTRACT

This chapter presents an up-to-date review of the latest researches into three heteroatom containing conducting polymer classes, i.e. polyaniline (PANI), polypyrrole (PPy) and polythiophene (PT).

The first one contains the heteroatom (nitrogen) placed into the main polymer chain as imine and amine nitrogen, the structure and properties of the polyaniline being strongly influenced by the presence of this atom. The existence of imine sites confers a specific feature for polyaniline unlike other semiconducting polymers, namely the possibility to be doped with protonic acids.

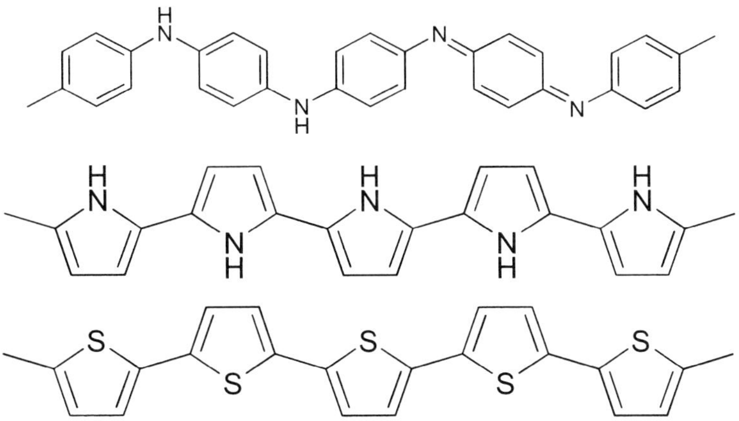

PANI

PPy

PT

[1] e-mail: grim@icmpp.ro.

The imine nitrogen is more basic and more easily to be protonate compared with the amine atom that can be protonated only at high acid concentrations. The last two polymers have a polyenic structure remembering of *trans –cisoidal* polyacetylene but the –NH- or –S- atom bridges stabilize the *cisoidal* conformation and polymers are very thermal and environmental stable, their properties being closer of arylene polymers. In the polyaniline class the emphasis will be centered on synthesis of nanostructures obtained by polymerization of aniline in different conditions and in presence of organic macrocyclics, organic polymers and inorganic materials. For polyheteroarylenes, the new exciting structures of oligomers and polymers, such as: tubular, macrocyclic, star-shape, hyperbranched and dendrimer, and their self-assembly will be shortly discussed. The applications of these polymer classes will be also briefly presented.

1. Introduction

The three types of heteroatom-containing polymers: polyaniline, polythiophene and polypyrrole, belong to the organic semiconducting and conducting polymer class and are materials of the future, they have a conjugated chain, a necessary condition for conduction, and after doping with oxidizing or reducing agents, or protonic acids, charge carriers responsible for electrical transport are created [1]. These are materials that combine solubility, processability, and flexibility of plastics with electrical and optical properties of metals and semiconductors. The molecular and supramolecular order is an important parameter because the structural defects limit mobility of the electrons and holes with unfavorable effects on the application of conducting polymers. The highest conductivity value for any known organic polymer (10^5 $S.cm^{-1}$), close to the conductivity of copper, was reported for oriented doped-polyacetylene [2], but this material is very unstable in environmental conditions. However, for the three mentioned above heteroatom-containing polymers, the highest conductivity values in doped form, reported until now, are much lower than that of oriented polyacetylene. Their energy gaps are wider (Table 1), but they have as an evident advantage that to be chemically and thermally stable materials with other interesting properties and applications.

These polymers can be used as active materials in thin film electronic devices becoming indispensable for development of low cost disposable, mechanically flexible and transparent plastic, paper electronics and opto-electronics having several advantages over their inorganic counterparts. They have many practical applications such as flat panel displays, photovoltaic cells, organic filed effect transistors (OFET), electroluminescent devices (LED), solar cells [9,10], etc. The first organic transistor on plastic substrate was reported in 1990 and this is based on vacuum-evaporated films of α-sexithiophene, a linear thiophene oligomer [11,12]. Their deposition as thin films with any desired form can be performed at room temperature from solution by simple processes as spin coating or drop casting but also by photolithography or ink-jet printing, or vacuum deposition of solid sample, all methods allowing fabrication of *sophisticated* architectures devices with large area [13]. The vacuum–evaporating techniques are generally used for small molecules and oligomers with well-defined structure that can be deposited in a pure form and their properties can be studied in correlation with the structure.

The chemical structure of the conjugated polymers offer several advantages as the responsive basis for chemical and biological (DNA sensing) detection schemes based on optical and electrochemical methods, they being very sensitive to minor perturbations.

Table 1. Structures and electrical properties of heteroatom-containing conjugated polymers as compared with polyacetylene

Polymer	Structural unit	Year [a)]	E_g (eV)[b)]	Conductivity (S. cm^{-1})[c)]
Polyacetylene (-CH=CH-)$_n$	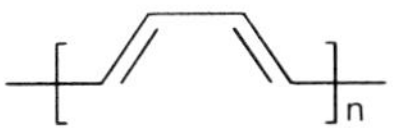	1977 [3]	1.5 [7]	10^5
Polypyrrole $(C_4H_3N)_n$	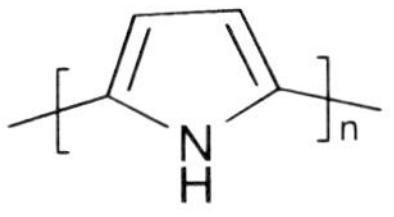	1979 [4]	3.1	10^2-$7.5.10^3$
Polythiophene $(C_4H_2S)_n$	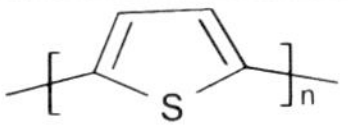	1981 [5]	2.0	10-10^3
Polyaniline $(C_6H_5N)_n$	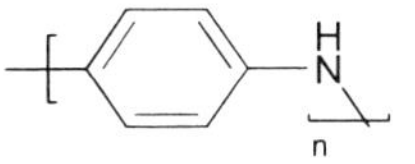	1982 [6]	3.2	30-200 (1000)[8]

a) year when it was reported first synthesis of the polymer as an electroactive material;
b) energy gap as difference between conduction and valence band;
c) the highest values reported until now.

Water-soluble polypyrrole and polythiophene have been synthesized by attaching charged pendant groups onto the conjugated polymer backbone, water solubility being a prerequisite for interaction with biological substrates [14].

2. Polyaniline

Among conducting polymers, polyaniline (PANI) is one of the most studied polymers during the last years due to a combined sum of factors: a cheap and easy to prepare monomer (aniline), the polymer can be synthesized both in aqueous or organic solutions, is environmental stable and has a lot of commercial applications [15-18]. It has to be pointed that PANI is unique between other conjugated polymers. Its metallic state can be obtained besides the using of common *p* or *n* dopants and also (and more important) by "*acid-doping*" using protonic inorganic or organic acids (HX). The doping process takes place by protonation of the polymer chain at imine atoms of emeraldine base form that then induces an internal redox reaction in which an electron is transferred from each p-phenylenediamine to p-quinodiiminium unit and formation of bipolarons (dications) and polarons (cation-radicals) after geometrical relaxation and redistribution of the charge on the chain (Scheme 1). In the emeraldine salt (conducting form of PANI) there is one unpaired electron per two aniline units and all aniline units are in the same intermediate oxidation state with all nitrogen atoms equivalent.

When as organic acid a macromolecular compound containing acidic functions (-COOH, $-SO_3H$, $-PO_3H_2$) is used then a composite material is obtained. A simple transformation of the basic form (emeraldine base) by treatment in dilute acid increases electronic conductivity by nine orders of magnitude, similar with transformation of semiconducting polyacetylene to a conducting material by doping with oxidant or reducing agents [3].

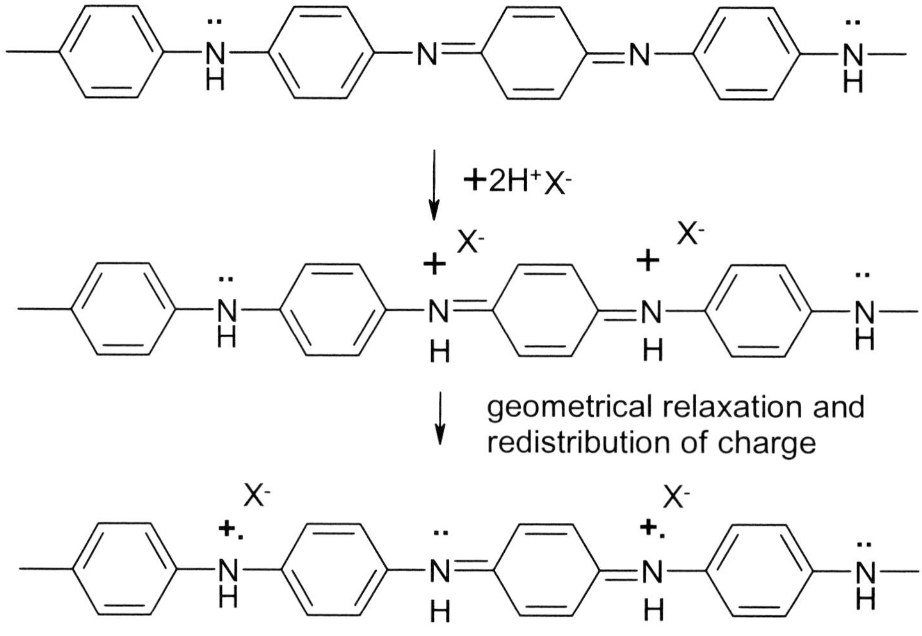

Scheme 1. The doping process of polyaniline with protonic acids.

Another specific property of polyaniline is the dependence of its conductivity on the content of water (or another solvent) bound to the polymer backbone.

Polyaniline is a "p-type" material, which indicates that the charge transport takes place mainly due to holes. Typical conductivity of polyaniline is 1-2 $S.cm^{-1}$, which can be increased to 1000 $S.cm^{-1}$ in the metallic state [8].

PANI can be synthesized by: (a) chemical and (b) electrochemical method, using water or organic solvents and polymer can result both as powder or thin film with colors strongly depending on the experimental conditions and doping state. The first way uses chemical oxidants such as peroxidisulfates [$K_2S_2O_8$, $(NH_4)_2S_2O_8$], bicromates ($K_2Cr_2O_7$), cerium salts ($CeSO_4$), H_2O_2, KIO_3, $KMnO_4$, Lewis acids ($FeCl_3$), etc, in acidic conditions or also at pH =7 or basic. The general synthesis procedure is described in literature [19-21].

The monomer/oxidant ratio can vary from 1:1 to 1:3 and this ratio determines the oxidation state of polyaniline chain, from all reduced form (leucoemeraldine) to all oxidized state (pernigraniline) (Scheme 2). Transitions between these redox forms are accompanied by dramatic changes of optical properties: colour changes from transparent to blue, violet and green. Transitions are relatively fast, about 0.1 μs for 0.1 μm film thickness, show high reversibility without damage of the film. This property can be used to build smart windows. The ideal emeraldine base (EB) oxidation state of PANI contains two types of structural units: benzenoid and quinoid in a 3:1 ratio with an equal number of alternating amine and imine repeating units and can be doped by protonation with acids in a two step process. Protonation occurs preferentially at imine nitrogen when a polysemiquinone radical cation (emeraldine salt-ES form) (Scheme 1) is obtained. The protonation can continue at high pH values at amine nitrogens but these protonated amine nitrogen atoms are unstable and even under a dynamic vacuum they lose the dopant.

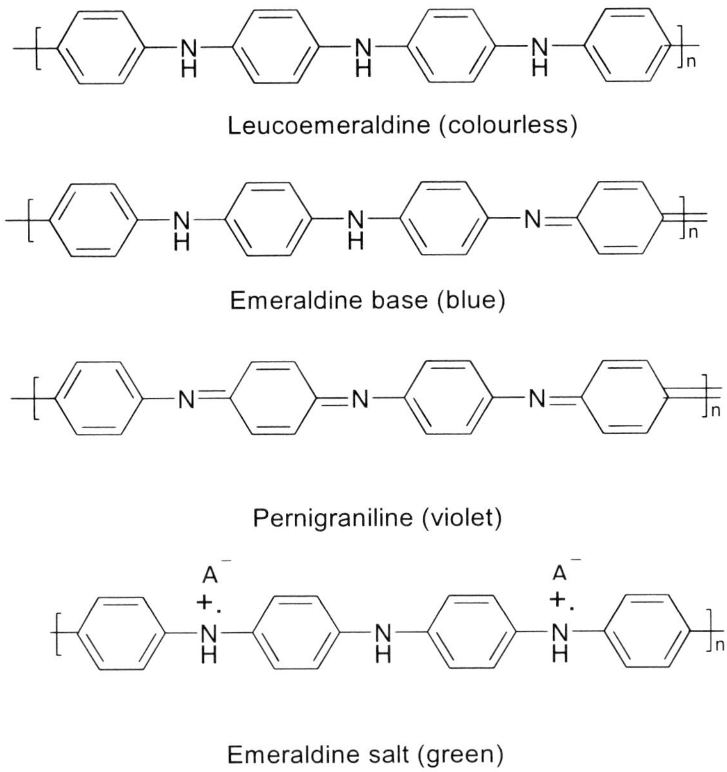

Scheme 2. Chemical structures of neutral forms of polyaniline and protonated emeraldine.

For the oxidation of aniline to emeraldine base (the best conducting form obtained after doping) for every mole of aniline at least 2.5 mole of oxidant are necessary. The reaction is carried out in acidic conditions using inorganic or organic acids: HCl, HBr, HI, HF, H_2SO_4, CH_3COOH, CCl_3COOH, $CH_3C_6H_4SO_3H$, $HClO_4$.

The anions can be included as counterions in doped form. When polymerization of aniline is carried out in presence of polymeric acids (i.e., poly- acrylic and methacrylic acids, poly styrenesulfonic acids or their copolymers, etc) blends and composites are obtained with many applications as metal protecting coatings [17,22]. The oxidation of aniline is an exothermic process. The polymerization can be performed at room temperature or lower temperature, the temperature decreasing favoures formation of polymers with higher molecular weight and less defects.

The second method, the electrochemical polymerization, can be carried out in an electrochemical cell with two or three electrodes, in potentiometric or amperometeric mode. The advantage of the method consists in the possibility to obtain thin films directly from polymerization while chemical method allows synthesis of the polymer powders in high quantities.

Based on kinetic studies Wei et al. [23], have proposed a similar mechanism for chemical and electrochemical polymerization of aniline that was critically reviewed by Gospodinova and Terlemezyan [18]. Anilinium cation-radicals, with three possible resonance forms [**2a-c**], are formed through chemical or electrochemical oxidation of aniline (**1**) in acidic conditions (Scheme 3).

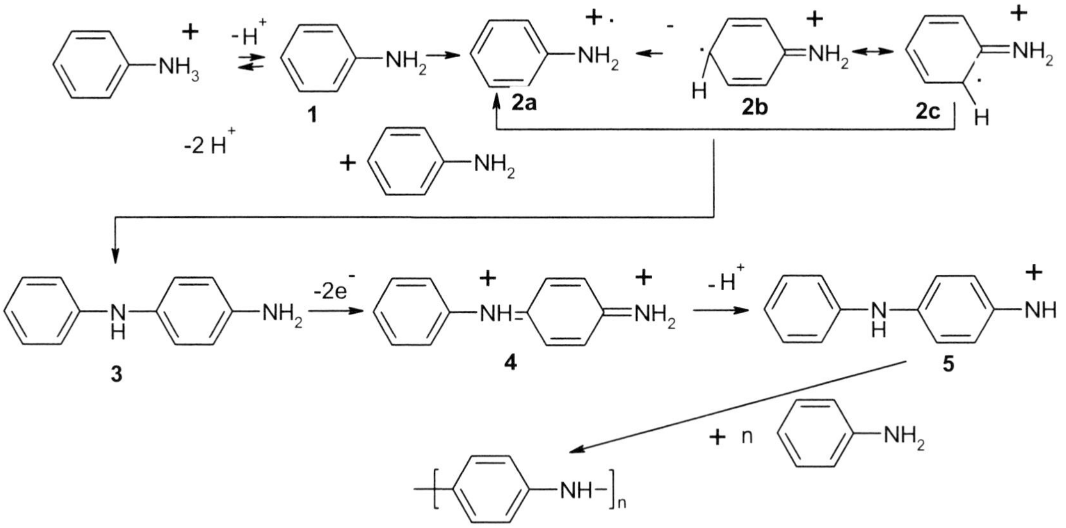

Scheme 3. Synthesis of polyaniline by chemical or electrochemical oxidation of aniline.

In the first steps of the reaction, oxidation is accompanied by formation of a dimer, p-amino diphenylamine (**3**), besides the secondary inert products, benzidine and N,N'-diphenylhydrazine, formed by *head-head* or *tail-tail* coupling of two cation-radicals. Oxidation of aniline takes place at an oxidation potential (E ~ 0.9 V versus standard calomel electrode) higher than those of dimers (E ~ 0.4-0.5 V), trimers, and so on. Therefore cation-radicals of each of these species instantly react with aniline to form dimers, trimers, and finally polyaniline. Aniline oligomers are easier oxidized then aniline monomer and they are fastly oxidized via a two-single electron step to quinoidal form dication (**4**) that reacts *via* electrophilic aromatic substitution with aniline monomer and oligomers. The chain propagation reaction goes on until all monomer is consumed, and by repeating the sequence of reactions leads to final polyaniline. The polymerization reaction is mainly directed to *para*-position of phenyl rings leading to a linear polyaniline chain. However, it can't be excluded the presence of some percents of *ortho* positions branched structures in the polymer chain.

From kinetic point of view, the aniline oxidative polymerization is a second order reaction; R_p=k[P][M], where R_p is the rate of polymer formation, k is the reaction constant, [P] and [M] are concentrations of polymer and aniline, respectively.

Water soluble and self-dopable PANIs have been synthesized by (co)polymerization of aniline derivatives containing sulfonate or carboxylate groups (*ortho*- and *metha*-aminobenzene sulfonic acid, *ortho*- and *metha*-aminobenzoic acid) as substituents on the aromatic ring [24-26]. The second way for introducing water solubility property consists in the incorporation of sulfonate groups onto the polymer backbone either by sulfonation of PANI-emeraldine base with fuming sulfuric acid [27, 28] or by deprotonation a polyaniline base and reacting with a sultone (1,3-propanesultone), the sulfonate group being positioned at phenyl ring or attached at amine nitrogen [29]. All polymers are self-dopable materials and have conductivities of ~ 0.1 S.cm-1, values which are independent of pH.

Chiral conducting polymers have attracted interest because of their potential applications for chiral separation, surface-modified electrodes, biological sensors, lasers, and chiral PANI can be synthezied by enantioselective polymerization of aniline in presence of an optically active dopant [30,31] or by post-protonation of polyaniline-emeraldine base with (+) or (-) 10-camphorsulfonic acid [32-34].

2.1. All-Organic Magnetic Polyanilines

Even the investigations are in early stage, there is an increasing interest in synthesis of purely organic materials with magnetic properties (organic magnets). Organic materials with magnetic properties have been reported until now but they are composites of an inorganic magnetic material dispersed into an organic polymer. The first *all*-organic magnetic materials studied were based on polyacetylenes containing as side substituents stable radicals or their precursors. Another class of polymers where magnetic properties can be induced is polyaniline with 1,3-disubstituted aromatic rings as spin-coupling units, poly-m-aniline (Scheme 4).

A condition that these materials have to fullfile is a precise regularity of the polymer chain to assure a long-range ferromagnetic coupling of the unpaired electrons. Polyanilines with *meta* substituted aromatic rings represent a target polymer for this scope because they can have stable radical sites with short ferromagnetic coupling units. If polyaniline (PANI) with 1,4-phenylene rings can be synthesized starting from aniline by oxidative polymerization, poly-*m*-anilines are not accessible by this direct approach. However, there are other methods: Ullman condensation reaction using cooper iodide as catalyst [35,36] and palladium–catalyzed amination reaction [37, 38] can be successfully used for synthesis of linear and branched *m*-polyanilines. Homopolymerization of an AB monomer (3-bromaniline) or polycondensation of 1,3-diaminobenzene with 1,3-dibromobenzene using palladium-based catalysts led to a linear poly-m-aniline (Scheme 5). When an AB_2 monomer (3,5-dibromaniline) was used, a 1,3,5-hyperbranched polyaniline could be obtained (Scheme 6) [39].

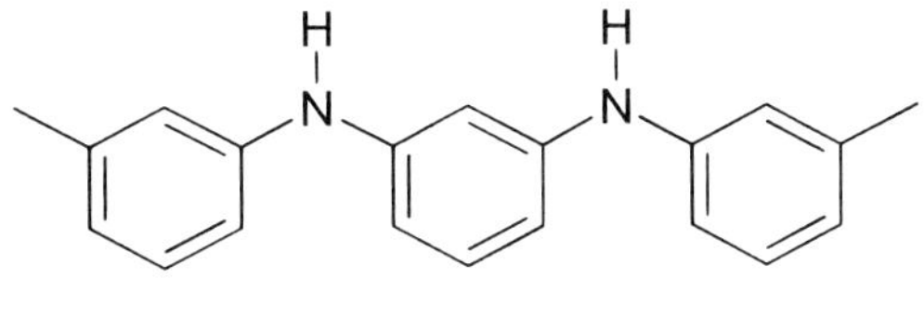

Scheme 4. Structure of poly-*m*-aniline.

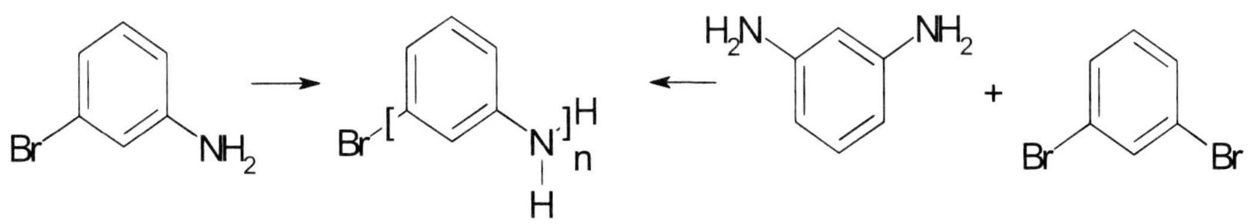

Scheme 5. Synthesis of linear poly-m-aniline by palladium-catalyzed amination reaction

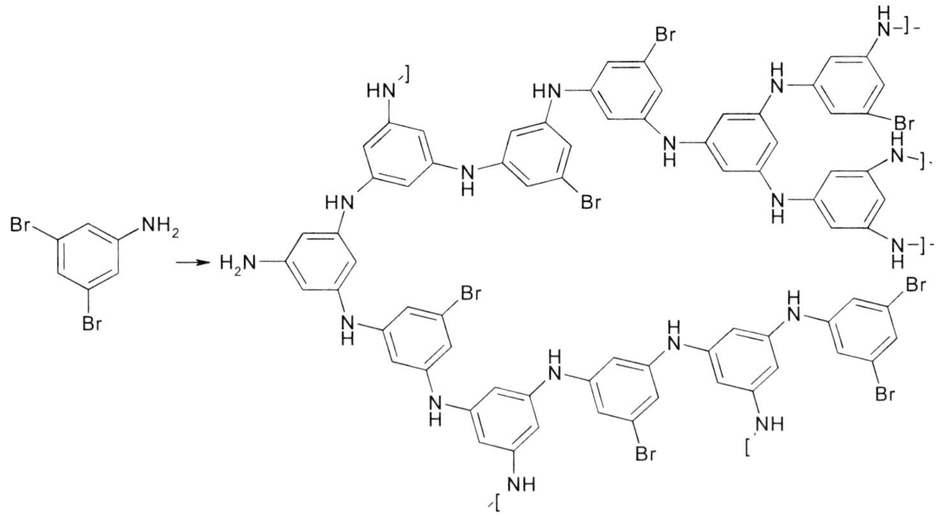

Scheme 6. Synthesis of hyperbranched polyarylamines by palladium-catalyzed homopolycondensation of 3,5-dibromaniline.

The linear and hyperbranched polymers can be either chemically or electrochemically oxidized to triplet radical-cations and their presence and intensity was studied by EPR spectra [40]. Even these studies are in an incipient stage, perspectives for synthesis of an *all*-organic magnetic material with interesting applications are very promising.

3. Polythiophene and Polypyrrole

Structurally, polythiophene (PT) and polypyrrole (PPy) (and also less studied polyfuran) (Scheme 7) remember of trans-cisoidal (t-c) polyacetylene where two hydrogens from adjacent monomer units are replaced with heteroatom bridges (-S- ,–NH- or -O-).

The heteroatom has a beneficial effect on polymer properties. Unlike polyacetylene that is a very unstable thermal and oxidative polymer, PT and PPy are environmentally stable materials that can be preserved for long time in ambiental conditions. Chronologically, PT and PPy constitute the third polymer class, after polyacetylenes and polyarylenes, in which was achieved the high electrical conductivity. Also, PPy was the first conjugated polymer obtained by electrochemical polymerization by Diaz et al [4] by electrooxidation of pyrrole using a wide variety of solvents. Thiophene has a stronger aromatic character than pyrrole due to lower electronegativity of the sulphur bridge compared to nitrogen, the lone electron pair of heteroatom being more disposable for conjugation. On the other part, pyrrole is a weak base and has a small pK_a value (-0.27) as compared with that of aniline ($pK_a \sim 4.63$) that has direct influences on their polymerization and structure of polymers. Both polymers can be synthesized starting from thiophene and pyrrole by chemical and electrochemical oxidative polymerization or from their functional derivatives by metal-catalyzed polycondensations. The oxidation potential versus a standard calomel electrode (SCE) is 2.07 V for thiophene, 1.2 V for pyrrole and 0.9 V for aniline. Therefore the number of oxidants for chemical polymerization of thiophene is more limited as compared with pyrrole and aniline.

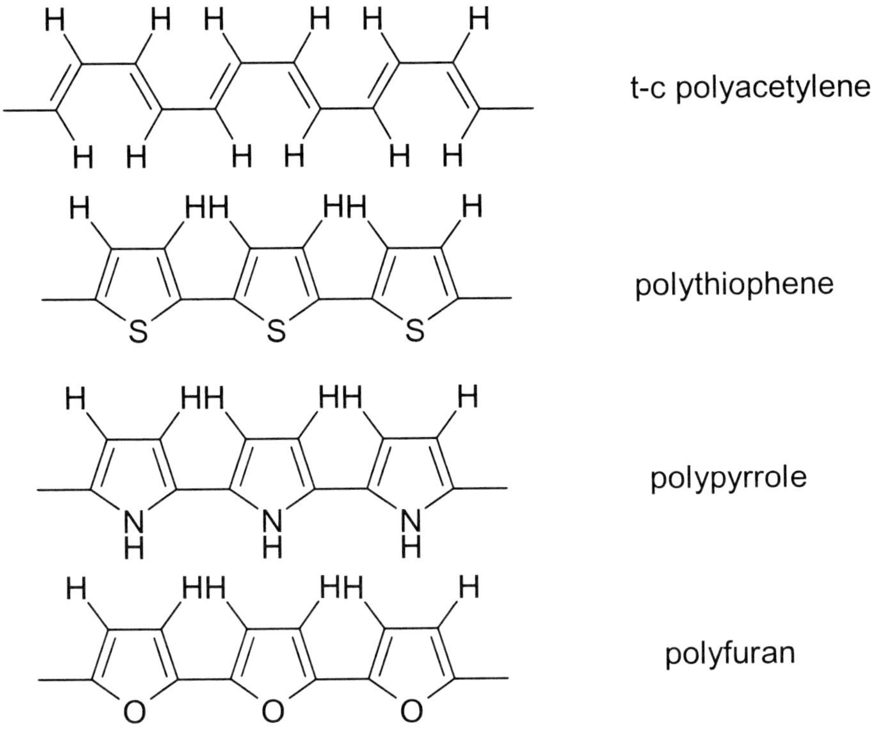

Scheme 7. The structures of conjugated polymers with heteroatom bridges.

The accepted mechanism for the chemical polymerization may be described as follows: upon oxidation of the monomer cation-radicals having three resonance forms are formed (Scheme 8). Two cation-radicals can recombine by radical sites yielding to a dication dimer that becomes electrical neutral species loosing two protons.

The dimer continues the reaction sequence by one-electron oxidation and coupling with cation-radicals of monomers (in the first stages of polymerization) or with cation-radicals of oligomers after monomer consumption. The chain propagation is interrupted by precipitation of the growing polymer chain.

Electropolymerization involves a similar mechanism but active species, cation-radicals, are electrogenerated at electrode (electrochemical process). Then radical-radical coupling reactions (chemical process) and electrochemical reoxidation lead to polymers. The all process can be described as an $E(CE)_n$ mechanism in which monomer and oligomers oxidations are electrochemical (E) processes and radical-radical couplings are chemical (C) processes. Unsubstituted PT and PPy are insoluble and infusible materials due to rigid conformation of the conjugated polymer chain. The substitution with longer or branched alkyl substituents (alkyl > C_6) allows obtaining soluble and processable polymers. In electrochemical polymerization, both PT and PPy are obtained as doped materials with approximately the same oxidation degree (about 0.3 charges per monomer unit) and charge-compensating counterions from electrolyte, while neutral polymers are obtained by dedoping.

The oxidation potential of thiophene is higher than of pyrrole. Therefore the number of chemical oxidants used in chemical polymerization of thiophene is limited as number and anhydrous $FeCl_3$ is the most used. Pyrrole and its derivatives can be polymerized using also ammonium and potassium peroxidisulfate and $FeCl_3.nH_2O$ with different hydratation grades.

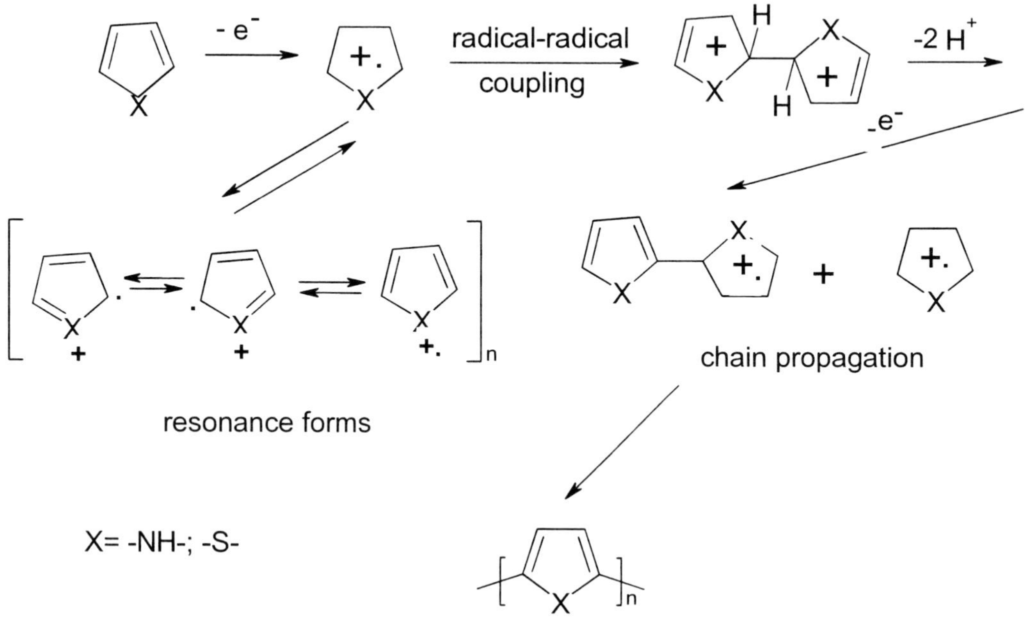

Scheme 8. The synthesis of polypyrrole and polythiophene by chemical oxidative polymerization.

4. Well-Defined Conjugated Oligomers

4.1. Linear Oligomers

Linear conjugated oligomers having well-defined structure and length, with rigid-rod form can be used as organic molecular wires in molecular electronics to connect nanodevices. They have lengths from 1 to 100 nm and width from 0.3 nm on up and are considered the future materials for replacing silicon in nanoelectronics. Tour et al. [41] have synthesized 2,5-thiophene ethynylene oligomers by interactive divergent-convergent method with thioester groups at one or both ends. By *in situ* deprotection of the ends, sulphur atom enables to bond covalently to metal surface of electrode and establish an electrical connection between molecular wire and electrode. Ethynyl units are used to link heteroaromatic rings because they confer a linear and rod-like shape. For instance, in Scheme 9 is presented an oligomer having a length of 10 nm but the length, side substituent and the number and nature of the ends, can be changed by synthesis.

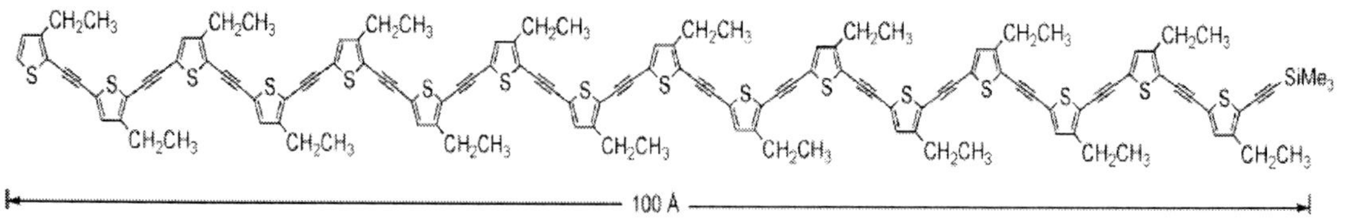

Scheme 9. Structure of a linear thiophene ethynylene oligomer of 10 nm length.

4.2. Cyclic Oligomers

Kaufman et al. reported in 1978 synthesis of thiophene-based macrocyclic compounds containing four and six thiophene rings (Scheme 10) by lithiation of 2,2'-bithiophene at 3,3' positions and cyclization by homocoupling reaction in the presence of copper (II) catalyst [42].

The cyclic tetramer, tetra[2,3-thienylene], can be considered a derivative of cyclooctatetraene with a nonplanar structure and most probable in bath conformation. This was studied as possible building block for construction of molecular electromechanical actuators (*molecular muscles*) [43] and double helices [44]. The extension of macrocycle size by one extra bithiophene unit yields the rigid hexa [2,3-thienylene] with three twisted *s*-trans bithiophene units. In solid state, X-ray studies show the arrangement of macrocycles, such as tubular channels are formed. Also, this compound can be imagined as a monomer and by polymerization should result in a porous and cross-linked material [45].

π-Conjugated macrocycles containing heteroarylene rings are of interest as bricks for modular building blocks of the new materials and supramolecular systems. Some macrocyclic structures containing thiophene and ethynylene or thiophene and ethynylene + vinylene linkages between them have been synthesized and characterized [46]. Macrocycles have a size and shape-persistent, circular and planar form with 1.8 until 6 nm inner cavities and 3.3-7.5 nm outside molecular diameters and can host alkanes of proper dimensions.

Macrocycles with thiophene and bis ethynylene linkages (Scheme 11) are precursors for *all*-thiophene macrocyclics because diyne linkages can react with Na_2S to form new thiophene rings. Bauerle et al. [47-50] have been reported the synthesis and characterization of conjugated giant macrocyclics based on *all*-oligothiophene (cyclo[n]thiophenes) with 8, 12, 16 and 18 thiophene rings (Scheme 12).

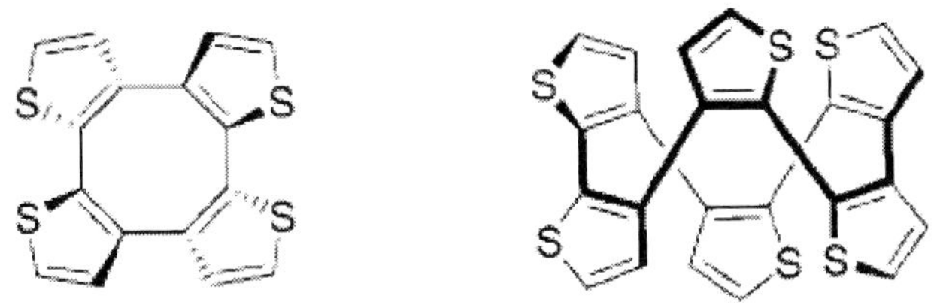

Scheme 10. Structure of cyclic thiophene-based tetramer and hexamer.

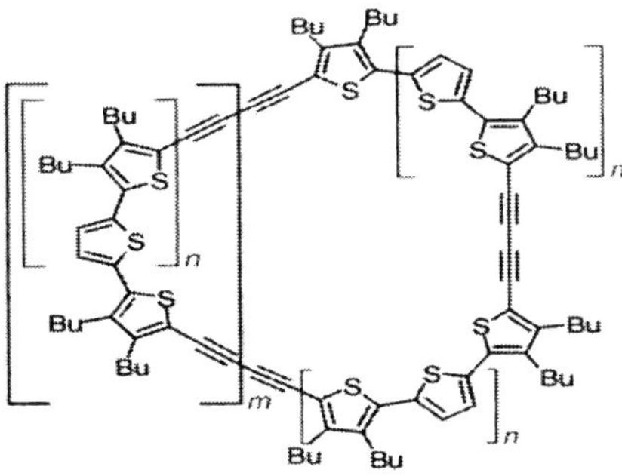

n=1or 2 ; m=0, 1 or 2

Scheme 11. (Continued on new page.)

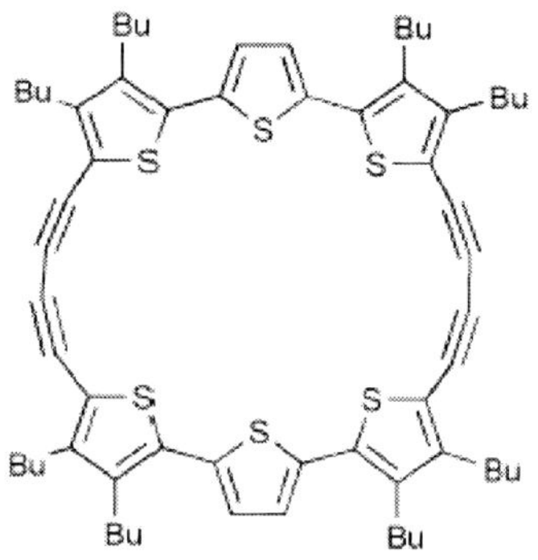

n=1; m=0

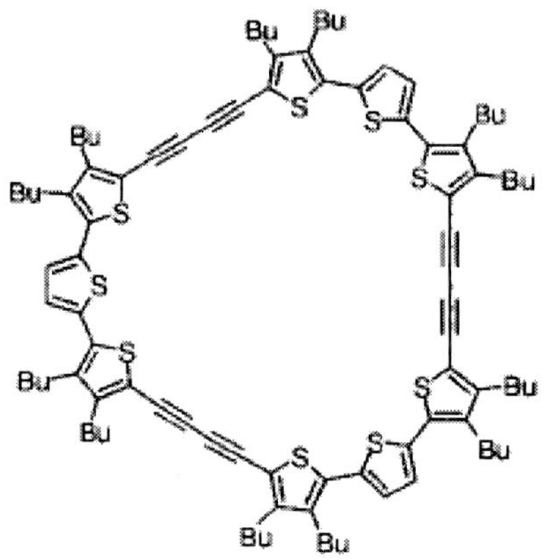

n=1; m=1

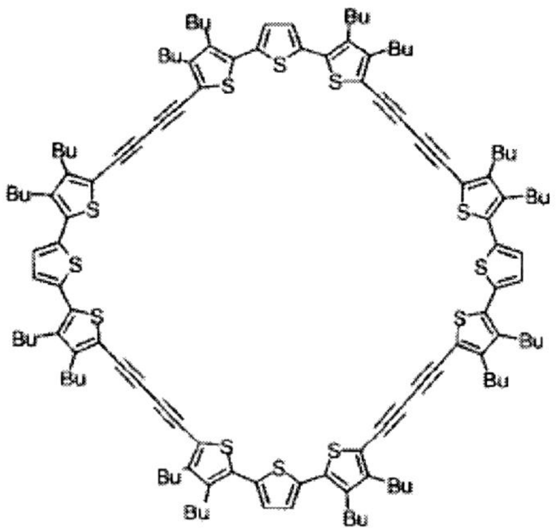

n=1; m=2

Scheme 11. Structure of cyclic thiophene bis ethynylene oligomers.

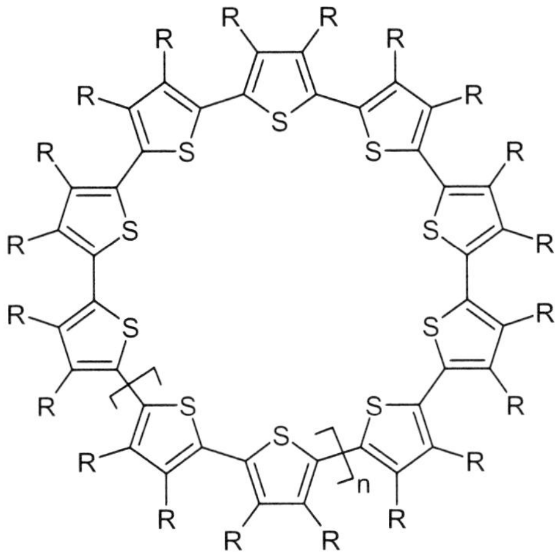

R is H or n-butyl and n=0, 2, 4, 5

Scheme 12. Structure of *all*-thiophene-based cyclic oligomers

All π-extended cyclic oligoheteroarylenes with large π-circles or disk shape can to self-assemble in solution due to strong π- π, X- π, X-X interactions (X = S, NH) and solvophobic effects. This solution preaggregation can be extended in solid state to one-, two- or three dimensional arrangement and 3D nanostructures, such as nano- or micro- scale wires, tubes, dots, ribbons, fibers, etc. for supramolecular electronics are obtained [51].

4.3. Conjugated Dendrimer and Branched Structures Containing Heteroatoms

Dendrimers, chemical compounds with a regular and highly branched and three-dimensional architecture are developed in the last years. This architecture has advantages to offer (macro)molecules with a well-defined form and structure, a high chemical purity and degree of order, are characterized by a polydispersity of one and have a tree-like structure. Their structure can be obtained using laborious step-by-step synthesis in a convergent or divergent methodology. These materials are very important for applications were a high mobility of charge carriers is asked: solar energy conversion, organic field-effect transistors and organic light emission diodes. On the other hand, the hyperbranched structure has randomness within the branching and a larger dispersity in the molecular weight but is more easily accessible and can be obtained in one step process using AB_2 or A_2+B_3 monomer pairs. In a dendrimer structure some key units can be observed: (a) the core, (b) the dendrons that contain also branching points, and (c) surface groups (optionally). Therefore, thiophene or pyrrole rings in a dendritic structure can be positioned at the core or within the branching and/or linking groups or at their surface. Dendrimers containing thiophene or pyrrole as core, branching sites or surface groups are reviewed [52-54], therefore in the following only *all*-thiophene dendrimers will be shortly discussed.

The first *all*-thiophene dendrimer synthesis was reported by Advincula et al. in 2002 [55] using a convergent synthetic sequence and metal-catalyzed reactions. The three dendrons (1-3) (Scheme 13) have synthesized and used as building units for construction of dendrimers but brick of all structures is dendron 1 obtained from 2,3-dibromothiophene (an AB_2 monomer). Dendrimers 4, 5 and 8 have synthesized by direct coupling of dendrons 1-3 through a $BuLi/CuCl_2$ method while dendrons 6 and 7 have obtained by palladium-catalyzed coupling reaction of 4,5,4',5'-tetrabromo-[3,3']thiophene and tributylstannyl derivatives of 2 and 3 (Stille reaction).

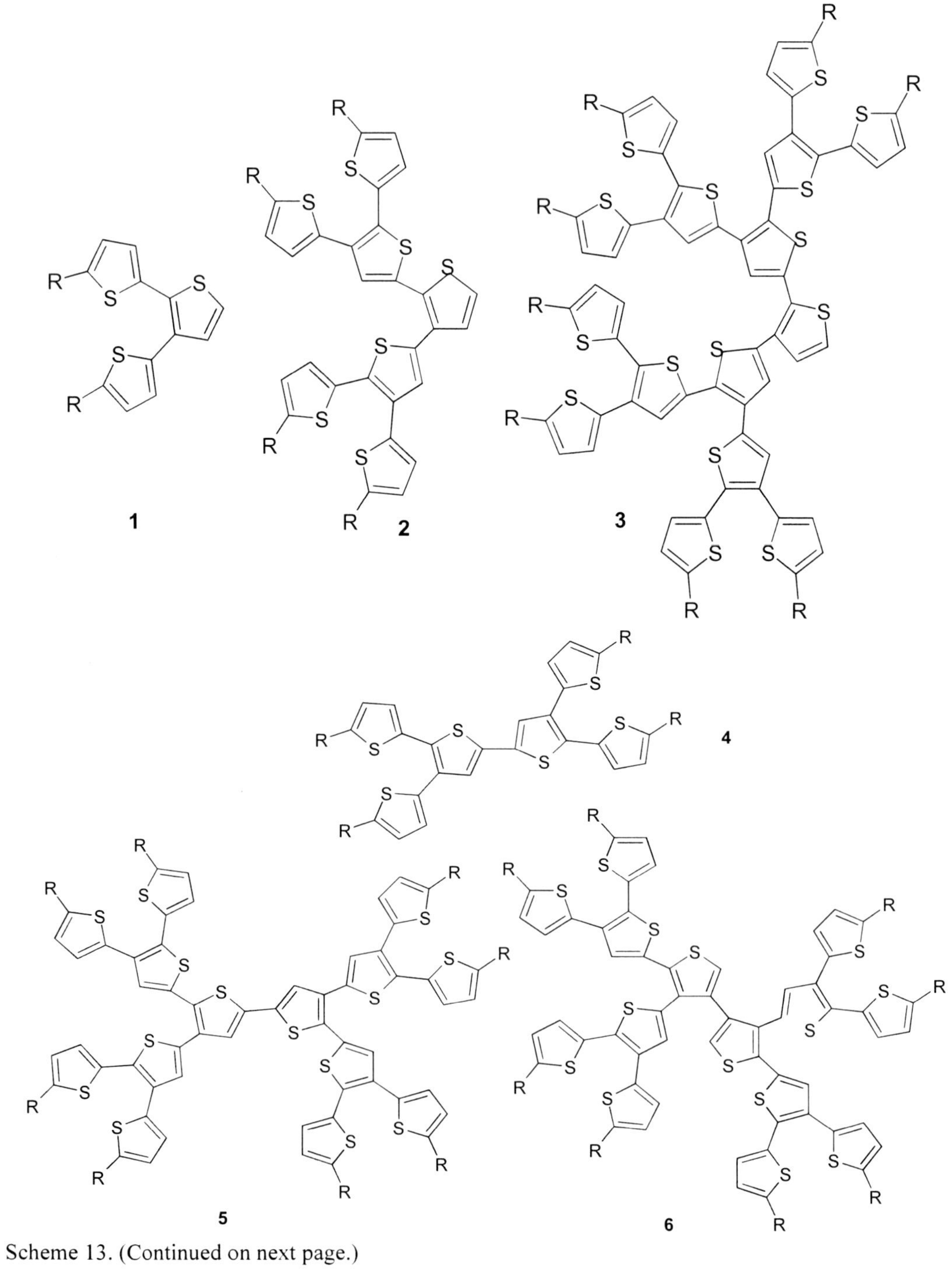

Scheme 13. (Continued on next page.)

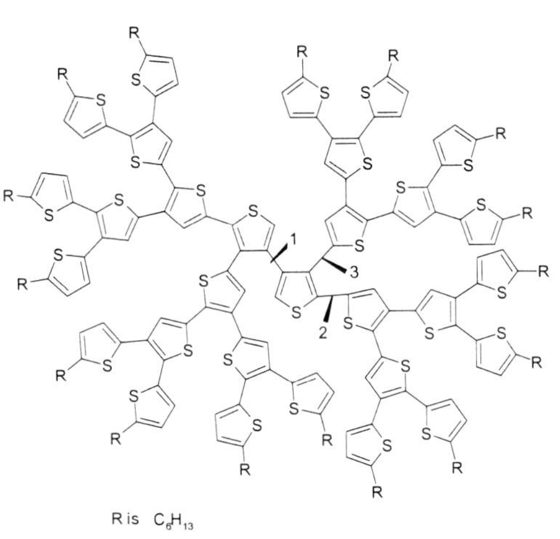

7

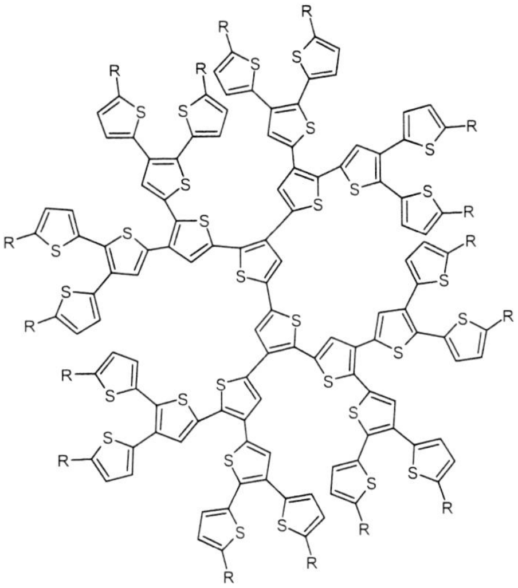

8

Scheme 13. Structure of all-thiophene-based dendrons and dendrimers.

Both dendrons and dendrimers are highly soluble in organic solvents (due to long alkyl tails) and can self-organize in supramolecular assembly (2D - crystals, nanowire and nanoparticles) depending on the dendrimer structure, substrate nature and experimental conditions [56].

The coupling reaction of the tributylstannyl derivatives derived from 1 and 2 (Scheme 13) with bromine tri- or tetrasubstituted benzene derivatives (9 and 10) was achieved by Mitchell et al [57] by Stille reactions and dendrimers containing 12 (11), 16 (12), 24 (13) or 32 (14) thiophene, benzene core and C_6 alkyl substituents are obtained (Scheme 14).

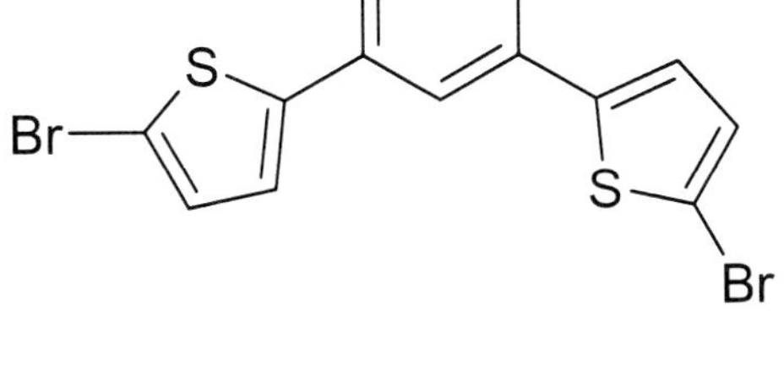

9

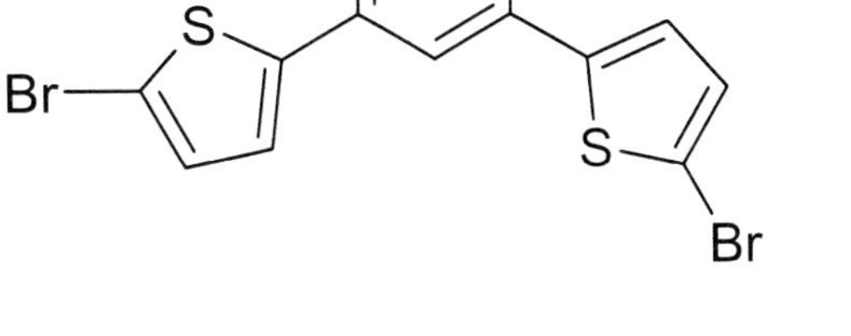

10

Scheme 14. (Continued on next page.)

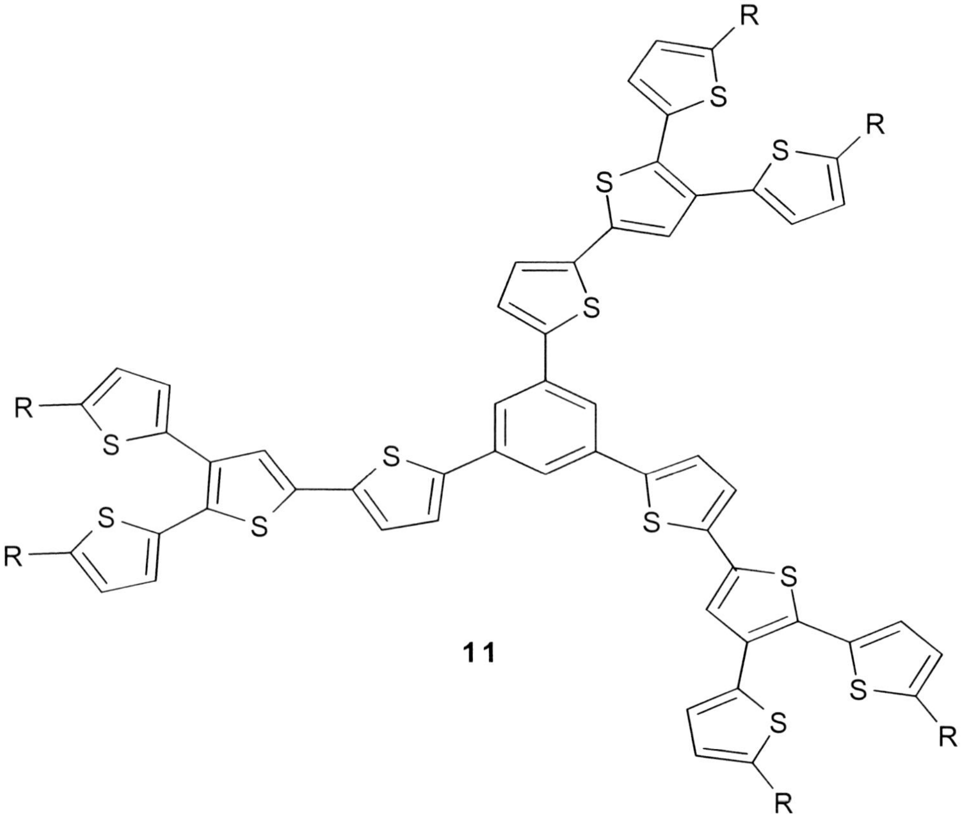

11

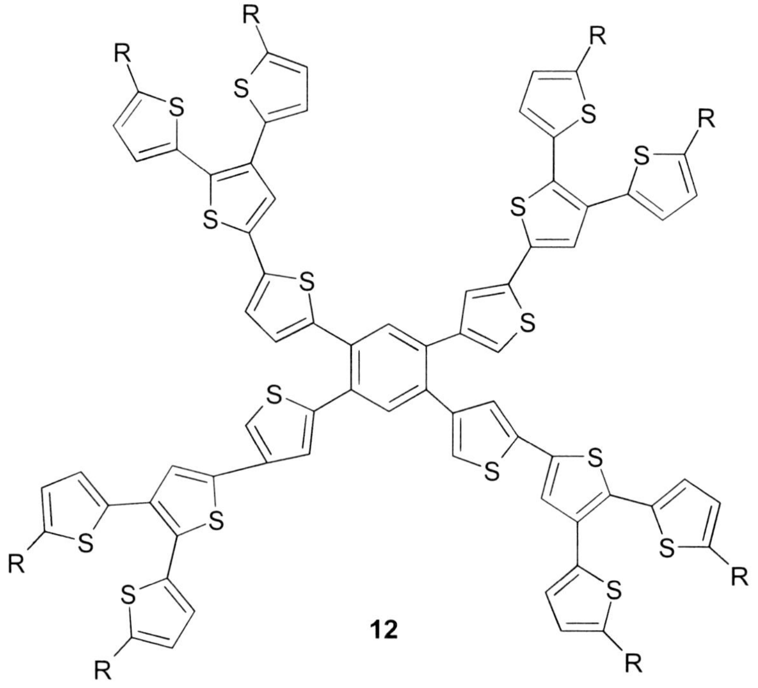

12

Scheme 14. (Continued on next page.)

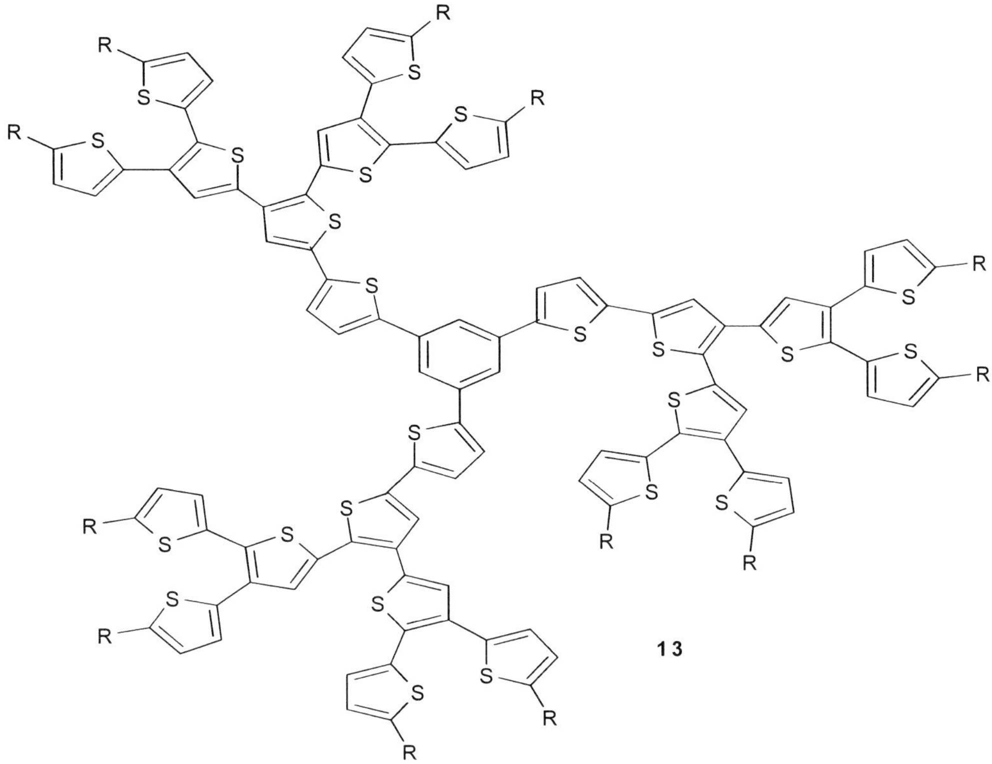

13

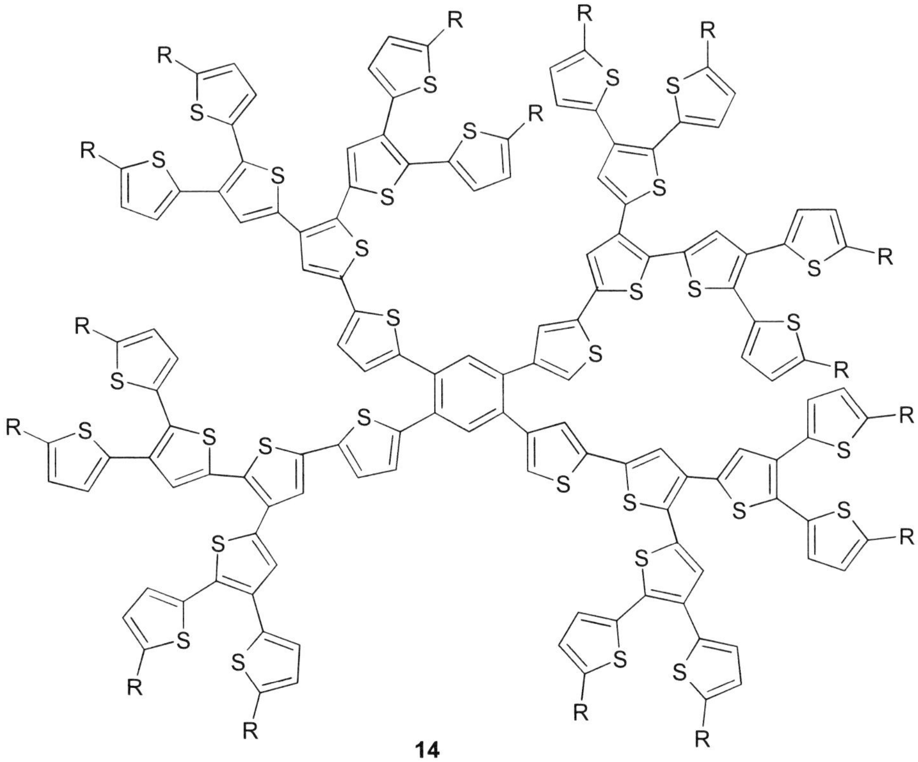

14

Scheme 14. Thiophene dendrimers having 1,3,5-tri-and 1,2,4,5-tetra-substituted benzene rings as core.

Other less compact dendrimer structures have builded by Negishi et al using 1,3,5-trisubstituted benzene as branching units and linear oligothiophenes with four and nine thiophene rings as dendrons containing phenyl or hydrogen at periphery [58]. OT is linear oligothiophene with 4 or 9 thiophene rings and at periphery H or phenyl and Δ is the branching site (1,3,5-trisubstituted ring) (Scheme 15).

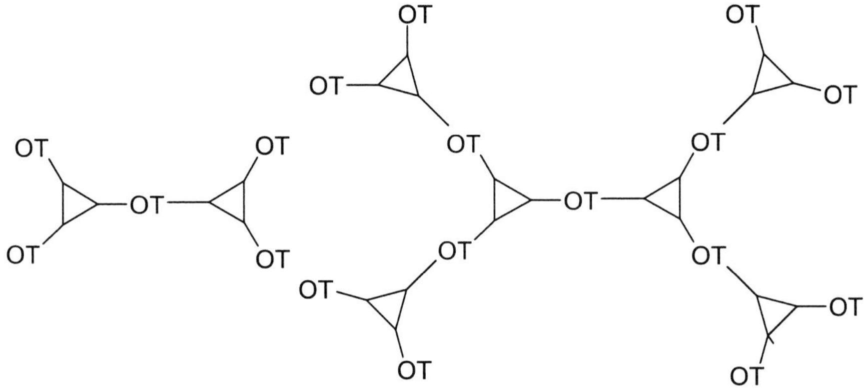

Scheme 15. Dendritic structures with oligothiophenes

Self-aggregation of these dendritic structures is attributed to π-π interactions of the two-dimensionally extended conjugation compounds and was evidenced by NMR spectroscopy and optical methods. Only a hyperbranched polythiophene structure has been synthesized until now by nickcl-catalyzcd homopolycondcnsation rcaction starting from 2,3-dibromothiophene, according to Scheme 16 [59].

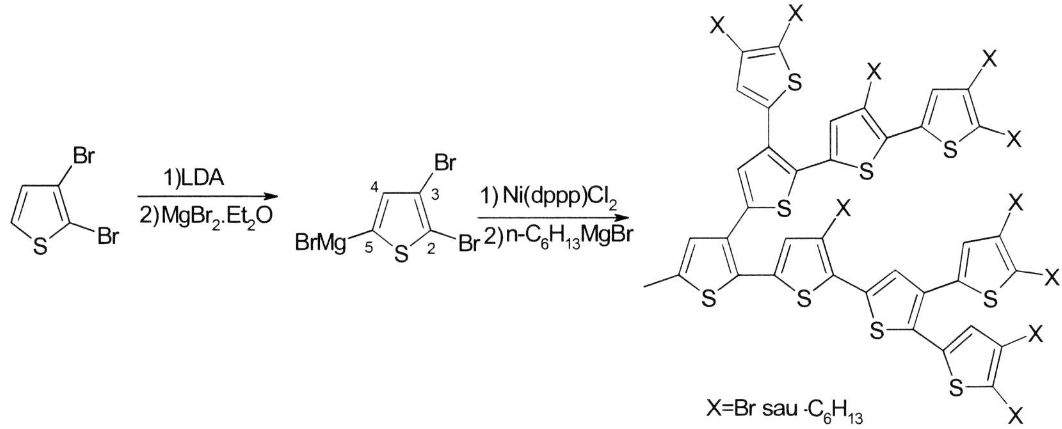

Scheme 16. Synthesis of hyperbranched polythiophene by organometallic route.

Molecular weights (M_n) for polymers having bromine and/or hexyl groups at periphery varies between 2500-3900 with polydispersity degrees of 1.6-1.7.

In all dendrimers and branched structures, thiophene rings can exist as 2,3/2,5 or 3,5-linked group and only in the last situation the conjugation is interrupted. The presence of fragments having different conjugation lengths make the film to absorb solar light on a broad absorption domain, all system can be considered a light- anntena harvesting for application in solar cells.

4.4. Thiophene and Pyrrole-Based Star Oligomers

Star-shaped oligoheteroarylenes with three (or more) arms have been synthesized starting from functional cores of type: 1,3,5-tri-, 1,2,4,5-tetra-benzene or 4,4',4''-triphenyl benzene, [60] 2,4,6-trisubstituted 1,3,5-triazine [61,62], 4,4',4''-triphenylamine [63,64], benzo[1,2-b:3,4-b':5,6-b''] trithiophene [65,66] and 10,15-dihydro-5H-diindeno[1,2-α,1',2'-c]fluorine (truxene) [67,68]. The arms are oligothiophenes linked by α position of thiophene, with one, two or three thiophene moieties and H, n-hexyl or an electron acceptor group [-CH=C(CN)$_2$] at periphery (Scheme 17).

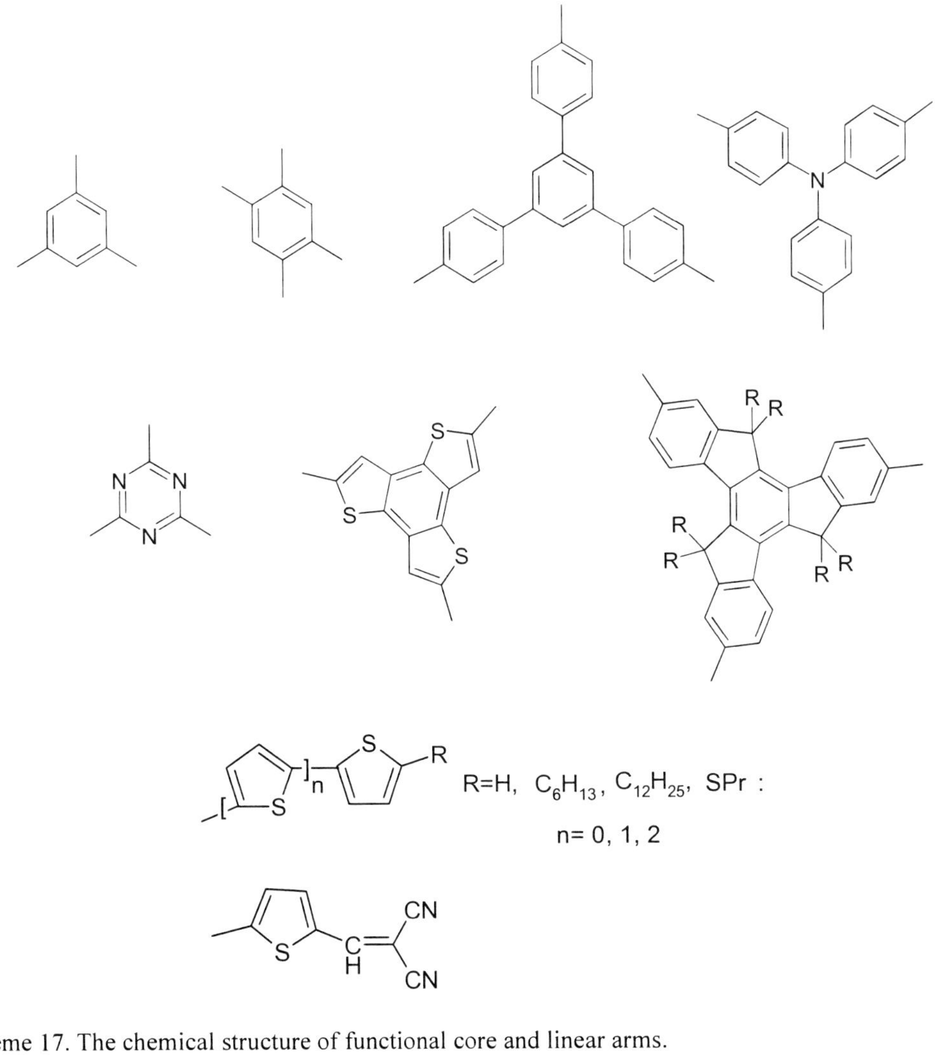

Scheme 17. The chemical structure of functional core and linear arms.

All-star oligomers are more soluble than their linear counterparts and have studied for application as active materials in organic field effect transistors and solar cells where high mobilities for charge carriers are necessary. In the case of the star-shaped oligomers, the charge carrier mobility in the correspondingly prepared films is achieved if the molecules are composed in such a way that side chains consisting of conjugated aromatic or heteroaromatic systems are attached in direct conjugation to a central aromatic or heteroaromatic ring so that the total molecule acquires an octupolar structure. This octupolar structure permits an effective π-π interaction of the molecules with one another in a manner such that stacking of a plurality of molecules along an imaginary axis (central ring) can take place and stacks can interact with one another by intermeshing of the side chains. The electronic properties of the materials are determined both by the arrangement of the molecules in a layer and by the molecular design.

5. Heteroatom-Containing Conjugated Polymers with Pseudorotaxane Architecture as Insulated Molecular Wires

Rotaxane and pseudorotaxane architecture in macromolecular chemistry, in which a linear macromolecular chain inserted inside the cavity of several macrocyclic molecules has received much attention in the last years due to their unique and aesthetically structure as well as potential use as method for improving the polymer solubility [69-72]. A conducting polymer with rotaxane architecture can be considered as an insulated molecular wire having as core one-dimensional conjugated macromolecular chain that can transport charge carriers and as shell many macrocyclic compounds threaded on it (Figure 1).

As general rule, macromolecular rotaxane structures are obtained mainly by three ways: (a) monomer polymerization or polycondensation in the presence of a macrocyclic compound, (b) threading of a preformed linear polymer chain through macrocycles, and (c) polymerization or polycondensation of a rotaxane monomer.

As macrocyclic compounds, until now only cyclodextrins and cucurbiturils were successfully threaded onto heteroatom-containing polymers.

Cyclodextrins are cyclic oligosaccharides consisting of six, seven or eight α-(1,4)-linked D-glucopyranose units and a torus-shaped structure having hydrophobic cavities with depths of 0.5 nm and diameters of 0.45 nm for α-CD, 0.7 nm for β-CD and 0.85 nm for γ-CD, respectively [73].

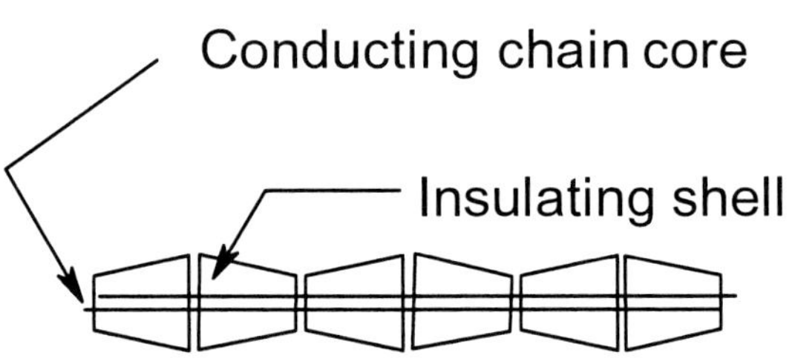

Figure 1. Structure of an insulated organic molecular wire.

They have the advantage to be soluble in water and some protic polar solvents and form inclusion complexes with a large variety of hydrophobic guests of inorganic and organic nature because of their polar hydrophilic outer shell and relatively hydrophobic cavity.

Cucurbit[6]uril, or CB[6], is another macrocyclic compound obtained by acid-catalyzed condensation reaction of glycoluril and formaldehyde [74-76]. Its name is derived from the Latin word for pumpkin (cucurbitaceae) because the structure of the molecule is like an open-ended pumpkin with a hydrophobic cavity having an inner diameter of ~ 0.5 nm, accessible through two polar and identical carbonyl fringed portals. It can form stable complexes with guest molecules, in particular protonated amines [77]. Rotaxane and pseudorotaxane structures containing CB[6], and its higher homologues [n= 7,8] as hosts have been synthesized and studied in the last time [78-84]. However, unlike ciclodextrins (CD) or other macrocylic hosts, the CB[n] compounds were used more less, probably due to their very low solubility in common organic solvents, they being soluble only in strongly acidic aqueous solutions. Therefore CB[n] are able to be used in chemical polymerization of aniline in acidic conditions.

The first members of the two macrocyclic compounds, α-CD and CB[5] have inside diameters of 0.45- 0.5 nm and inclusion of aniline, thiophene or pyrrole monomers is difficult to take place, therefore the most studies have used higher homolgues of CD and CB.

There are only few reports about synthesis of conducting polymers with pseudorotaxane structure. Thus, Lacaze *et al.* [85] have studied the electrochemical polymerization of 2.2'-bithiophene hosted inside of hydroxypropyl- β -CD; the reaction was carried out in water where the complex is soluble. A soluble and low molecular weight polythiophene was obtained in these conditions. Harada *et al.* [86] have also obtained low molecular weight polythiophene (5000 daltons) included in macrocycles as β -CD and 2,6-dimethyl- β -CD by chemical oxidative polymerization ($FeCl_3$ as oxidant) of inclusion compounds: 2,2'-bithiophene/β -CD and 2, 2-bithiophene/2,6-dimethyl- β -CD in aqueous solution. On the other hand, pyrrole and 3,4-ethylenedioxythiophene with α -, β - or γ –CD or randomly dimethylated α -, β - or γ –CD form 1:1 molar ratio crystalline adducts, with a herringbone fashioned cage-type crystal structure. The chemical oxidative polymerization of adducts in water, using $K_2S_2O_8$ or $FeCl_3$ as oxidants have led to insoluble polymers; during the polymerization reaction the adduct is decomplexed, CDs remain in solution and the net insoluble conjugated polymers are obtained [87]. Conducting pseudorotaxanes were also obtained by including of polythiophene or poly(3-methyl)thiophene in the cavity of β –CD [88].

Polyaniline/ β –CD pseudorotaxane was synthesized by two ways: (a) by chemical polymerization of inclusion adduct aniline/ β –CD [84] or N-phenyl-1,4-phenylenediamine/ β –CD in aqueous solution using ammonium persulfate [89]; (b) by inclusion of polyaniline emeraldine base into β –CD in aqueous solution [90-92] (Scheme 18). It was observed that aniline forms with β –CD an inclusion complex of 2:1 molar ratio and its oxidative polymerization can lead to a pseudorotaxane structure only if the reaction was carried out in neutral media [89]. In aqueous acidic media the dissociation of aniline/ β –CD inclusion complex in free partners is favourized because aniline that exists predominantly as aniline cation with a highly hydrophilic $NH_3^{(+)}$ group has the tendency to come out of the β–CD cavity and free aniline is polymerized to its emeraldine salt form.

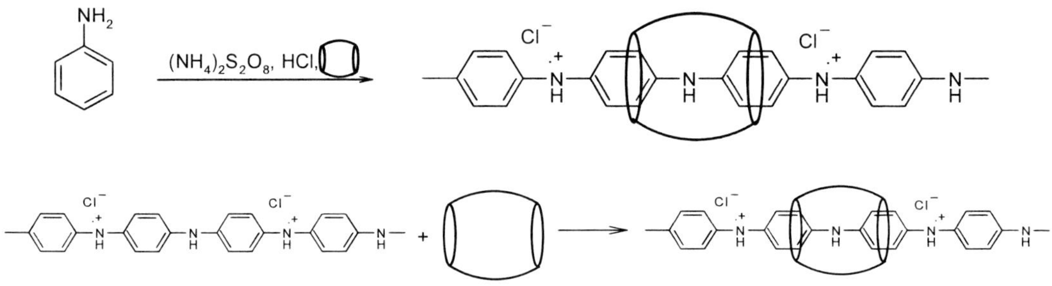

Scheme 18. Synthesis of polyaniline-rotax-CB[6] by (a) chemical polymerization of aniline/CB[6] inclusion adduct, and (b) encapsulation of polyaniline emeraldine salt in CB[6].

Unlike CDs which preferentially bind neutral guest molecules, CBs[n] bind preferentially guests with positive charges. In the aniline polymerization, anilinium cations are responsible species in the growing step. Therefore, CB[6] can form more stable complexes with aniline than β-CD. Pseudopolyrotaxane polyaniline/ CB[6] have been synthesized also by two ways: (a) chemical oxidative polymerization of aniline in presence of CB[6] as host in acidic aqueous solution and using ammonium persulfate as oxidant and (b) by threading of preformed polyaniline chain through inner cavity of CB[6]. The way (b) is entropically unfavorable but is enthalpically favourized by noncovalent interactions. Polyaniline is the first conducting polymer with rotaxane structure synthesized by using cucurbiturils as macrocyclic compounds [93].

Other heteroatom polymers such as full aromatic polyazomethines having a rotaxane architecture have been synthesized by polycondensation of aromatic diamine/aromatic dialdehyde pairs in the presence of α –and β –CD [94-97].

The threading of many macrocycles onto a single conjugated conducting polymer is accompanied by a coil-rod transition. As a result, the electronic conjugation can spread over the whole macromolecule. The interchain interactions and aggregation are avoided by keeping the chains at a distance determined by the thickness of macrocyclics. However, the encapsulated polymer can not be doped because the dopant molecules have not access to conjugated chain to reduce or oxidize it.

6. Heteroatom Polymers Nanostructures

Conjugated polymers can form nanoscale structures down to the molecular scale that can be used in nano-electronics and -optoelectronics. The limit of silicon manufacturing technology is thought to be tenths of micrometers wide because the silicon's band structure disappears when silicon layers have just a few atoms thick, such as for further miniaturization, molecular–scale materials have to be produced. Therefore, conducting polymer based- nanostructures such as: (nano) fibriles, sticks, cylinders, tubules, wires, belts, particles, have received much attention in the last years due to their fascinated architectures and interesting physical properties that open new application directions. Conducting polymer nanostructures offer the possibility to build high-density electronic devices with superior performance being more responsive than bulk polymers to external stimuli due to a high interfacial surface area between polymer and its environment, important for applications in sensing or catalysis. Nanostructured conducting polymers in sensor applications have a greater sensitivity, faster time response relative to its conventional counterpart due to higher effective surface area and short penetration depth for target molecules [98]. For instance, nanostructures of conducting polymers can have electronic conductivities of over an order of magnitude when compared to powder or thin film samples. These nanoscopic structures can be builded by template method or directly, due to the predisposition of conducting polymers for self-assembly in fibrilar morphology. The conjugated polymer chain has an anistropic growth, shows anisotropic properties and can be considered as one-dimensional (1-D) conductor. It has to be pointed out that polyacetylene, the first and best conducting polymer structure, was synthesized still from 1974 having nanofibrilar morphology [99]. Thus, by polymerization of acetylene using $Al(C_2H_5)_3/Ti(OC_4H_9)_4$ as a catalyst, thin or thick films or powders are obtained and polymer morphology is dependent on catalyst concentration. At catalyst concentrations higher than 10 mmole/l, adherent films are obtained because the number of grown-up fibrils is enough to allow their interweaven. At lower concentrations the number of fibrils is low and they fall down at reactor bottom before to be tangled and polymer results as a powder. The fibrils width is between 20-30 nm having an undefinite length.

There are several general methods to shape polyaniline, polypyrrole and polythiophene into nanostructures such as wires, fibers, tubules, etc usefull for nanoscale molecular electronic devices. These can be classified in:

- physical methods; electrospinning on conductive substrates [100], mechanical stretching [101], dip-pen writing [102] or drawing [98]. However, these methods can be applied for conducting polymer submicrofibers but nanofibers are more difficult to obtain. Moreover these methods can be used only for soluble or melt polymers but their number is reduced.
- chemical methods are considered an approach more important because nanostructures can be grown in the polymerization process. Therefore such methods can be applied for all insoluble polymers and the assembly of conducting polymers might be directed by templates or by itself.

The first chemical methods have used templates of "hard" or "soft" nature.

a) The pores of a nanoporous polymeric membrane (track-etched polycarbonate) [103-105] or channels of zeolites [106] and anodized porous aluminium oxide [107,108] were used as hard templates in chemical polymerization of monomers. The hard template method proposed by Martin [103,109] has advantage that the length and diameter of fibers and tubes can be precisely controlled by the selected porous membrane (membranes with pores diameters down to 10 nm and pore densities of 10^9 pores/cm^2 are now commercially products) and can be applied both for conducting polymers, metals, semiconductors or other materials. The disadavantages of the method is necessity of a porous material that has to be removed after nanofibres synthesis when these nanostructures may be destroyed or form undesirable aggregated structure when released from the template. Nanotubes of polypyrrole, poly(3,4-ethylenedioxythiophene) and polyaniline have synthesized by this way. Nano-tubules of PPy of 50-200 nm diameters, [110, 111] or nanotubes/nanofibers of poly(3,4-ethylenedioxythiophene) and polyaniline [112] with diameters of 75-150 nm were synthesized by electrochemical method using the pores of track-etched polycarbonate membranes as hard templates. PT has synthesized using a micro- or nanoporous membrane as a template so that each tubule is grown up within membrane pores (i.e., polycarbonate) and every tubule is isolated from neighboring tubules for prevention of aggregation. However, after removing of membrane, nanotubules can aggregate in a disordered material. The pores of the membrane play the role of beaker, similar as inner cavity of cyclodextrins and cucurbiturils in other cases. In the first steps, the polymer nucleates and grows on the pore walls due to solvophobic and electrostatic interactions between growing polymer chain and membrane walls. Depending on polymerization time, hollow tubules with thin or thick walls and solid fibriles for extreme case (long polymerization times) are obtained. Macromolecular chains are alignment, have less kinks and defects and as a consequences a higher electronic conductivity [103,105].

b) Another method uses soft templates as structural directing molecules. As soft templates, micelle of surfactants [113-115], multiblock copolymers, polyelectrolytes [116] nanowire seeds [117], aniline oligomers [118] and bulky dopants [119,120], liquid crystalline phases [121], reverse microemulsion have been used.
c) The synthesis of *na*nostructures in the absence of any templates but carrying the polymerization in very controlled condition. Thus, the chemical oxidative polymerization of aniline or pyrrole using ammonium peroxidisulfate in presence of protonic acids, such as naphthalene sulfonic acid [115], azobenzene sulfonic acid [122], p-toluenesulfonic acid [123] camphor sulfonic acid [124] or even inorganic acids (HCl, H_2SO_4, HBF_4, H_3PO_4) [125] can be used to obtain nanostructures. In all these cases, the dopants play the role of soft templates [115,125,126]. Organic protonic acids having a hydrophilic group ($-SO_3H$) and a lipophilic part (aromatic ring), react with aniline to form salts that adopt micellar structures. Dimensions of micro/nanostructures can be controlled from the size of micelle that is dependent on the molar ratio aniline/protonic acid. In the case of inorganic acids, anilinium cations formed between aniline and acids can also adopt a micelle form that is template for aniline polymerization. The size of nanostructures is dependent on the dopant structure and reaction conditions (temperature, aniline/acid ratio). It was observed that the diameter of tubes decresed from micro- (650 nm) to nanoscale (75 nm) with increasing the molar ratio between protonic acid and aniline from 0.5 to 4. [127].

Interfacial polymerization is another method to create nanostructures without the need for templates or even functional dopants [128]. In the case of interfacial polymerization of aniline, this is dissolved in organic phase while oxidant is in aqueous acidic solutions. The polymerization takes place at the phase boundary of an immiscible aqueous/organic biphasic system. Once formed, the hydrophilic polyaniline-emeraldine salt migrates in the water layer where termination occurs, hence the nanofibers are not subject to further polymerization and overgrowth the dimensions. The role of interface is only to separate the primary nanofibers from secondary overgrowth. In the absence of the interphasic system, a rapidly mixed reaction, when the two solutions containing monomer and oxidant are rapidly mixed and stirred [129], can lead to pure nanofibers with a relatively uniform distribution. As a general conclusion, is possible that nanofiber morphology to be obtained in the first polymerization stadies of aniline, pyrrole and thiophene due to one-directionally growing of the polymer chain but final structure can be affected by the secondary growth of polymers and agglomeration.

Microtubules of PANI and poly (o-toluidina) can also be obtained in absence of template-membrane, by in situ polymerization and doping. β-Naphthalene sulfonic acid is used as a dopant. The diameter microtubules, estimated by SEM method are between 0.8- 6 μm while wall thicknes was 50-100 nm [130, 131].

7. Heteroatom Polymers-Metal Nanocomposites

Homogeneous dispersed Au nanoparticles in bulk PANI have been synthesized by polymerization of aniline using as oxidant and doping agent a gold salt [132]. Polyaniline nanofibers and nanotubes decorated with noble metals, Au, Ag, have been synthesized by γ-irradiation of aniline/free radical oxidant/noble metal salt ($AgNO_3$, $HAuCl_4$). The dimensions of metal particles are dependent of metal precursor/aniline ratio [133]. Chemical polymerization of aniline in presence of $HAuCl_4$ and camphorsulfonic acid [134] can also allow synthesis of gold decorated PANI nanofibers. Polyaniline fibers have diameters of 50-100 nm and lengths of 1-3 μm while metal particles have dimensions between 2-25 nm. The electrical conductivities are up to 50 times higher than that of polyaniline nanofibers. Polyaniline in the undoped state can interact by imine centres with many inorganic oxids, sulfide, nitride, etc, to form hybrid organic/inorganic materials and then metal nanoparticles can be grown by a carefully controlled redox process [135-137]. This combination between conducting polymers and inorganic nanocrystals offers a new and exciting direction for electronic devices with complementary behavior between organic and inorganic partners.

8. Applications

Heteroatom-containing conducting polymers are attractive materials for a number of advanced applications as they posses electronic, optoelectronic and magnetic properties similar to metals (they are also known as *organic metals*). In the same time they have classical polymer properties; flexibility, processability, low cost, low density, etc [1]. They have well-defined electrochemical activity, stability in air, and aggressive media. Therefore these plastics can be used as components of electrochemical, electronic and photonic devices [9-12].

Polyaniline, polypyrrole and polythiophene and their derivatives can be used as the component of electricity dissipative films, electromagnetic shields, IR absorbers, as the element of electrochemical batteries, printed circuit boards, supercapacitors, electro-optical windows, field effect transistors, organic light emitting diodes, transparent conducting films [138,139]. Thin film transistors formed with active layers of thiophene –based oligomers of well-defined structure have field-effect carrier mobilities of 0.1 to 0.5 $cm^2/V.sec$, values considered useful in many electronic applications [11]. It also can be applied as the electron-proton conducting electrode, or proton conducting solid electrolyte in fuel cells and other devices. Because physical properties of conducting polymers are strongly modified over several orders of magnitude in response to interactions with species from environment they can be used as active thin films for sensors and biosensors [140-142 and references therein]

Polyaniline and other materials show excellent anticorrosion properties and are used as coatings for steel. The most significant achievement is the application of these polymers as components of field effect transistors and polymer light emitting diodes.

Metal nanoparticle-decorated nanofibers, nanotubules can find applications for digital non-volatile flash memory device with an important impact on the future of information technology [143] or catalysts for fuell cell applications or metal-catalyzed organic reactions

9. Conclusions

The great research effort to explore these polymer types (polypyrrole, polythiophene and polyaniline) has been stimulated by both their simple chemical structures and synthesis methods, polymers are thermally and environmentally stable materials, that can be doped with oxidizing or reducing agents (or in the case of polyaniline, using also inorganic or organic acids) to conducting materials and due to their interesting applications. All polymers are synthesized starting from cheap and easy accessible monomers by chemical and electrochemical oxidative polymerization. Unfortunately, all doped unsubstituted polymers are insoluble and infusible materials. The introduction of long and flexible pendant substituents on the main chain, or encapsulation of the conjugated polymers inside of many organic macrocyclics, has improved solubility. Well-defined linear, cyclic and star-shape oligomers, mainly based on thiophene rings, are other very interesting materials with aesthetically shapes and applications in solar cells, molecular wires for nano-electronics, materials for OLEDs and OFETs. They have advantage to be free of defects therefore they have high carrier mobilities and can be deposited as very thin films in controlled conditions by vacuum evaporating techniques or from solution. In the polyaniline class, many efforts are directed for obtaining of polymer nanostructures, such as: nano-fibriles, tubules, sticks, belts, particles, etc, very important for sensing and biosensing applications. Hybrid nanocomposites: polyaniline/ inorganic oxids, polyaniline/ metal particles or polyaniline/mesoporous or layered materials are other very investigated materials due to their interesting properties and applications.

Acknowledgements:

The author thanks for financial support by Projects CEEX-MATNANTECH 10/2005.

References

[1] Skotheim, T. A.; Reynolds, J. R. eds., *Handbook of Conducting Polymers, 3-rd edition* CRC Press, 2007.

[2] Naarman, H.; Theophilou, N. *Synth. Met.* 1987, 22, 1-8.

[3] Chiang, C. K.; Fincher, C. R.; Park, Y. W.; Heeger, A. J.; Shirakawa, H.; Louis, E. J.; Gau, S. C.; MacDiarmid, A. G. *Phys. Rev. Lett.* 1977, 39, 1098-1101.
[4] 4 (a) Diaz, A. F.; Kanazawa, K. K.; Gardini, G. P. *J. Chem. Soc. Chem. Commun.*1979, 635-638; (b) Kanazawa, K. K.; Diaz, A. F.; Gill, W. D.; Grant, P. M.; Street, G. B.; Gardini G. P.; Kwak, J. F. *Synth. Met.* 1979/1980. 1, 329-336.
[5] Yamamoto, T.; Sanechika, K.; Yamamoto, A. *J. Polym. Sci. Polym. Lett. Ed.* 1980, 18, 9-12.
[6] MacDiarmid, A. G.; Epstein, A. *Faraday Discuss. Chem. Soc.* 1989, 88, 317-332.
[7] Eckhardt, H.; Shacklette, L. W.; Jen, K. Y.; Elsenbaumer R. L. *J. Chem. Phys.* 1989, 91, 1303-1315.
[8] Lee, K., *Nature* 2006, 441, 65-68.
[9] Walzer, K.;Maening, B.; Pfeifer, M.; Leo, K. *Chem. Rev.* 2007, 107, 1233-1271.
[10] Murphy, A. R.; Frecht, J. M. J. *Chem. Rev.* 2007, 107, 1066-1096
[11] Garnier, F.; Horowitz, G.; Peng, X. H.; Fichou, D. *Adv. Mater.* 1990, 2, 592-594.
[12] Peng, X. Z.; Horowitz, G.; Fichou, D.; Garnier, F. *Appl. Phys. Lett.* 1990, 57, 2013-2015.
[13] Menard, E.; Meitl, M. A.; Sun, Y.; Park, J. U.; Lee, D. J.; Nam, Y. S.; Jeon, S.; Rogers, J. A. *Chem. Rev.* 2007, 107, 1117-1160.
[14] Pinto, M. R.; Schanze, K. S. *Synthesis* 2002, 9, 1293-1309.
[15] Lux, F. *Polymer* 1994, 35, 2915-2936.
[16] Kang, E. T.; Neoh, K. G.; Tan, K. L. *Prog. Polym. Sci.* 1998, *23*, 277-324.
[17] Anand, J.; Palaniappan, S. A.; Sathyanarayana, D. N. *Prog. Polym. Sci.*, 1998, 23, 993-1018.
[18] Gospodinova, N.; Terlemezyan, L. *Prog. Polym. Sci.*, 1998, 23, 1443-1484.
[19] Chiang, J. P.; MacDiarmid, A. G. *Synth. Met.* 1986, 13,193-205.
[20] Armes, S. P.; Miller, J. P. *Synth. Met.* 1988, 22, 385-393.
[21] Focke, W. W.; Wuck, G. E.; Wei, Y. *J. Phys. Chem.* 1987, 91, 5813-5818.
[22] Pud, A.; Ogurtsov, N.; Korzhenko, A.; Shapoval, G. *Prog. Polym. Sci.* 2003, 28, 1701-1753
[23] Wei, Y.; Tang, X.; Sun, Y. *J. Polym. Sci. Part A: Polym. Chem.* 1989, 27, 2385-2396
[24] Fan, J.; Wan, M.; Zhu, D. *J. Polym. Sci. Part A: Polym. Chem.*1998, 36, 3013-3019.
[25] Chen, S. A.; Hwang, G. W. *J. Amer. Chem. Soc.* 1994,116, 7939-7940.
[26] Nguyen, M. T.; Kasal, P.; Miller, J. L.; Diaz, A. F. *Macromolecules*,1994, 27, 3625-3631.
[27] Yue, J.; Epstein, A. J. *J. Amer. Chem. Soc.* 1990, 112, 2800-2801.
[28] Yue, J.; Wang, Z. H.; Cromack, K. R.; Epstein, A. J.; MacDiarmid, A. J. *J. Amer. Chem. Soc.* 1991, 113, 2665-2671.
[29] Chen, S. A.; Hwang, G. W. *J. Am. Chem. Soc.* 1995, 117, 10055-10062.
[30] Majidi, M. R.; Kane-Maguire, L. A. P.; Wallace, G. G. *Polymer* 1994, 35, 3113-3115.
[31] Innis, P. C.; Norris, I. D.; Kane-Maguire, L. A. P.; Wallace, G. G. *Macromolecules* 1998, 31, 6521-6528.
[32] Havinga, E. E.; Bouman; M. M.; Meijer E. W.; Pomp, A.; Simenon, M. M. J. *Synth. Met.*, 1994, 66, 93-97.
[33] Majidi, M. R.; Kane-Maguire, L. A. P.; Wallace, G. G. *Polymer* 1995, 36, 3597-3599.
[34] Ashraf, S. A.; Kane-Maguire, L. A. P.; Majidi, M. R.; Pyne, S. G.; Wallace G. G. *Polymer* 1997, 38, 2627-2631.

[35] Wienk, M. M.; Janssen R. A. *J. Am. Chem. Soc.* 1997, 119, 4492-4501.
[36] Wienk, M. M.; Janssen, R. A. *J. Am. Chem. Soc.*1996, 118, 10626-10628.
[37] Hartwig, J. F.; *Acc. Chem. Res.* 1998, 31, 852-860.
[38] Wolfe, J. P.; Wagaw, S.; Marcoux, J. F.; Buchwald, S. L. *Acc. Chem. Res.* 1998, 31, 805-818.
[39] Nikolaos, S.; Rachel, E.; Wardand, T.; Tra Y. Meyer, *Macromolecules* 1998, 31, 3158-3161.
[40] Galecka, M.; Wielgus, I.; Zagorska, M.; Pawlowski; M.; Kulszewicz-Bajer I. *Macromolecules* 2007, 40, 4924-4932.
[41] James, D. K.; Tour, J. M. *Top. Curr. Chem.* 2005, 257, 33-62, and references therein
[42] Kauffmann, T.; Greving, B.; Kriegesmann, R.; Mitschker, A.; Woltermann, A. *Chem. Ber.* 1978, 111, 1330-1336.
[43] Marsella, M. J.; Reid, R. J. *Macromolecules* 1999, 32, 5982-5984.
[44] Marsella, M. J.; Kim, I. T.; Tham, F. *J. Am. Chem. Soc.* 2000, 122, 974-975.
[45] Marsella, M. J.; Yoon, K.; Tham, F. *Org. Lett.* 2001, 3, 2129-2131.
[46] Nakao, K.; Nishimura, M.; Tamachi, T.; Kuwatani, Y.; Miyasaka, H.; Nishinaga, T.; Iyoda, M. *J. Am. Chem. Soc.* 2006,128, 16740-16747.
[47] Fuhrmann, G.; Debaerdemaeker, T.; Bauerle, P. *Chem. Commun.* 2003, 948-949.
[48] Osteritz, E. M.; Bauerle, P. *Adv. Mater.* 2001,13, 243-246.
[49] Fuhrmann, G.; Kromer, J.; Bauerle, P. *Synt. Met.* 2001, 119, 125-126.
[50] Kromer, J.; Carreras, I. M.; Fuhrmann, G.; Musch, C.; Wunderlin, M.; Debaerdemaeker, T.; Osteritz, E. M.; Bauerle, P. *Angew. Chem. Int. Ed.* 2000, 39, 3481-3486.
[51] Hoger, S. *Shape Persistent Acetylenic Macrocycles for Ordered Systems* in Acetylenic Chemistry, F. Diederich, P. J. Stang, R.R. Tykwinski, Eds., Wiley-VCH: Weinheim, Germany, 2005.
[52] Inoue, K. *Prog. Polym. Sci.* 2000, 25, 453-571.
[53] Vogtle, F.; Gestermann, S.; Hesse, R.; Schwierz, H.; Windisch, B.; *Prog. Polym. Sci.* 2000, 25, 987-1041.
[54] Gao, C.; Yan, D. *Prog. Polym. Sci.* 2004, 29, 183-275.
[55] Xia, C.; Fan, X.; Locklin, J.; Advincula, R. C. *Org. Lett.*, 2002, 4, 2067-2070.
[56] Xia, C.; Fan, X.; Locklin, J.; Advincula, R. C.; Gies, A.; Nonidez, W. *J. Am. Chem. Soc.* 2004, 126, 8735-8743.
[57] Mitchell, W. J.; Kopidakis, N.; Rumbles, G.; Ginley, D. S.; Shaheen, S. E. *J. Mater. Chem.* 2005, 15, 4518-4528.
[58] Negishi, N.; Ie, Y.; Taniguchi, M.; Kawai, K.; Tada, H.; Kaneda, T.; Aso, Y. *Org. Lett.* 2007, 9, 829-832.
[59] Xu, M. H.; Pu, L. *Tetrahedron Lett.* 2002, 43, 6347-6350.
[60] Ponomarenko, S. A.; Kirchmeyer, S.; Elschner, A.; Huisman, B. H.; Karbach, A.; Drechsler, D. *Adv. Funct. Mater.* 2003, 13, 591-596.
[61] Cherioux, F.; Guyard, L.; Audebert, P. *Chem. Commun.* 1998, 2225-2226.
[62] Cherioux, F.; Guyard, L.; Audebert, P. *Adv. Mater.* 1998, 10, 1013-1018.
[63] Carvine, A.; Roquet, S.; Leriche, P.; Aleveque, O.; Frere, P.; Roncali, J. *Chem. Commun.* 2006, 1416-1418.
[64] Carvine, A.; Roquet, S.; Aleveque, O.; Leriche, P.; Frere, P.; Roncali, J. *Chem. Mater.*, 2006, 18, 2584-2590.

[65] Nicolas, Y.; Blanchard, P.; Levillain, E.; Allain, M.; Mercier, N.; Roncali, J. *Org. Lett.* 2004, 6, 273-276.
[66] Bettignies, R.; Nicolas, Y.; Blanchard, P.; Levillain, E.; Nunzi, J. M.;. Roncali, J. *Adv. Mater.* 2003, 15, 1939-1943.
[67] Sun, Y.; Xiao, K.; Liu, Y.; Wang, J.; Pei, J.; Yu, G.; Zhu, D. *Adv. Funct. Mater.* 2005, 15, 818-822.
[68] Pei, J.; Wang, J. L.; Cao, X. Y.; Zhou, X. H.;.Zhang, W. B. *J. Amer. Chem. Soc.* 2003, 125, 9944-9945.
[69] Nepogodiev, S. A.; Stoddart, J. F. *Chem. Rev.*1998, 98, 1959-1976.
[70] Raymo F. M.; Stoddart, J. F. *Chem. Rev.* 1999, 99, 1643-1664.
[71] Sauvage, J. P.; Dietrich-Buchecker, C. Eds. Molecular Catenanes, Rotaxanes and Knots, 1999, Wiley-VCH, Weinheim.
[72] Kim, K.; Selvapalam, N.; Oh, D. H. *J. Incl. Phenom. Macrocylic Chem.* 2004, 50, 31-36.
[73] Szejtli, J. Cyclodextrins and Their Inclusion Complexes, Akademiai Kiado, Budapest, 1982.
[74] Behrend, R.; Meyer, E.; Rusche, F. *Liebigs Ann. Chem.*1905, 1, 339-340.
[75] Freeman, W. A.; Mock, W. L.; Shih, N. Y. *J. Am. Chem. Soc.* 1981,103, 7367-7368.
[76] Day, A.; Arnold, A. P.; Blanch R. J.; Snushall, B. *J. Org. Chem.* 2001, 66, 8094-8100.
[77] Lee, J. W.; Samal, S.; Selvapalam, N.; Kim,H. J.; Kim, K. *Acc. Chem. Res.* 2003, 36, 621-630.
[78] Tuncel D.; Steinke, J. H. G. *Macromolecules* 2004, 37, 288.-302
[79] Whang, D.; Jeon, Y. M.;. Heo J.; Kim, K. *J. Am. Chem. Soc.* 1996,118, 11333-11334.
[80] Whang D.; Kim, K. *J. Am. Chem. Soc.* 1997,119, 451-452.
[81] Whang, D.; Park, K. M.; Heo, J.; Ashton P.; Kim, K. *J. Am. Chem. Soc.* 1998,120, 4899-4900.
[82] Park, K. M.; Kim, S. Y.; Heo, J.; Whang, D.; Sakamoto, S.; Yamaguchi K.; Kim, K. *J. Am. Chem. Soc.* 2002, 124, 2140-2147.
[83] Tan, Y.; Choi, S. W.; Lee, J. W.; Ko Y. H.; Kim, K. *Macromolecules* 2002, 35, 7161-7165.
[84] Isobe, H.; Sato, S.; Nakamura, E. *Org. Lett.* 2002, 4, 1287-1289.
[85] (a) Lagrost, C.; Lacroix, J. C.; Aeijach, S.; Jouini, M.; Chane-Ching K. I.; Lacaze, P. C. *Chem. Commun.* 1998, 489-490; (b) Lagrost, C.; Lacroix, J. C.; Aeijach, S.; Jouini, M.; Chane-Ching K. I.; Lacaze, P. C.: *J. Chim. Phys.*1998, 95, 1208-1212.
[86] Takashima, Y.; Oizumi, Y.; Sakamoto, K.; Miyauchi, M.; Kamitori S.; Harada, A. *Macromolecules* 2004, 37, 3962-3964.
[87] Storsberg, J.; Ritter, H.; Pielartzik, H.;. Groenendaal, L. *Adv. Mater.* 2000, 12, 567-569.
[88] Yamaguchi, I.; Kashiwagi K.; Yamamoto, T. *Macromol. Rapid Commun.* 2004, 25, 1163-1166.
[89] Grigoras, M.; Conduruta, G. D. *J. Incl. Phenom. Macrocylic Chem.* 2006, 54, 101-107.
[90] Yuan, G. L.; Kuramoto N.; Takeishi, M. *Polym Adv. Technol.* 2003, 14, 428-432.
[91] Shimomura, T.; Yoshida, K. Ito, K.; Hayakawa, R. *Polym. Adv. Technol.* 2000, 11, 837-839.
[92] Yoshida, K.; Shimomura, T.; Ito, K.; Hayakawa, R. *Langmuir* 1999, 15, 910-913.
[93] Grigoras, M.; Conduruta, G. D. *Rev. Roum. Chim.* 2006, 51. 987-992

[94] Simionescu, C. I.; Grigoras, M.; Farcas, A.; Stoleru, A. *Macromol. Chem. Phys.* 1998, 199, 1301-1306.
[95] Farcas, A.; Grigoras, M. *High Perf. Polym.* 2001, 13, 149-158.
[96] Farcas, A.; Grigoras, M. *High Perf. Polym.* 2001, 13, 201-210.
[97] Farcas, A.; Grigoras, M. *Polym. Int.* 2003, 52, 1315-1320.
[98] Li, H.; Kameoka, J.; Czaplewski, D.A.; Craighead H. G. *Nanoletters* 2004, 4, 671-675.
[99] Ito, T.; Shirakawa, H.; Ikeda, S. *J. Polym. Sci. Polym. Chem. Ed.* 1974, 12, 11-20.
[100] Reneker D. H.; Chun, I. *Nanotechnology* 1996, 7, 216-223.
[101] He, H. X.; Li, C. Z.; Tao, N. J. *Appl. Phys. Lett.* 2001, 78, 811-813
[102] Xu, P.; Kaplan D. L. *Adv. Mater.* 2004, 16, 628-633
[103] Martin, C. R. *Chem. Mater.* 1996, 8, 1739-1746.
[104] Parthasarathy, R. V.; Martin, C. R. *Chem. Mater.* 1994, 6, 1627-1632.
[105] Martin, C. R. *Acc. Chem. Res.*1995, 28, 61-68.
[106] Wu, C. G.; Bein, T. *Science* 1994, 264, 1757-1759.
[107] Martin, C. R. *Adv. Mater.*1991, 3, 457-459.
[108] Wang, C. W.; Wang, Z.; Li M. K.; Li, H. L. *Chem. Phys. Lett.* 2001, 341, 431-434.
[109] Martin, C. R. *Science* 1994, 266, 1961-1966.
[110] Park, J. G.; Lee, S. H.; Kim, B.; Park, Y. W. *Appl. Phys. Lett.* 2002, *81*, 4625-4627.
[111] Park, J. G.; Kim, B.; Lee, S. H.; Park, Y. W. *Thin Solid Films* 2003, *438-439*, 118-122.
[112] Duvail, J. L.; Rétho, P.; Garreau, S.; Louarn, G.; Godon, C.; Demoustier-Champagne, S. *Synth.Met.* 2002, *131*, 123-128.
[113] Wei, Z.; Zhang, L.; Yu, M.; Yang, Y.; Wan, M. *Adv. Mater.* 2003, 15, 1382-1385.
[114] Li, G.; Zhang, Z. *Macromolecules* 2004, 37, 2683-2685.
[115] Wei Z.; Zhang Z.; Wan, M. *Langmuir*, 2002, 18, 917-921.
[116] Liu, J. M.;Yang, S. C. *J. Chem. Soc. Chem. Commun.* 1991, 1529-1530.
[117] Zhang, X.; Goux, W. J.; Manohar, S. K. *J. Amer. Chem. Soc.* 2004, 126, 4502-4503.
[118] Li, W. G.; Wang, H. L. *J. Amer. Chem. Soc.* 2004, 126, 2278-2279.
[119] Wei, Z. X.; Wan, M. X. *J. Appl. Polym. Sci.* 2003, 87,1297-1301.
[120] Qiu, H. J.; Wan, M. X.; Matthews, B.; Dai, L. M. *Macromolecules* 2001, 34, 675-677.
[121] Huang, L. M.; Wang, Z. B.; Wang H. T.; Cheng, X. L.; Mitra, A. Yan, Y. X. *J. Mater. Chem.* 2002, 12, 388-391.
[122] Huang, K.; Wan, M. X. *Chem. Mater.* 2002, 14, 3486-3492.
[123] Yang, Y. S.; Liu, J.; Wan, M. X. *Nanotechnology* 2002, 13, 771-774.
[124] Zhang L. J.; Wan, M. X. *Nanotechnology* 2002, 13, 750-755.
[125] Zhang, Z. M.; Wei, Z. X.; Wan, M. X. *Macromolecules* 2002, 35, 5937-5942.
[126] Shen, Y.; Wan, M. *J. Polym. Sci.; Part A: Polym. Chem.* 1999, 37, 1443-1449.
[127] Huang, J.; Kaner, R. B. *J. Amer. Chem. Soc.* 2004, 126, 851-855.
[128] Huang, J.X.; Kaner, R. B. *Angew. Chem. Int. Ed.* 2004, 43, 5817-5821.
[129] Huang, J.; Wan, M. X. *J. Polym. Sci.; Part A: Polym. Chem.* 1999, 37, 151-157.
[130] Wan, M. X.; Li, J. C. *Polym. Adv. Technol.* 2003, 14, 320-325.
[131] Kinyanjui, J. M.; Hatchett, D. W.; Smith, J. A.; Josowicz, M. *Chem. Mater.* 2004, 16, 3390-3398.
[132] Pillalamarri, S. K.; Blum, F. D.; Tokuhiro, A. T.; Bertino, M. F. *Chem. Mater.* 2005, 17, 5941-5944.
[133] 134. Feng, X.; Yang, G.; Xu, Q.; Hou, W.; Zhu, J. J. *Macromol. Rapid Commun.* 2006, 27, 31-36.

[134] 135. Godovsky, D. Y.; Varfolomeev, A. E.; Zaretsky, D. F.; Nayana Chandrakanthi, R. L.; Kundig, A.; Weder C. *J. Mater. Chem.* 2001, 11, 2465-2469.
[135] 136. Kan, J.; Zhou, S.; Zhang, Y.; Patel, M. *Eur. Polym. J.* 2006, 42, 2004-2012.
[136] 137. Li, L.; Jiang, J.; Xu, F. *Eur. Polym. J.* 2006, 42, 2221-2227.
[137] 138. Gurunathan, K.; Vadivel Murugan, A.; Marimuthu, R.; Mulik, U. P.; Amalnerkar, D. P. *Materials Chemistry Physics* 1999, 61, 173-191.
[138] 139. Saxena, V.; Malhotra, B. D. *Curr. Appl. Phys.* 2003, 3, 293-305.
[139] 140. Bai, H.; Shi, G. *Sensors* 2007, 7, 267-307.
[140] 141. Adhikari, B.; Majumdar, S. *Prog. Polym. Sci.* 2004, 29, 699-766.
[141] 142. Tyler McQuade, D.; Pullen, A. E.; Swager, T. M. *Chem. Rev.* 2000, 100, 2537-2574.
[142] 143. Tseng, R. J.; Huang, J. X.; Ouyang, J.; Kaner, R. B.; Yang, Y. *Nanoletters* 2005, 5 1077-1080.
[143] 144. Rajesh, B.; Ravindranathan Thampi, K.; Bonard, J. M.; Mathieu, H. J.; Xanthopoulos N.; Viswanathan B. *Electrochem. Solid-State Lett.* 2004, 7, A 404-A 407.

In: Advances in Functional Heterochain Polymers
Editor: Maria Cazacu
ISBN: 978-1-60456-598-0

Chapter 5

DEVELOPMENTS ON PHOTOPOLYURETHANES AND ITS VARIOUS APPLICATIONS

Emil C. Buruiana, Tinca Buruiana*[1] *and Violeta Melinte
Romanian Academy, Petru Poni Institute of Macromolecular Chemistry, 41 A Gr. Ghica Voda Alley 700487, Iasi

ABSTRACT

This review aims to address the main issues of polyurethanes containing different chromophores chemically incorporated in the macromolecular backbone with focus on ionic polymers with photoisomerizable or photolabile structures. First, aspects of chemistry and photochemistry of the polyurethanes are presented, such as isocyanate, polyols, chromophoric derivatives, prepolymers and segmented elastomers as well as their properties induced by the presence of chromophore (for instance, azobenzene). Then, some specific applications of the photopolyurethanes and the corresponding ionic forms are described.

Keywords: *polyetherurethane ionomers, azobenzene, triazene, photoisomerization/ photochromic, photolabile units*

1. INTRODUCTION

Interest in the synthesis and characterization of new photopolymers has grown in recent years due to their unusual properties and potential applications in the optic and optoelectronic field, biomedical devices, microlithography, and so on [1-4]. Consequently, the literature data are referring to various polymeric structures bearing chromophore moieties, among these the polyurethanes having an important role [5-7]. Polyurethanes (PU) are linear block copolymers, typically assembled of statistically alternating soft (SS) and hard (HS) segments. Their versatile properties have made them promising candidates for many applications in

[1] e-mail: tbur@icmpp.ro.

various markets where tough, durable, elastomeric materials are required. These polymers offer design freedom not found in other materials due to a unique and broad-ranging portfolio of characteristics including outstanding abrasion resistance, strength and toughness, chemical and hydrolysis resistance, fungal growth resistance, low temperature flexibility, clarity, colourability, and a wide processing window in many plastic procedures. Their adaptable physical properties are usually attributed to the microphase-separated structure deriving from the thermodynamic incompatibility between the soft and hard segments [8-10]. Although an almost unlimited variety of chemical compositions is possible, the soft segments are frequently oligomeric diols of polyether or polyester type (Mw: 500-3000), while the hard segments are the result of the diisocyanate chain extended by low molecular weight diols. An interesting development of this field pioneered by Dieterich (1970), Cooper (1973) and further, by Hwang (1981) was the insertion of small amount of ionic functionality into polymers to yield polyurethane ionomers [11-13]. Though the polymers contain a minority of ionic groups, below 250 meq/100 g [14] or a more constrained range between 10-120 meq/100 g polymer [15], their properties are strongly influenced by the presence of ionic charges, which tend to aggregate *via* Coulombic interactions into so-called "ionic domains" promoting thus phase separation and hard domain cohesion. Depending on the charge nature (positive, negative or both) placed within the polymeric backbone, pendant to the chain or only at the chain ends (telechelic systems), all types of ionomers, namely cationomers, anionomers and zwitterionomers have received constantly much attention [16-21]. Considered now one of the most fast developing and active branches of polyurethane chemistry, such polymers combine the well-known properties of the parent polymers with those derived from the existence of ionic structures in the macromolecular chain, offering a unique opportunity to form aqueous dispersions in the absence of any external emulsifier. The qualities of aqueous dispersions, notably their excellent adhesion at many polymeric and glass substrates and film-forming ability at ambient temperature, have recommended them as suitable for modern ionomer adhesive and coating applications [22, 23]. Since they were first reported in 1970, a broad variety of polyurethane ionomers has been synthesized and their specific properties have been reviewed [24, 25].

Chromophoric polymers exhibit a change of their absorption and/or fluorescence characteristics in response to a thermal [26, 27], chemical [28], optical [29] or other [30] stimulation and are useful for a large range of technologies [31, 32]. Incorporation of photochromic molecules into polymeric materials leads to photoresponsive systems, whose properties can be manipulated by light. Numerous optical phenomena emerged and generated intensive research, the most exploited among them being the linear absorbance and refractive index changes in solid systems which utilize ordering effects leading to photoinduced birefringence and dichroism, principally studied in azo polymers. Therefore, the potential applications of these photopolymers as reversible optical data storage, diffraction, and holography have stimulated a great deal of experimental and theoretical work, which extended the use of different photochromic architectures [33, 34].

On this line, it was very interesting to introduce variable chromophores in polyurethane ionomers, recognized for their distinctive functional properties induced by the existence of ionic groups associated in ionic aggregations, which act as cross-links in order to give photoactive polymeric materials. Using a versatile synthesis methodology, various monomeric and polymeric materials carrying chromophoric functionalities (azoaromatic, nitroaromatic, quinone, anthryl, triazene) covalently or electrostatically attached to the hard segments of the polymeric chain were prepared and studied. The aim of this review is to give a brief description of the polyurethane ionomers with cationic groups and will refer to our own work carried out in the last ten years on cationomers bearing different chromophoric structures with focus on the specific role of the chromophore in adjustment of polymer properties.

2. Polyurethane Cationomers

2.1. General Considerations on their Synthesis and Properties

For the synthesis of polyurethane ionomers various methods have been tried out, but with a few exceptions [35-39], the ionic groups have been introduced in the hard segments of polyurethane chains. Based on the urethane chemistry, such polymers were obtained during the polymerization by use of a chain extender with preformed ionic or ionogene groups. The first elastomeric cationomers have been prepared by the well-known procedure of using a chain extender with tertiary amino group and subsequent modification with an ionisable component. The most common approach has been to use *N*-methyldiethanolamine [40-48] as low molecular diol with tertiary amine functionality, followed by a quaternization of the aliphatic tertiary nitrogen atoms with various acids, alkyl halides or dimethyl sulfate.

Cooper et al. [49,50], Varma et al. [51], have been designed new polyurethane cationomers with pendant ammonium groups using as chain extender short diols of ionic type or diols with trialkylamine groups, quaternized latter. Polymer morphology and some properties were investigated as a function of ionic content, neutralizing anion and alkyl group length. Polyurethane cationomers have also been prepared by using viologen diol [52] or a soft segment based on polybutadiene carrying quaternary ammonium salt [53, 54].

Other ionic polymers based on ionene chemistry allowed incorporation of the non-covalently attached cyclic species on each "paraquat" hard segment and consequently the preparation of segmented ionic "rotaxane" [55]. Thus, polyetherurethane ionenes with viologen units have been prepared by Menschutkin reaction from di-tertiary amines and dihalides and then examined in terms of photochromic activity [56]. Moreover, some polyurethane ionenes with good antimicrobial response were obtained in these conditions [57]. A different set of polymers containing polyurethane-polyvinylbenzylchloride blocks or pyridinium rings was also reported [58-60].

Extensive studies have been performed on the structure-properties relationship in the polycations revealing and explaining in the same time their segmental structure and peculiarity to be utilized as coatings and adhesives in the form of water dispersions [61-63]. The effects of ionomeric architecture both on the polymer solution [64, 65] and film properties [66-70] were detailed investigated. Indeed, due to the propensity of the electrostatic charges to strongly interact in polar dilute solutions, these polymers have similarities with polyelectrolytes [71]. In the same time, the ionic groups segregations into microdomains with formation of physical (Coulombic) crosslinks as well as the preparation method and thermal history influences the elastomeric characteristics of the polymers. The property profile of ionomeric films obtained from urethane-acrylate cationomer aqueous dispersions was exploited in applications including coatings, adhesives or impregnation formulations [61, 72].

In contrast with the scientific and commercial interest focused on polyurethane ionomers, a few works concerning insertion of the chromophore structures in the polymeric chains, as a way to diversify the traditional applications, were reported in the literature [73-76]. Such materials included in their structure dyes [73-75] or mesogenic azobenzene moieties [76].

In the last decade, our laboratory has developed a new family of polyurethane cationomers containing quinone, anthryl, nitroaromatic and/or azoaromatic chromophores anchored to the ionomeric backbone [77-84]. Insertion of some photoactive units that elicit optical response in such polymers could result in an enhanced knowledge of their peculiarities, opening the way to design new materials for possible practical optical devices.

Inspired by the studies performed on classical polyurethanes bearing azo chromophores and implicitly on the obtaining of mecano(photo)chromic, liquid-crystal, nonlinear optical or photocleavable materials [85-88], it was considered usefully to investigate the synthesis and photobehavior of some polyurethane cationomers with azobenzene attached to polymeric chains. As firstly demonstrated, use of photochromic azobenzene labels in the main chains of single-phase polyurethanes, as a sensitive molecular probe azoaromatic structures provided information on the aging time of the polyurethane backbone related to the loss of free volume, its distribution and its change with aging [89,90].

2.2. Polyurethane Cationomers with Pendent Azobenzene Chromophores

2.2.1. Synthesis of Azo Cationomers with Chromophore Anchored to the Hard Segment

As was described in our previous papers [80,84,91-93], azoaromatic units containing cationomers may have a variety of chain structures introduced by means of new azo chain extenders, azo quaternizing agents or azo polyol components (Table 1).

Table 1. Chemical structures of the precursors used in the synthesis of azo polyurethane cationomers

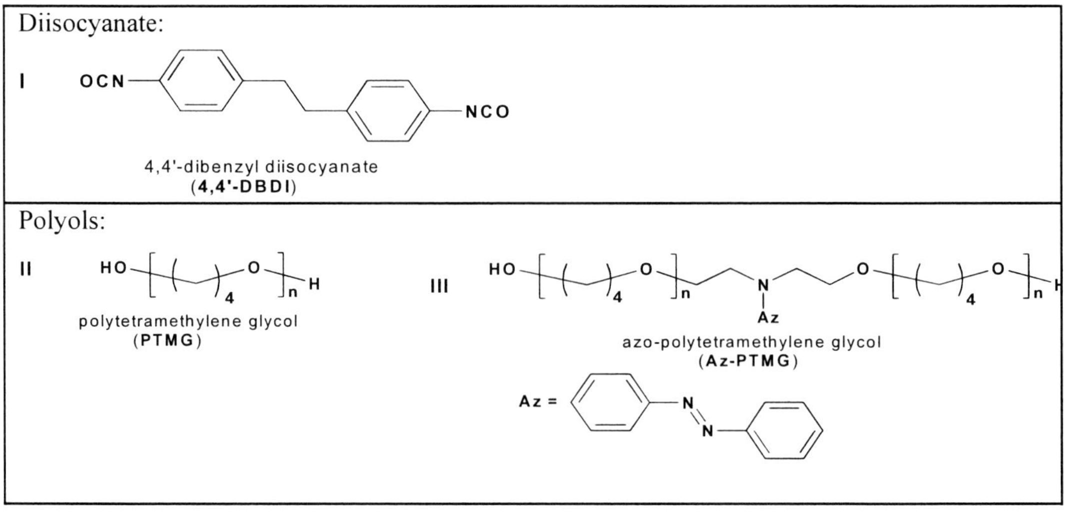

Chain extenders:

IV

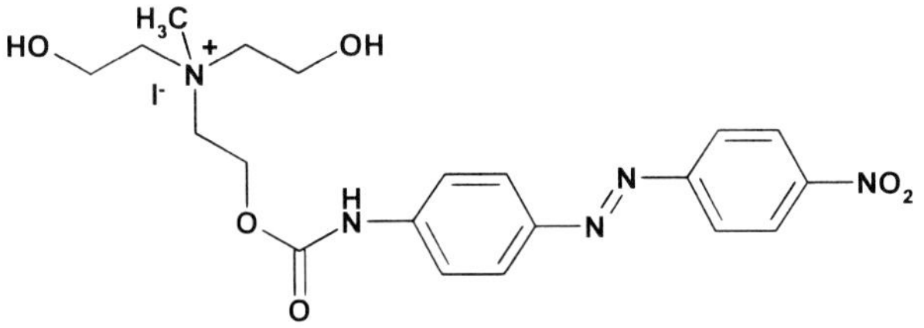

2-ethyl-N-methyldiethanolamoniu-N[4(nitrophenyldiazenyl)phenyl] aminoformate iodide

V

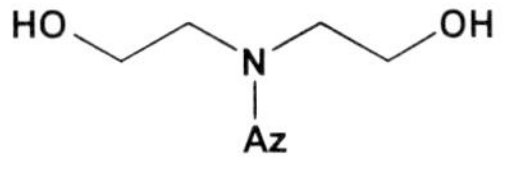

N-di-(β-hydroxyethyl)-4-amino azobenzene

VI

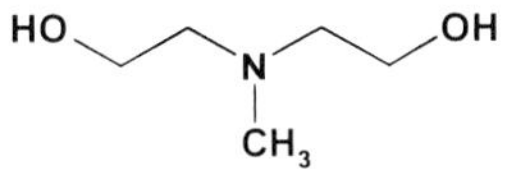

N-methyldiethanolamine

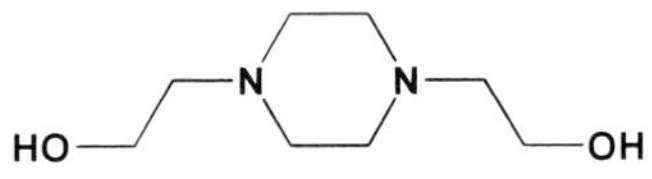

VII

N,N-di-(β-hydroxyethyl)piperazine

VIII

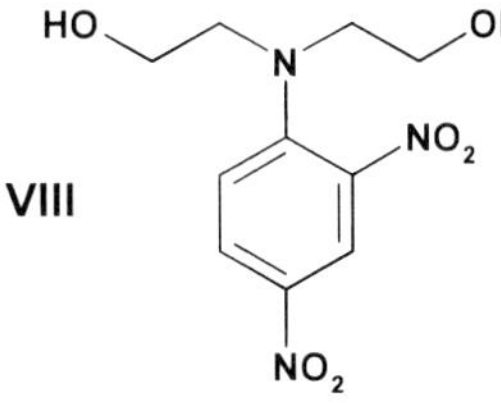

N-(2,4-dinitrophenyl) diethanolamine

Quaternizing agents:

IX

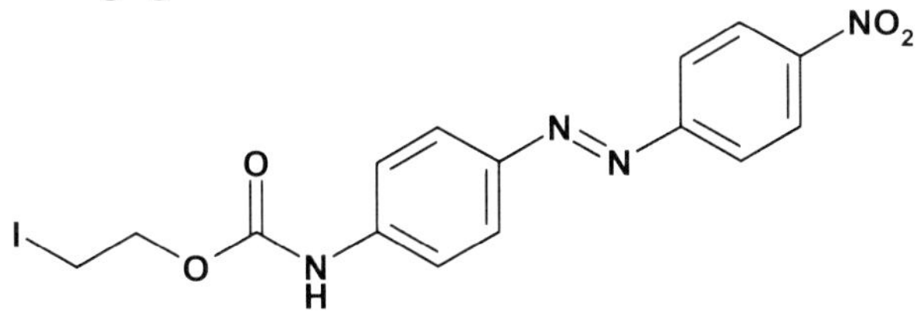

2-iodoethyl-N-[4(4-nitrophenyl-diazenyl)phenyl]aminoformate

XI

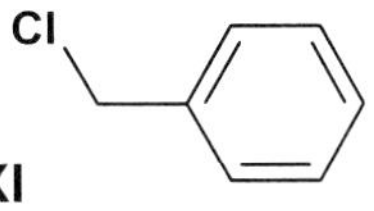

benzyl chloride

X

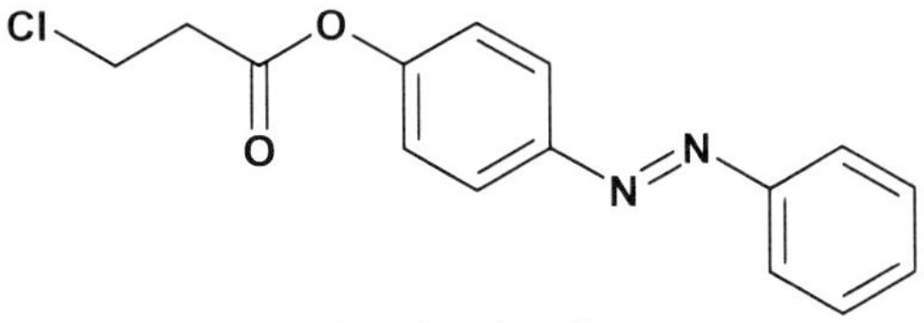

4(3-chloropropionyloxy)azobenzene

CH_3I

XII

methyl iodide

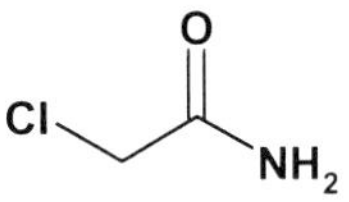

XIII

chloroacetamide

Azo salts:

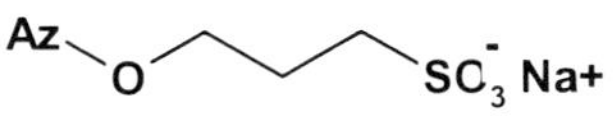

3(p-oxyazobenzene)propane sulfonic acid sodium salt

XIV

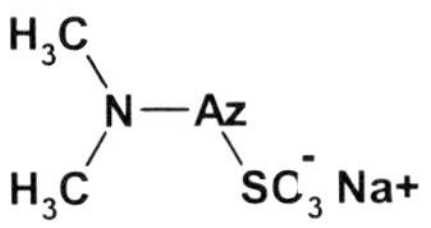

methyl orange sodium salt

XV

Some of the most highly investigated polymers were elastomeric cationomers based on poly(tetramethylene glycol) of 1000 or 2000 average molecular weights (PTMG), 4,4'-dibenzyl diisocyante (4,4'-DBDI) and different chain extenders of ionic or non-ionic type (Table 2).

Table 2. Structural characteristics of azo polyetherurethane cationomers

Polymer	Azo-derivative (cf.Table1)	Molar ratio (polyol/ NCO/OH)	Ionic halogen cont. (meq/100g)	Azo cont.* (wt %)	λ_{max} (nm) DMF	Film	Ref.
Az-PUC1	IV	1/3/2	49.23	28.59	452	440.0	[90]
Az-PUC2	IX	1/3/2	48.35	28.08	450	440.5	[79,91]
Az-PUC3	X	1/3/2	50.35	13.92	451	440.5	[92]
Az-PUC4	V	1/3/2	28.56	8.50	415.0	412.5	[83]
Az-PUC5	V	1/2/1	16.42	5.00	408.0	406.0	[83]
Az-PUC6	III	1/3/2	51.93	7.24	410	405.5	[77]
Az-PUC7	III	1/3/2	65.52	9.13	-	407	[77]
Az-PUC8	XIV	1/2/1	33.57	9.29	351	345	[81]
Az-PUC9	XIV	1/2/1	4.95	15.79	350	347	[80]
AzPUC10	XV	1/3/2	47.09	23.50	423	434.5	[95]

* - calculated from the ionic halogen content

Therefore, the use of azo ionic diol (IV) as chain extender for the prepolymer based on PTMG (II) and 4,4-DBDI (I), was a direct route to afford azo polyetherurethane cationomers (Az-PUC1) with *para*-nitroazoaromatic chromophore located in the quaternary ammonium structure, as shown in Scheme 1 [91].

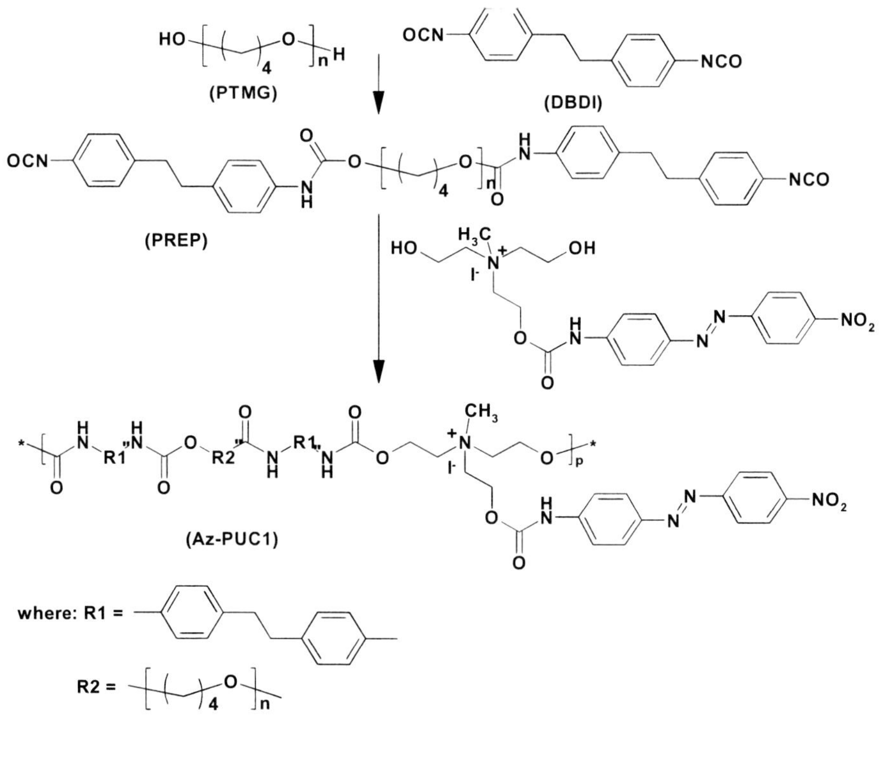

Scheme 1. Synthesis of Az-PUC1

Another technique widely used to introduce azo chromophore is based on quaternizing tertiary amine (VI), previously introduced in polyurethanes by means of azo derivative (IX, X), when the resulting polycations (Az-PUC2, Az-PUC3) contain *para*-nitroazoaromatic or azoaromatic groups attached to quaternary nitrogen atoms [80,92,93]. A general scheme may be described for the Az-PUC3 synthesis (Scheme 2).

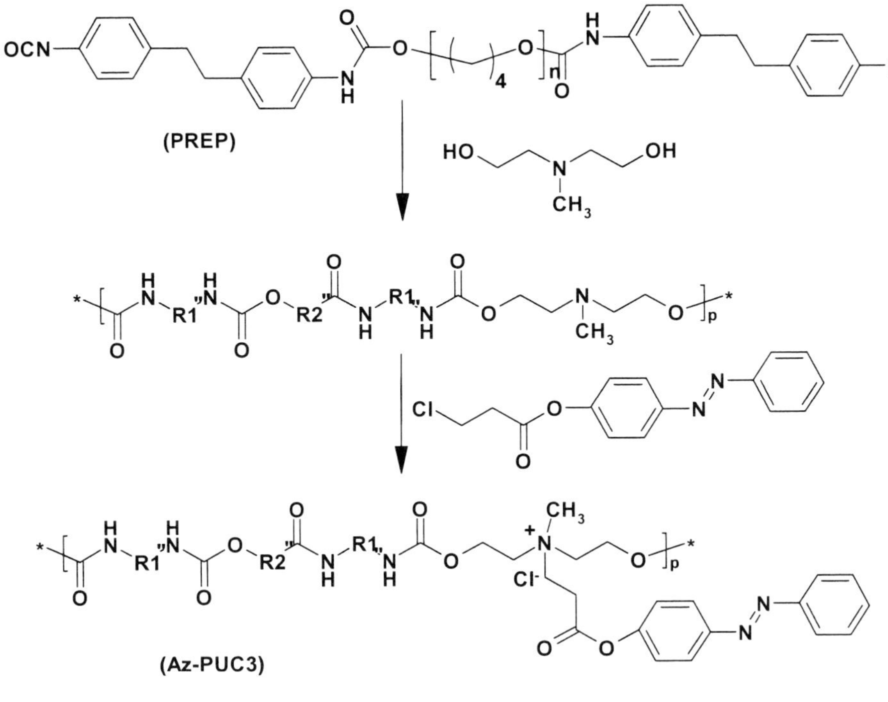

where R1 and R2 are identical with those from scheme 1

Scheme 2. Synthesis of Az-PUC3.

In a similar manner, the use of a mixture of chain extenders, such as azo-diol (V), and NMDA, followed by quaternizing with benzyl chloride led to azo cationomers (Az-PUC4, Az-PUC5), schematically represented in Figure 1 [84]. Both polymers have the azoaromatic chromophore placed in the non-ionic fragment of the hard segments but differ by the concentration of ionic groups.

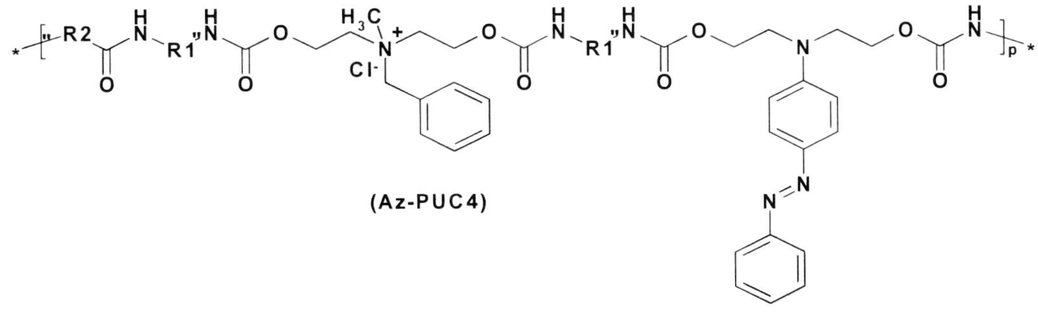

Figure 1. Schematic representation of Az-PUC4.

Comparative photoreactivity of the azo chromophores in all above mentioned polycations exposed to UV irradiation was explored in four studies [80, 84, 91, 92]. The obtained results indicated that the photoresponsiveness of the chromophore would be directly influenced by the structure and attachment point of the azoaromatic unit in ionomeric chains, as will be shown lately. In the same time, another question was suggested: what happens when the chromophore is located into flexible chain of the ionomeric backbone ? A concrete answer is given bellow.

2.2.2. Polyurethane Cationomers with Azobenzene in the Soft Segment

More recent study was conducted to find the appropriate conditions for the incorporation of azoaromatic groups into polyether component, such as PTMG, widely used in polyurethanes synthesis. The attention was directed toward the preparation of new azo PTMG that contains one azobenzene side group and its use as starting diol in the designing polyurethane cationomers [78]. The key intermediate was the preparation of 4-*N,N*-di(β-tosyloxyethyl)aminoazobenzene (1) by esterification of 4-*N,N*-di(β-hydroxyethyl)-aminoazobenzene with *para*-toluenesulfonyl chloride (TsCl) in the presence of pyridine. Another derivative assumed activation of the hydroxyl group from PTMG with a group susceptible to a nucleophilic attack by the oxygen atom of the alkoxide moiety. Thus, the reaction of PTMG of 1000 average molecular weight or of shorter block length (Mw = 650) with sodium hydride, used in an appropriate amount to transform only one hydroxyl group in an alkoxide structure, produced mainly monoalkoxide-PTMG (2 or 3). Since the above formed ditosylate contains the tosyloxy leaving groups (TsO^-), it can serve as alkylation agent for the latter polyether with one terminal alkoxide functionality [94], so that coupling can occurs only at one end of the PTMG molecule. Therefore, using the coupling reaction between the active ester (1) and monoalkoxide-PTMG (2 or 3), two blocks of polyether were connected by means of azo aromatic derivative. It is expected that the formation of the dialkoxide derivate in a small amount would be also possible, reason for that multiple azo units may be incorporated within the soft segment. Obviously the subsequent purification provided the azo diol with one azo chromophore, whose structure was further confirmed. By this simple procedure, two azo polyether diols of average molecular weight 2239 (Az-PTMG1) and 1542 (Az-PTMG2) respectively, bearing one azobenzene side group as coupling moiety between two symmetric flexible segments were firstly produced. The synthetic approach to Az-PTMG1 is sketched in Scheme 3.

Synthesis of azo polyurethane cationomers (Az-PUC6, Az-PUC7) developed by us, was similar to the formation of other polymers, differing only by the starting polyol, which contains one azoaromatic side group between two flexible blocks having same length and composition. One of the polyurethane cationomers (Az-PUC6) based on Az-PTMG1, 4,4'-DBDI and NMDA, further quaternized with chloroacetamide, is outlined in Figure 2. Under such circumstances, polycations with azobenzene in the soft segment and quaternary ammonium groups in the hard block might be obtained as thin films with good optical qualities.

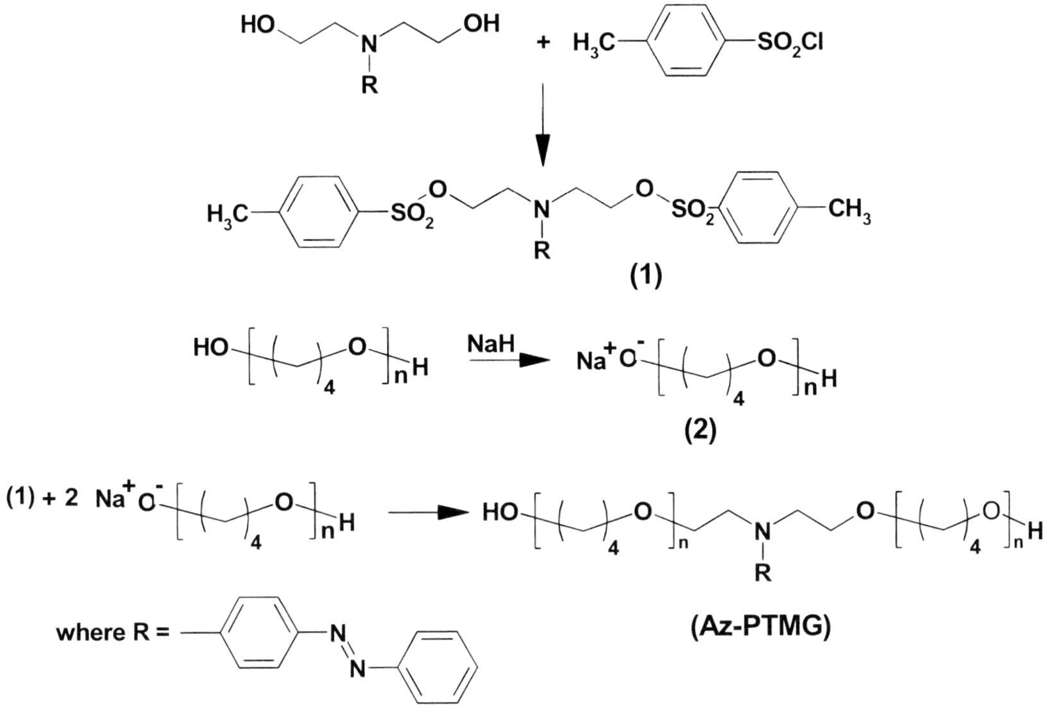

Scheme 3. Synthesis of Az-PTMG

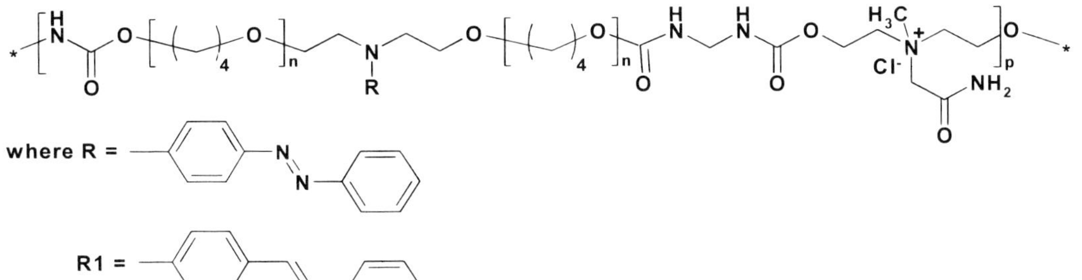

Figure 2. Chemical structure of Az-PUC6.

2.2.3. Polyurethane Cationomers with Azobenzene in the Field of Ionic Structure

In order to investigate more closely the influence of the ionic groups on photoreactivity of the chromophore, azo cationomers bearing piperazine ring and azoaromatic units as counterions of the ammonium quaternary structure were prepared expecting that they will be able to be used as photochromic materials [81]. Synthesis of these polymers implied the use of *N,N'*-bis(β-hydroxyethyl)piperazine as chain extender and its quaternization with methyl iodide. Taking into consideration the polycation ability to complexe the chromophore by a simple route, such as the ionic exchange reaction between polymer and 3(*p*-

oxyazobenzene)propane sulfonic acid sodium salt, the electrostatic incorporation of different concentration of azo groups was possible (Scheme 4).

This facile reaction which generates azo functionality in the field of quaternary structure was suggested by decreasing of ionic iodine content from the starting polycation up to a certain value, simultaneously with increasing of the sulfur content. Polycations thus formed differ from the pure cationomer only by inclusion of the azoaromatic ionic groups instead of the commonly used anion (iodine). Two different azo polymers (Az-PUC8, Az-PUC9) with an ionic exchange degree of 52.7 wt.% and 92.31 wt.%, respectively, were synthesized in order to investigate the effect of chromophore content on the photophysical properties related to chemical composition of the polymers.

In the same connection, by using methyl orange (as sodium salt) in this ionic exchange reaction, polyurethane cationomers with azo-dye as counterions for the quaternary ammonium groups were also prepared [95]. Structure of azo-dye polycation (Az-PUC10) is presented in Figure 3.

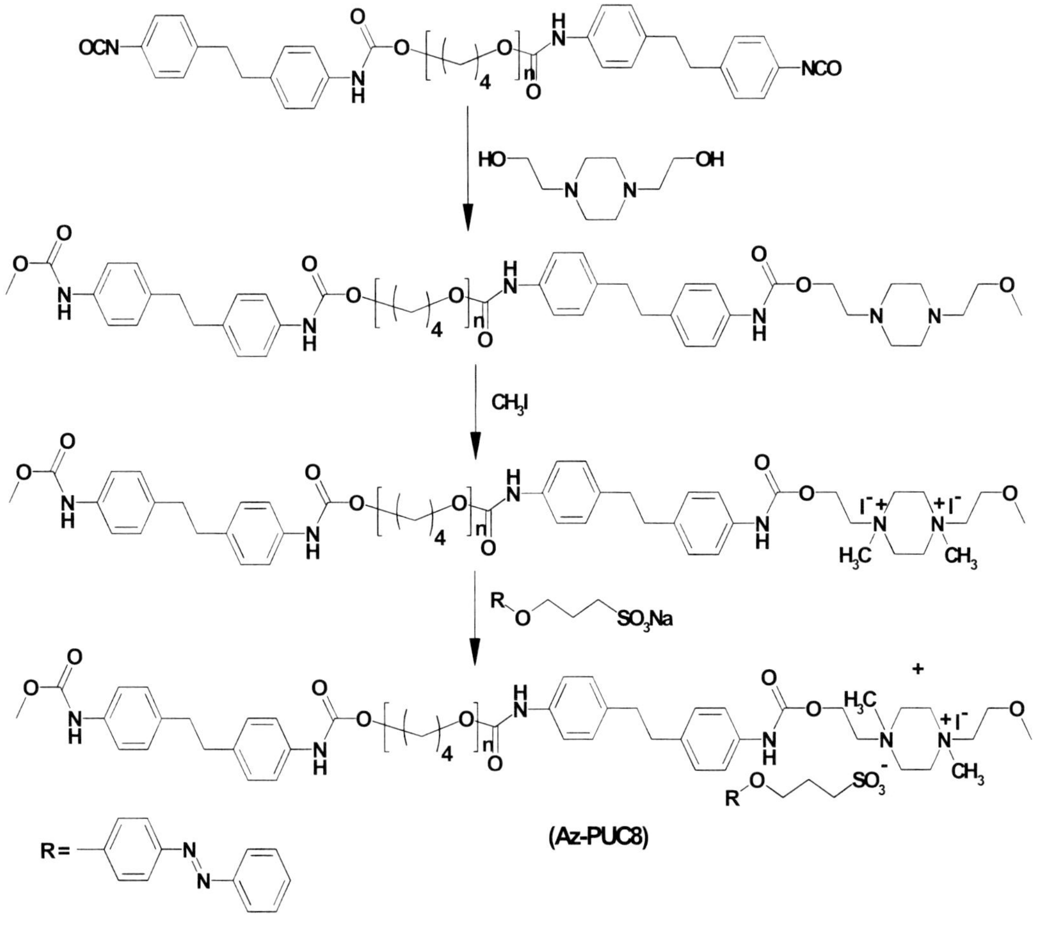

Scheme 4. Synthesis of Az-PUC8.

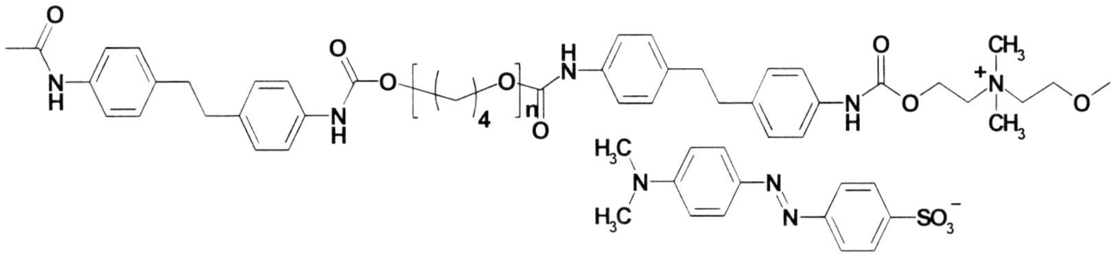

Figure 3. Chemical structure of Az-PUC10.

The experimental investigations allow us to confirm that all synthesized azo polyetherurethane cationomers (Az-PUC1-10) were soluble in common solvents like THF, CH_3OH, DMF, DMAc and DMSO. Solutions from these red polymers, as resulted from reaction, were coated onto a glass substrate and dried to form coloured and flexible polymeric films having typical soft elastomeric behaviour characteristic for the conventional polyurethanes. Moreover, the optical clarity of these films is suitable for various applications. Structure of all polymers was determined by elemental analysis, IR, ^{1}H-NMR and UV-vis absorption spectroscopy. The results of elemental analysis were in good agreement with the calculated values for the expected structure of polymers, the amount of ionic groups being estimated from the ionic halogen content. GPC analysis performed on some precursors indicated the absence of low or high molecular weight tails and reasonable number-average molecular weights. Such polymers have a variety of interesting features worthily of study. First, they contain different azobenzene groups in varying proportions which provide an unusual opportunity to approach photobehavior of these chromophores in solution and in film state. Part of the difficulty in treating azo polymers derives from the interchain interactions of polyurethane ionomers *via* Coulombic forces. In the following, the photoresponsive properties of the azo polycations investigated by photoirradiation with UV light will be discussed.

2.3. Photobehavior of Azo Polyurethane Cationomers in Dilute Solutions

2.3.1. Trans-Cis Photoisomerization of the Azo Chromophore in DMF Polymer Solution

It is well known that a large number of chromophores used in polymer synthesis may undergo diverse photochemical reactions when irradiated by UV light [96]. Since we had considered that the insertion of photosensitive chromophores into ionomeric chains might be a very useful and versatile technique for modification of the polymer structure, it was interesting to follow if some polymer properties may be controlled by means of photoirradiation. In order to enlighten these processes, photochemical experiments on all azoaromatic chromophores covalently or electrostatically bound to polymeric chains were done using as irradiation source, a high pressure Hg lamp equipped with a proper filter. The photoresponse of the azo chromophores was investigated by electronic absorption spectroscopy and evaluated comparatively with the corresponding azo models. In general, no large difference was observed between the photobehavior of azo polymers and analogous low molecular weight derivatives, excepting the isomerization rates, always higher in the latter case.

Normally, in every derivative the azoaromatic chromophores are in the thermodynamically more stable *trans* configuration. Photoexcitation of this structure induces a structural change of the *trans* conformer to *cis* isomer [97]. Once the molecule is in the *cis* form, it often recovers to the initial *trans* configuration by either the back thermal reaction or inverse photoisomerization cycle.

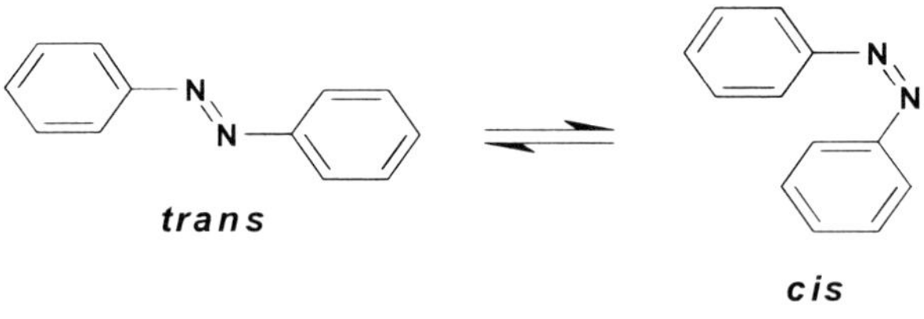

In our case, the main feature of the azo polycations arises from the photoprocesses induced by UV irradiation, with major influence on the structural properties of the materials, accompanied by reversible or irreversible modification of the polyurethane properties, as will see later. Thus, depending on the chromophore nature, its location and abundance, the determination of the isomerization characteristics of the azo moieties indicated significant differences in the rate of isomerization and photoprocess evolution. These factors were studied by the UV-vis spectral changes of azo chromophores under different exposure times, as shown in the next figures. In every azo polycation, upon the UV exposure of polymer solution in DMF, the intensity of long wavelength π-π* absorption band characteristic to the *trans* isomer decreased gradually with irradiation times and only in some cases a photostationary state is possible to reach [80, 91-93]. The modifications in the electronic absorption spectra were employed to examine the involved kinetics. The kinetic ratios for the irradiated sample (A_t) to the non-irradiated one (A_0) are given in Figure 4 (plot a) as a function of irradiation time for Az-PUC1, Az-PUC2 and Az-PUC3 in DMF solutions.

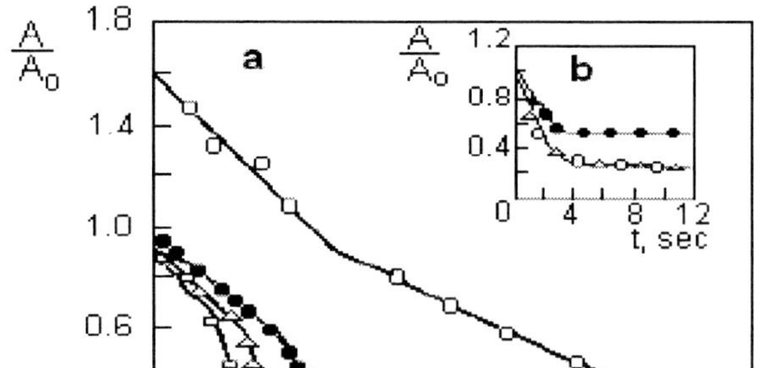

Figure 4. Decrease of relative absorbance (A/A_0) as a function of irradiation time for *trans-cis* photoisomerization in DMF solution at 298 °K, plot a: Az-PUC1 (•), Az-PUC2 (□), Az-PUC3 (o), Az-PUC10 (Δ); plot b: Az-PUC4 (o), Az-PUC5 (Δ), Az-Pr4 (•). *Reprinted with permission of [Ref. 80], Copyright (2002) – Elsevier.*

As can be seen from these plots, the *trans-cis* photoisomerization of *para*-nitroazoaromatic or azoaromatic groups covalently attached at the quaternary structure of polymeric chains requisite long irradiation periods. This was caused by steric restrictions imposed to the photochrome by the rigid spacer groups as well as by the inter- and intrachain electrostatic interactions of the charged polymers. Cooperative effects involved in strong interactions, generated by the nature of the disubstituted azo chromophore with electron-withdrawing and electron-releasing group (Az-PUC1, Az-PUC2) lead to a *trans-cis* isomerization in the first stage of the photoprocess. The appearing of the isosbestic points in this first stage suggested that only the *trans-cis* photoisomerization is the single photochemical change occurred and no side reactions are produced. A consequence of the irradiation is the producing of the photobleaching effect in the second stage but the question about the exact structure of the resulting products, remain open. Evolution of the photoprocess may be considered as a two step phenomenon in which the exposure to UV irradiation first excites a ground state of chromophore to an excited *trans* isomer, accompanied in its relaxation by a stable *cis* state. Reactions between species can further lead to an irreversible photobleaching. A similar photobehavior was also evidenced for some NLO polyurethanes and urethane-urea copolymers [98].

Unlike *para*-nitroazo chromophore, the *trans-cis* photoisomerization of azoaromatic groups placed in the same quaternary structure (Az-PUC3) with attainment of the photostationary state after 120 min. of irradiation was evidenced. As a result, the molar fraction of *cis* isomer in this state was about 0.78. From these observations one can easily see a strong tendency for favourable interactions of these latter groups that are brought into neighbouring positions which facilitates the *cis* isomer formation even when the azo groups are irradiated for a longer time. Such behaviour of the azo derivative provided information about the intrachain interactions between azo ammonium quaternary groups and the local environments around the chromophores, which, in turn, can be related to polymer structure in solution.

Comparatively with the above polycation, a synthetic mixture based on polyurethane cationomer and azobenzene (1:1 azo/tertiary amine groups) reached the photostationary state in shorter time (about 20 sec). In their pioneering work on azo polyurethanes, Sung [89,90] and Eisenbach [99] investigated the kinetics of the *trans-cis* photoisomerization and the thermal *cis-trans* isomerization of the azobenzene labels comprising hard segments and revealed that the relatively facile photoprocess reflected the structure, morphology and segmental mobility of the surrounding polyether(ester)urethanes in dilute solution. They reported that for more than 75% of the course of the reaction there is no difference in the photobehavior of the dilute solution and polymeric film. As the reaction attains the photostationary state, its composition in *cis* conformer is higher in solution (0.8 *cis* isomer) than in film (0.62 *cis* isomer).

The difference in the isomerization of azo derivative as ammonium group into the ionomeric architecture, whose properties are at least in part, governed by electrostatic considerations is illustrated by the photoprocess dynamics influenced by the strong interaction between polycation having a local charge density and counterion around it.

One of the ways to overcome a part of this drawback affected mainly by restrictive mobility of the azo chromophore may be the photoisomerization of azoaromatic groups placed in the non-ionic fragment of the hard segments (Az-PUC4, Az-PUC5) [84]. Indeed, during careful irradiation of the polymer solutions the *trans-cis* photoisomerization of azo chromophore occurred and the time required for attaining photostationary state considerably decreased at about 30 sec. (Fig 4, plot b). Under these conditions, the molar fraction of *cis* isomer in the equilibrium state was of 0.55. At first glance, one might think that increased sensitivity to UV irradiation of this last polycation is a direct consequence of the higher conformational mobility of the polymeric chains with azo groups as non-ionic fragment of the hard segments. Comparison of the photobehavior of Az-PUC4 with that of non-ionic polyurethane precursor showed that the presence of ammonium quaternary structure besides chromophoric groups on the same polymer chain slightly retards the attaining of photostationary state. It is logically to expect that the *trans-cis* photoisomerization of azoaromatic chromophore located in the polyether flexible segment to be a faster process than in the case of the chromophores placed in the hard segment. Consequently, upon UV irradiation of the Az-PUC6 and Az-PUC7 solutions, the *trans-cis* photoisomerization of azoaromatic groups, after about 5 sec, was interpreted as arising from the specific polymer structure [78]. The effect of higher mobility of the azo chromophore controlled by both local chain stiffness imposed by pendant inserted azobenzene and overall chain length on the macromolecular conformations are well reflected in the isomerization behaviour in part, due to the more efficient incorporation of the azo groups within the soft microdomains.

Photobehavior of 3(*p*-oxyazobenzene)propane sulfonate electrostatically incorporated as anion around the quaternary structure from the ionomeric backbone was relatively similar [81]. From the comparison of the data concerning evolution of the photoprocess, the notable difference in the photoresponse of polymeric solutions may be primarily attributed to the chromophore structure and its ionic bonding. Following the progress of the *trans-cis* photoisomerization of azoaromatic group in the polyurethane cationomer (Az-PUC8) with irradiation time (Figure 5, plot a), a rapid reaction with reaching a photostationary state in only 20 sec was observed. In the equilibrium state, the value of *cis* isomer was estimated at 0.8. On the other hand, well defined isosbestic points (308 and 415 nm) confirmed the presence of *cis* and *trans* isomers resulted by a single chemical phototransformation of the azo chromophore. The increase of photoisomerizable structure content in polymer chains (Az-PUC9) has not been favourably to a faster *trans-cis* photoisomerization in polymer solution.

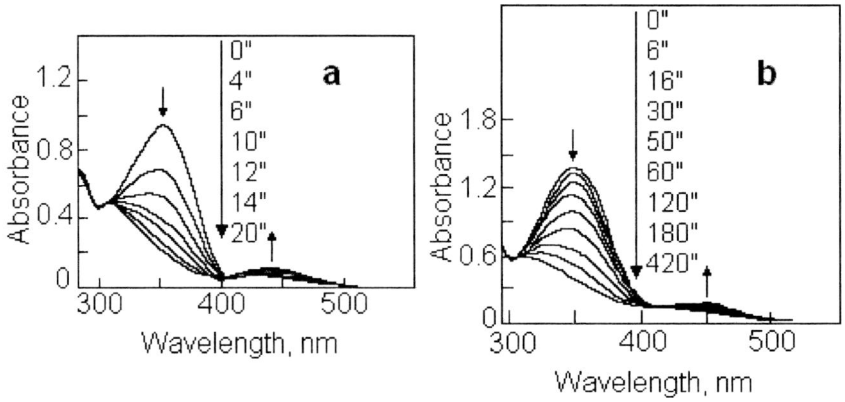

Figure 5. UV-vis absorption spectra monitoring the trans-cis photoisomerization of azo chromophores in polycations, plot a: Az-PUC8 in DMF solution; plot b: Az-PUC8 in film. Reprinted with permission of [Ref. 80], Copyright (2002) - Elsevier

By contrast, a significant difference in the photobehavior of azo cationomer prepared by using methyl orange as photoactive chromophoric agent capable of interacting electrostatically with ammonium groups was noticed (Figure 4, plot a) [95]. As a consequence of azo-dye structure, photoisomerization of the dye-polycation (Az-PUC10) was detected only in the first step of the photoprocess, with a tendency to return to the initial state when irradiation was interrupted. On continued irradiation (over 20 min), colourless polymer solutions were obtained and this photobleaching effect was accompanied by a photodecomposition reaction of polyurethane cationomers. An important cause of the above photobehavior may arise from the chromophore nature, dissociated in polar solvent, its distribution and its orientation on the ionomeric chains. By employing an acid pH, the dissociation effect is suppressed and so the sulfobetaine structure of azo chromophore formed is more favourable to respond at the irradiation. The rather peculiar viscosity behaviour of Az-PUC10 in DMF solution may be mentioned at this point. Measurement of the reduced viscosity - as sensitive indicator of photoprocess evolution, showed that the polymer solution underwent viscosity modifications with the irradiation time, suggesting that the photodegradation of the polycation with azo-dye chromophore was occurred. This seems to be a feature for ionomeric backbone with an azo-dye electrostatically loaded. It was also remarked that the systematic shifts of the absorption maximum towards longer wavelengths for all chromophoric polymers, are apparently due to a microenvironmental polarity effect of the surrounding azoaromatic groups.

2.3.2. Cis-Trans Reaction of the Azo Chromophore in DMF Polymer Solution

The kinetics of *cis-trans* relaxation process of all azo chromophores in polymer solution was monitored by the changes of specific absorbances to the *trans* configuration, of the irradiated samples. A better understanding of the important factors affecting this process is essential for the development of new photochromic systems of soft elastomeric type. One of the key parameters that affected the rate of *cis-trans* isomerization was the temperature. For this reason, the back isomerization of *cis* isomers was studied by recovery at room temperature in dark or at temperatures between 50 and 90 °C, and by irradiation with 365 nm. In some cases (Az-PUC3, Az-PUC4, Az-PUC5, Az-PUC6, Az-PUC8, Az-PUC9), the return to the initial *trans* configuration state occurred but with different rates.

The thermal *cis-trans* isomerization was expressed by a first order rate equation:

$$\ln(A_\infty - A_0)/(A_\infty - A_t) = k\,t$$

where A_0, A_∞ and A_t are values of the absorbance at times t_0, t_∞ and t, respectively and k is the rate constant for the *cis-trans* isomerization [100]. Apparently, the differentiation between the photoresponse of the above azo polycations may be due to the different incorporation of the azoaromatic chromophore in polymeric chains. In spite of the strong interactions exercised by the quaternary ammonium structure bearing azoaromatic side group on the *cis* isomer once formed (Az-PUC3), the thermal recovery (68 °C) in DMF solution becomes possibly but requires high times, over 120 h [93]. At lower temperatures, the same reaction necessitates more long times. For example, after a standing at room temperature of about 200 h, only 10% *cis* form was capable to return in its *trans* configuration. This is not surprising because the relaxation process assumes the reorientation of azobenzene molecules in different constrained environment of the ionic matrix. The understanding of the time dependence of this photoprocess is more difficult because several structural factors including a few highly polar ionic sites in an organic hydrophobic neighbourhood may perturb the reorientation of azobenzene units. In contrast with these structures, the return of the aforementioned synthetic mixture (at 68 °C) with a complete transformation of the *cis* isomer to initial *trans* form takes place within 48 h. This is obviously due to the weak chromophore-polycation interactions leading to *trans* isomer reformation. In this mode, the physical incorporation of the azo chromophore in the polyurethane ionomer matrix does not offer information on the ionomeric chain mobility.

An important difference in the return of *cis* isomer located far away from the ionic vicinity of the polymer chain, namely in the non-ionic fragment of hard segments (Az-PUC4, Az-PUC5) was evidenced. In both azo polycations as well as in their precursors was clearly observed an enhancement of the thermal recovery at higher temperature (relaxation time is about 7 min at 90 °C), suggesting a considerable mobility of the azo chromophore in solution (Figure 6, plots b). A quite comparable situation was also found in Az-PUC6 and Az-PUC7 containing lateral azobenzene in the polyether segments of the ionomeric chains.

Following the back isomerization process of Az-PUC8 and Az-PUC9 in DMF solution, the return of *cis* azo chromophore to the starting *trans* form was noticed in both polymers, but the thermal recovery required higher times (over 180 min at 72 °C). An increase of the azo photochromic groups concentration as anions around the quaternary structure (Az-PUC9) led to a decrease in the *cis-trans* isomerization rate (Figure 6, plot a). On the other hand, although they are in intimate contact into polar medium, the effect of ionic groups on the isomerization of azo chromophore is significantly smaller than those observed in the case of polycation (Az-PUC3) possessing azoaromatic unit covalently bound to the ammonium quaternary structure.

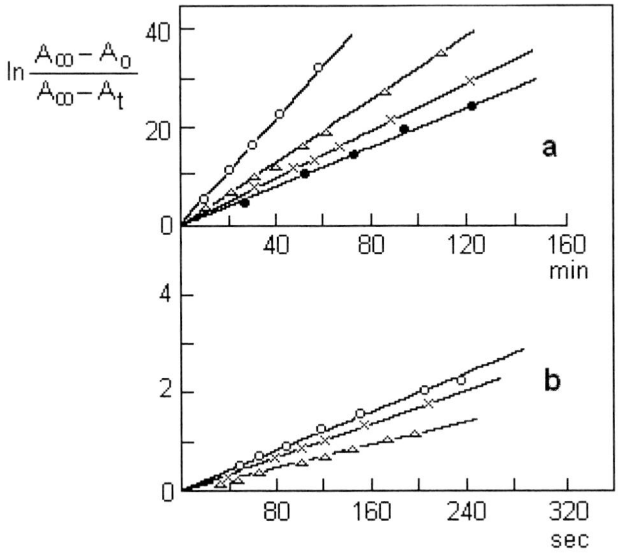

Figure 6. Plots for cis-trans isomerization of azo polyurethanes, plot a: Az-PUC8 (x), Az-PUC9 (•) in DMF solution; Az-PUC8 (o) and Az-PUC9 (Δ) in film; plot b: Az-PUC4 (Δ), Az-PUC5 (x) and Az-PR4 (o) in DMF solution. Reprinted with permission of [Ref. 80], Copyright (2002) – Elsevier.

In fact, an improvement of the recovery efficiency of the azoaromatic groups upon changing incorporation from covalently to electrostatically really occurs for this reason, due to a higher local mobility of the chromophores in the latter. It is reasonable to believe that besides the size of the azo moiety and the free volume around the azobenzene, the local mobility is controlled and by the ionic strength of the coupling interaction between the azo moiety and the polymeric backbone.

2.4. Photobehavior of Azo Polyurethane Cationomers in Thin Films

2.4.1. Trans-Cis Photoisomerization of the Azobenzene in Polymeric Films

The photophysical properties of the chromophores are more complex when the azoaromatic groups are studied in polymeric films. It is well known that the isomerization of the azo chromophores is substantially slow-down in polymer films than in solutions, as a direct consequence of their reactivity influenced by the local mobility of reactants and heterogeneity of reaction sites in the solid state [101]. The same rule also applies when the substrate itself is polyurethane. As mentioned, the earlier work of Sung [89,90] and Eisenbach [99] based on the molecular labelling techniques of polymer chains studied the *trans-cis-trans* photoisomerization of azobenzene attached to the hard segment of the amorphous, single-phase polyether(ester)urethanes. Their results showed that in polymer films the reaction was completely reversible but the quantum yield was reduced to about 30%, as compared to the analogous process occurred in solution. In contrast to this, distinct diversities were observed for the two-phase polyurethanes containing a similar chromophore in the hard segment or in the soft segment. Direct evidence of the reversible isomerization of the azo chromophore in the soft segment has been obtained. The ability to differentiate the photobehavior of the flexible and rigid portions of the polymeric films, and the effects of molecular mobility, widely suppressed within hard domain, on the isomerization process were demonstrated. The authors have also sustained as possible the *cis-trans* isomerization of azobenzene placed in hard segments explained in terms of no phase separation occurrence.

A straightforward extension of these results can be made for the polyurethane cationomers containing various azo chromophores in order to detect possible photochromic properties of some ionomeric films of soft elastomer type. Although still in their infancy, our current work has revealed some possibilities in the structural tailoring of such ionomers capable to assure good photoresponse of the azo chromophores in thin films.

Therefore, azo polycations are expected to behave in a different way compared to the non-ionic polymers with the same azo sequence. Consequently, it was confirmed that in thin films the isomerization rate and phototransformation degree of the azobenzene groups were affected by restrictions in movement of the chromophores, imposed by steric and shielding effects to the molecular chains. Moreover, these findings are considered to be a result of the electrostatic perturbation induced through the ammonium quaternary groups exerted by means of charge distribution in polymer backbone. Data achieved for the azo polymers, showed that the feasible level of *cis* isomerization was primarily affected by the attachment point of the azoaromatic derivative in polymeric chains, content and the incorporation technique, covalently or electrostatically. This was implicitly observed by comparing the *trans-cis* photoisomerization for Az-PUC3 with that of the polymers series, Az-PUC4, Az-PUC5, Az-PUC6-7 and Az-PUC8, Az-PUC9. The evolution of the photoprocess was followed in every thin polymeric film subjected to irradiation under the same conditions, and it was concluded that the absorption maximum corresponding to the π-π* transition, decreased much slower as the UV light exposure time increased. This tendency to diminish the photoresponse is the result of slower *trans-cis* photoisomerization of the azo units in film state with or without attainment of an equilibrium state. Therefore, the distinction in the *trans-cis* photoisomerization of the azo chromophores in azo polycations based films may be reasonably explained not only by the variability of azobenzene structure but also by the different mobility of the polymeric chains assisted by the ionic interactions and hydrogen bonds from polyurethanes.

When the photobehavior of Az-PUC3 film was compared to that of Az-PUC4 and Az-PUC5 films, it was observed that photoreaction required longer times of irradiation than those in solutions, with the mention that only the polymers Az-PUC4, Az-PUC5 (Figure7, plot a) reached the photostationary state. For example, the measurement of the relative absorbance A/A_0 indicated a value of 0.63 after 240 min of irradiation for PUC4 film, while in DMF solution, the same parameter was 0.45, after only ten seconds.

As in the case of the non-ionomeric precursors, the first photoprocess was attributed to the photoisomerization of azoaromatic groups, proved by the presence of two isosbestic points at 364 and 497 nm, respectively. Under these UV irradiation conditions, a photobleaching process starts in the polymer film after 150 min and it takes place very slowly as the reaction proceeds. By contrast, in the case of Az-PUC3, the strong electrostatic interactions of azo chromophores covalently attached to the quaternary structure greatly reduced the *trans-cis* photoisomerization of azo derivative in film state. The increase in the irradiation time up to 5h, lead to a transformation degree of the *trans* isomer of only 10%, suggesting a very limited mobility of the chromophore in thin film.

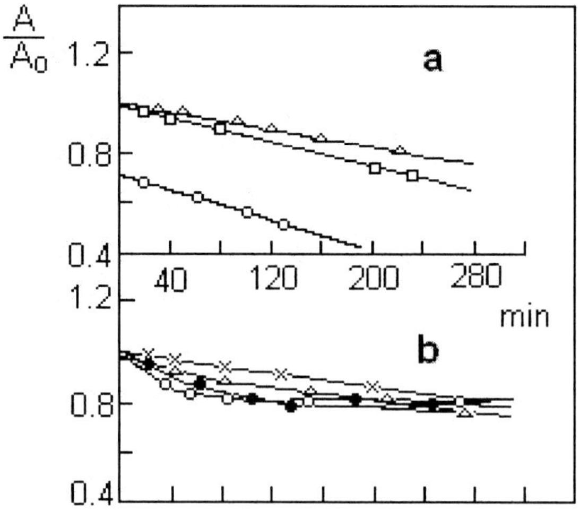

Figure 7. Dependence of relative absorbance of azo polymeric films on irradiation time, plot a: Az-PUC4 (□), Az-PUC5 (o), Az-PR4 (Δ); plot b: Az-PUC6 (x), Az-PUC7 (□), Az-PR6 (•), Az-PR7 (o). Reprinted with permission of [Ref. 80], Copyright (2002) – Elsevier.

At a lesser amount of azoaromatic groups, the changes appeared in the electronic absorption spectra are much more pronounced for Az-PUC5 film (Figure7, plot a). In this case, the intensity of the absorption band at 406 nm decreases faster during irradiation reaching a molar fraction of *cis* isomer of about 0.24 after 50 min of irradiation. For the same irradiation time, in the Az-PUC4 film approximately 0.09 *cis* form was measured. In the first polycation, the chromophore loading on the ionomeric chains was limited to a low level to avoid chromophore-chromophore interactions that often lead to a complex photobehavior. The process is dynamic in nature with a photobleaching effect observed in the time-dependent phototransformation data. The electronic absorption spectra of Az-PUC5 film exhibited no modifications by its standing at 120 ^{o}C for 405 min, the photobleaching process being irreversible in these conditions.

Azoaromatic chromophores located in the soft segment of polyurethane chain are assumed to have an interesting photobehavior [78]. In the following, we first studied some properties of the corresponding polymers of ionomeric (Az-PUC6, Az-PUC-7) or non-ionomeric type in thin films. Accordingly, one might reasonably expect that the *trans-cis* photoisomerization of the azo chromophore to be a faster reaction comparatively with the above discussed azo polycations. Indeed, as can be seen from the figure 7 (plot b), in the ionomer film the photostationary state is reached in only 30 sec of UV irradiation with a molar fraction of *cis* isomer of about 0.34. In the analogous non-ionic polyurethane film, it was found that absorption maximum corresponding to π-π* transition more quickly decreased as the UV light exposure time increased. This diminution is the result of the same *trans-cis* photoisomerization of azo unit, with attainment of an equilibrium state in 12 sec, when the molar fraction of *cis* form was about 0.25. Although the polyether matrix in which the azobenzene was incorporated can provide a better mobility to the chromophore, the composition of the photostationary state indicates that a high portion of *trans* isomer is unable to isomerize to the *cis* form, probably due to the energetically unfavourable local environments. As assumed, the *trans-cis* photoisomerization was particularly sensitive to the competitive interactions with other portions of the polymer chains, namely at the ionic interactions within the hard segment dispersed in the polyether phase. As a result, a difference in the photoisomerization rate of the azo aromatic chromophore in the ionic film (Az-PUC6, k = 7.7 x 10^{-2} s^{-1}) and in the analogous non-ionic film (PR, k = 35.5 x 10^{-2} s^{-1}) was found. The general photobehavior of the Az-PUC7 corresponds to that previously seen for Az-PUC6, with a small decrease in the photoisomerization rate of ionomeric film (Az-PUC7, k = 7.2 x 10^{-2} s^{-1}) as well as in its analogous precursor with identical soft chain (PR, k = 33.8 x 10^{-2} s^{-1}). Consequently, it should be noted that the observed deference between Az-PUC6 and Az-PUC7 might result from changes in the content of ionic groups and implicitly in the extent of physical crosslinking. In this context, we can imagine that the activity of precursor is a reflection of the matrix polarity, which allowed a more efficient arrangement of the chromophore units due to the lack of electrostatic interactions within the polymer. In fact, the enhancement of photoisomerization efficiency of the azo moieties by providing a lower barrier to photoreaction occurred for this reason. The appearance of isosbestic points at 362 nm and 470 nm is a clear indicator that in this case, the photoisomerization is the only chemical phototransformation suffered by polymeric chains.

Similar results were also reported for Az-PUC8 and Az-PUC9 with azobenzene attached as anion to the piperazine rings of ionomeric backbone. In thin film, the rapid *trans-cis* photoisomerization of azoaromatic chromophore from Az-PUC8 has also evidenced (Figure

5, plot b), the equilibrium state being reached in about 60 sec of irradiation with a molar fraction of *cis* isomer in the photostationary state of approximately 0.56. At a higher content of azobenzene chromophore (Az-PUC9) the isomerization reaction has not favoured. In the determination of the isomerization rate from *trans* to *cis* isomer in thin film, a decreasing in the rate value ($k = 10.8 \times 10^{-3}\ s^{-1}$) as the chromophore concentration increased in this latter polymer comparatively with Az-PUC8 ($k = 12.4 \times 10^{-2}\ s^{-1}$) was found. Moreover, it was evident that the difference in the photoisomerization rate of the chromophore in thin film comparatively with that found in solution is primarily due to the more restricted mobility of the photochrome in polymer film.

Studies reported so far have mostly concentrated on the *cis-trans* isomerization of azobenzene chromophore in polymer films based on all above described polyurethane cationomers. In general, the direct detection and characterization of the polymer photosensibility *via* UV measurements depended on the different local environments of the chromophores inserted in the polyurethane ionomer which give rise to a different photobehavior. With few exceptions, the thermal recovery of the azo chromophore in the cationomer matrix was not evidenced. This aspiration is important relatively to the photochromic phenomena which may be induced in ionomeric films of soft elastomer type, which are assumed to be of great interest in versatile applications. The principal driving force for the chromophores relaxation is a large increase in mobility and free volume, sufficiently to promote the return of the *cis* azobenzene structure to the initial *trans* form. These conditions are satisfied by a small number of polymers such as those containing azo chromophore in the soft segment of the ionomeric chains (Az-PUC6, Az-PUC7) or as anion in the field of ammonium quaternary structure (Az-PUC8, Az-PUC9). Evidence for this claim is that the spectral changes during the return of the chromophore can be easily observed in thin polymeric films.

2.4.2. Cis-Trans Isomerization of the Azobenzene in Polymeric Films

The kinetics of *cis-trans* relaxation process of azobenzene in the irradiated films, was followed by monitoring the changes in the absorbance specific to the *trans* isomer, appeared by keeping the films at 80 °C or at ambient temperature. In both cases, the chromophores returning to the initial *trans* configuration take place, but the rates differ. The first order plot for thermal *cis-trans* isomerization of azo chromophore in Az-PUC6 and Az-PUC7 films are illustrated in Figure 8.

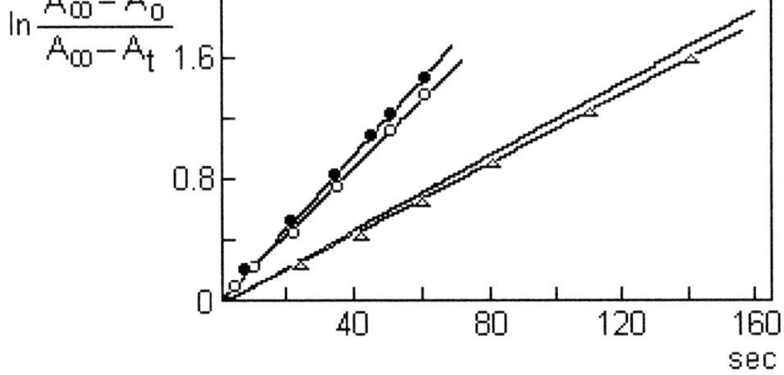

Figure 8. Plots for cis-trans isomerization of azo polyurethane films: Az-PUC6 (x), Az-PUC7 (Δ), Az-PR6 (•), Az-PR7 (o).Reprinted with permission of [Ref. 80], Copyright (2002) – Elsevier.

Although the Az-PUC6 and Az-PUC7 films contain azobenzene in the flexible segment, comparing the rate value found in ionomeric films (Az-PUC6: k = 29.17 x 10^{-3} s^{-1}; Az-PUC7: k = 25.9 x 10^{-3} s^{-1}) with that of the non-ionomeric films (PR6: k = 59.3 x 10^{-3} s^{-1}; PR7: k = 57.8 x 10^{-3} s^{-1}), is clear that the thermal recovery of the *cis* isomer to the *trans* form is also strongly influenced by specific effects such as ionic interactions from the hard domains whose microphase separation is not very well delimited. In contrast with the ionomeric films, whose *cis-trans* isomerization necessitated 300 sec., the same back reaction required about 60 sec. in the non-ionic films. Additionally, a slower recovery of the chromophores in both films at ambient temperature, over 200 min, was noticed too. The thermal recovery data indicated the role of the structural and steric environments surrounding the azobenzene groups inserted in the polymeric backbone. Since the fatigue behaviour of these films can be quite important, five cycles of *trans-cis* and *cis-trans* isomerizations were followed, but no modifications were detected in the electronic absorption spectra.

The same trend of chromophores to relax has been observed and in the case of polycations with azobenzene as counterions of piperazine rings (Az-PUC8, Az-PUC9). By comparing the rate value of Az-PUC8 film (k = 8.46 x 10^{-4} s^{-1}) with that found in solution (k = 3.81 x 10^{-4} s^{-1}), is clearly that the thermal recovery of the *cis* form to *trans* isomer is much faster (about 70 min) in polymer film (Figure 6, plot a). Such result suggested that in film state, the *cis* configuration of azobenzene structure is in a very constrained state. On the other hand, as a result of the increasing of azo chromophore concentration in Az-PUC9 film, the return to the initial *trans* state is a slower process, which needs over 100 min. Data for the rate value found in the latter polycation film (k = 5.2 x 10^{-4} s^{-1}) reflected the temperature-induced slower back reaction, comparatively with Az-PUC8 film. These results confirmed the photochromic nature of the thin elastomeric films prepared from azo cationomers with photoisomerizable azobenzene chromophore placed as anion around ionic groups. Hence, the employing of a certain azo counterion in the field of ammonium quaternary structure may result in a better *cis-trans* isomerization, while the use of an azo dye (methyl orange) may induce the appearance of a photodegradation reaction.

Our findings sustain that the covalent introduction of the azobenzene chromophore as side group in the polyether segments of a polyurethane backbone seems to be a decisive factor for the reversible photo(thermo)reactions. An alternative approach should be represented by the electrostatic incorporation of photoisomerizable azoaromatic structures into polyurethane cationomers as a convenient way for the inducing of photochromic properties in polymeric films of elastomer type, intended for special applications. We expect that in the near future these polymers to be developed and characterized in many other aspects concerning the polyurethane photochemistry. By selecting the measurement conditions and instrumentation, it is also possible to investigate selectively the chemical structure, molecular dynamics and the orientation of a specific chromophore in polymer films. Correlation of some photoinduced chain motions with mechanical properties should improve understanding of the polymer behaviour at the macroscopic level.

2.4.3. Effect of Some Additives on the Isomerization of the Azo Chromophore in Films

In terms of *trans-cis* isomerization and *cis-trans* relaxation of the azobenzene chromophore in film state, one can not always explicitly interpret the relative small content of *cis* isomer from the equilibrium state or inability of the *cis* isomer to return to the *trans* form. One of the major unanswered questions on this field is how the photoreactivity of azo groups in ionomeric chains would be influenced by the presence of small amount of photosensitive derivative. A convenient way to enhance the photostimulated activity of the polymer chains was the incorporation of sensitizers in polymer films, such as benzophenone or iodine vapours [83]. In a previous investigation, we reported that the nitropolyurethane with 2,4-nitroaromatic groups in polymer backbone was sensitive to UV irradiation [102]. Consequently, in order to improve the isomerization of azobenzene through physical interactions between an azo polycation and the nitropolyurethane in polymer film, we select 2,4-dinitro-polyurethane as sensitizer. Following the evolution of the photoprocess in Az-PUC5 films, it was found that the presence of both sensitizers, either nitropolyurethane or benzophenone does not enhance the dynamics of the photoprocess. Moreover, the photoprocess rate in azo cationomer-nitropolyurethane based film was reduced comparatively with that found in the absence of the sensitizer. Therefore, the adding of these sensitizers had a negligible effect on the photosensibility of azo cationomers. A somewhat different behaviour was evidenced by the exposure of the irradiated films at iodine vapours (for low exposure times, of only 1-2 min), known as catalyst of the *cis* - *trans* isomerization of the monomeric azobenzenes. The origin of this feature of the *cis* isomer may be explained by the photochemical formation of the iodine atoms, capable to lead to a radical intermediate, whose free rotation is possible. In Figure 9 is depicted the effect exerted by the iodine vapours on the behaviour of Az-PUC5 film, irradiated 170 min, when the ratio between A/A_0 is 0.54.

In this situation, a fast increase of the absorbance at 405 nm was noticed, the starting absorbance value being reached completely after about 265 sec. In the same time, a common polyurethane film based on PTMO, 4,4'-DBDI and NMDA, treated with iodine vapours exhibited an absorption band at 357 nm, assigned to some complexes formed between iodine and the aromatic structure. Such behaviour confirms that the *cis-trans* isomerization of Az-PUC5 film evidently proceeded. By reirradiation of the azo polycation films exposed to iodine vapours, the *trans* to *cis* isomerization takes place but with a slower rate and finally, a photobleaching effect was evidenced, too. It is reasonable to assume that the combinative effect of this treatment may affect the stability of the polymer.

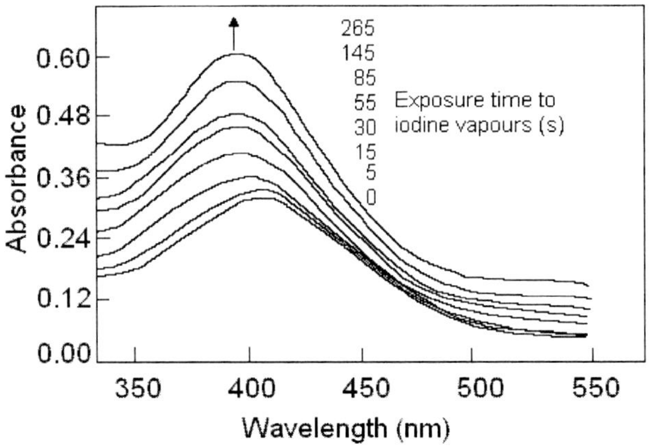

Figure 9. Influence of iodine vapors on the electronic spectra of UV irradiated thin film Az-PUC5. The exposition times to iodine are given in figure. Reprinted with permission of [Ref. 80], Copyright (2002) – Elsevier.

So, the data analysis led to some preliminary conclusions regarding the role played by sensitizer present into azo ionomer matrix. Clearly, it is not as easy to predict how specific structure of the sensitizer will influence the isomerization reaction of the azobenzene. The examination of a variety of sensitizers and methods might led to correlations that are less subject to the bias due to the specific probe dependent interactions and interpretation of such results based upon insufficient knowledge of the molecular architecture of ionomer films.

Our approach in the field of azo cationomers has been shown clearly to be flexible and adaptable to a variety of azo coupling agents and attachment methods of the chromophore. General properties of the azo cationomers described above should be just a starting point for their own area of interest as polyurethane elastomers. Already this field was marked by an enormous progress stimulated of environmental reasons and a current wide diversity of applications, new polyurethane ionomers are expected to be developed, as we improve overall properties and structural design.

2.5. Polyurethanes with Photosensitive Triazene Moieties

2.5.1. Triazene Diols and the Corresponding Polyurethanes

A modern trend in the design of photopolymers intended for optical applications is represented of the achieving of new materials highly photosensitive by means of active chromophoric structures covalently attached to the polymeric backbone. Recent developments in microlithography and photoresists technique prove that increasing demands in view of materials properties (structure dimensions, resolution, and sensitivity to the irradiation process) are the dominating aspects that qualify the performance of photoresist systems. The investigation of photosensitive materials through laser ablation as a powerful technique for direct patterning of the polymeric surfaces has recently imposed, because such a method necessitated only few processing steps comparatively to the standard lithographic techniques and additionally, eliminates the pollutant solvents [103-106]. Among these, the triazene sensitized polymers were the subject of extensive investigations since the triazene moiety owing to its high photosensitivity and spontaneous fragmentation into gaseous products is considered one promising candidate for excimer laser ablation lithography [107-109]. Therefore, various classes of polymers containing photolabile triazene groups (-*N*=*N*-*N*-) in the main or side chain, like polyesters, polyethers, polymethylmethacrylates, polysulfones or polytriazenes with high thermal stability and favourable photolytic sensitivity have been described in the literature [110-115].

From this perspective, the extending of the research on new polyurethane structures having triazene groups seems to be an interesting opportunity for a better understanding of structure/property/photoreactivity relationship that could be exploited in modern applications for such versatile polymers. Therefore, our group has approached the synthesis of triazene polyurethanes with different topologies in order to combine the sensitivity of the triazene structure with the particular properties of the polyurethanes [116-120].

A very important feature to be mentioned is that firstly the synthesis of triazene diols employed in the formation of polyurethanes was accomplished, following some classical reactions from the organic chemistry. The driving force in this field was represented by the obtaining of triazene monomers with various inductive effects exerted through the aromatic substitution pattern, well known for their influence on the further photobehavior of the monomers and polymers. Consequently, the synthesis of new triazene diols was performed through an electrophilic coupling between an aryldiazonium salt and a secondary amine, the structures of the obtained triazene (XVI-XX) being represented in table 3. In addition, for an enhancement of the polymers photosensitivity new bistriazene monomers which comprise two triazene units in the same molecule (XXI, XXII and XXIII) were also prepared.

Subsequently, the flexible polyetherurethanes with triazene moieties were prepared by a well know pathway in a two step process [116-118]. In the first stage was prepared a prepolymer consisting of PTMO and 2,4-TDI or MDI diisocyanate, which was further extended with the triazene diol used in tandem with *N*-methyldiethanolamine (NMDA), the molar ratio of the partners being 1/3/1/1 (Scheme 5).

Table 3. Triazene monomers used for the synthesis of photoresponsive polyurethanes

XVI 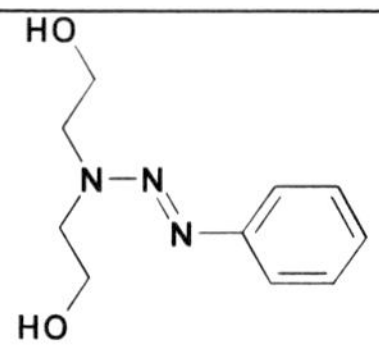 *1-phenyl-3,3-di-(2-hydroxyethyl) triazene-1*	**XVII** 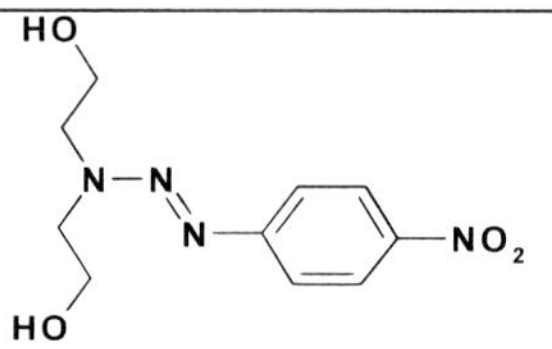 *1-p-nitrophenyl-3,3-di(2-hydroxyethyl)triazene-1*
XVIII 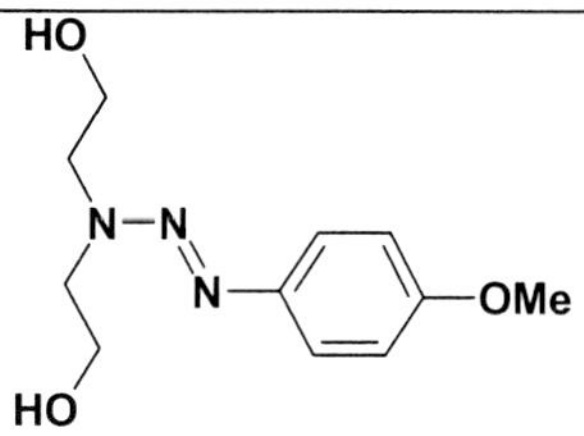 *1-p-methoxyphenyl-3,3-di(2-hydroxyethyl)triazene-1*	**XIX** 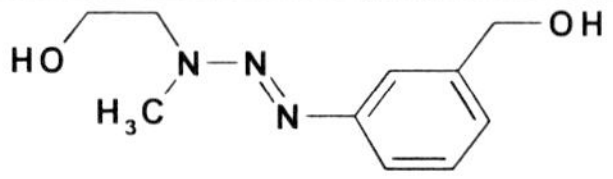 *1-(m-hydroxymethyl) phenyl-3-(2-hydroxyethyl)3-methyltriazene-1*
XX 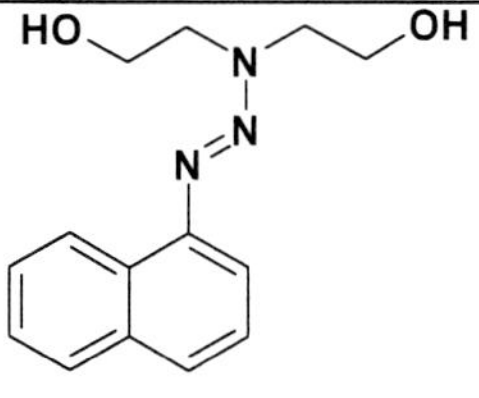 *1-(α-naphthyl)-3,3-di-(2-hydroxyethyl) triazene-1*	**XXI** 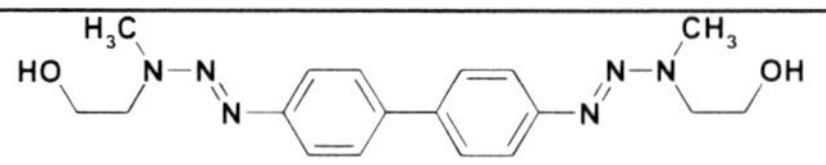 *1,1'[4,4'-diphenyl]-3,3'-di(β-hydroxyethyl, methyl)-bistriazene*
XXII 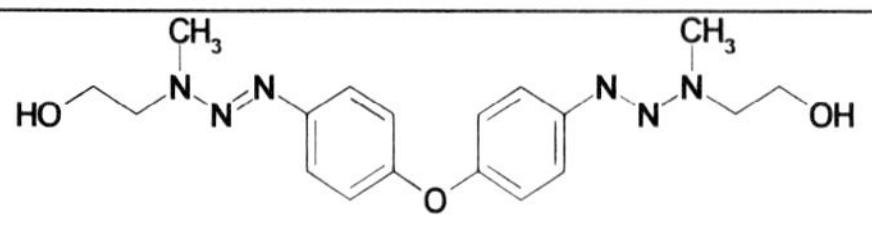 *1,1'[4,4'-diphenylether]-3,3'-di(β-hydroxyethyl methyl)-bistriazene*	**XXIII** 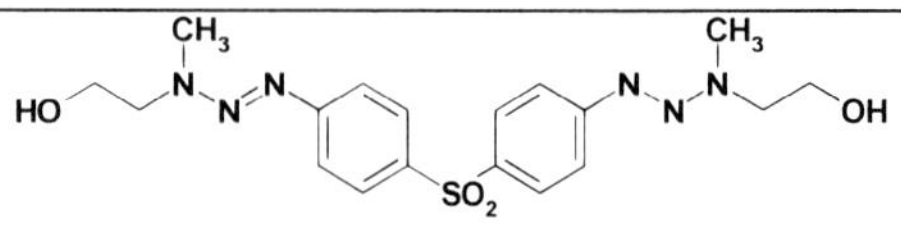 *1,1'[4,4'-diphenylsulfone]-3,3'-di(β-hydroxyethyl, methyl)-bistriazene*

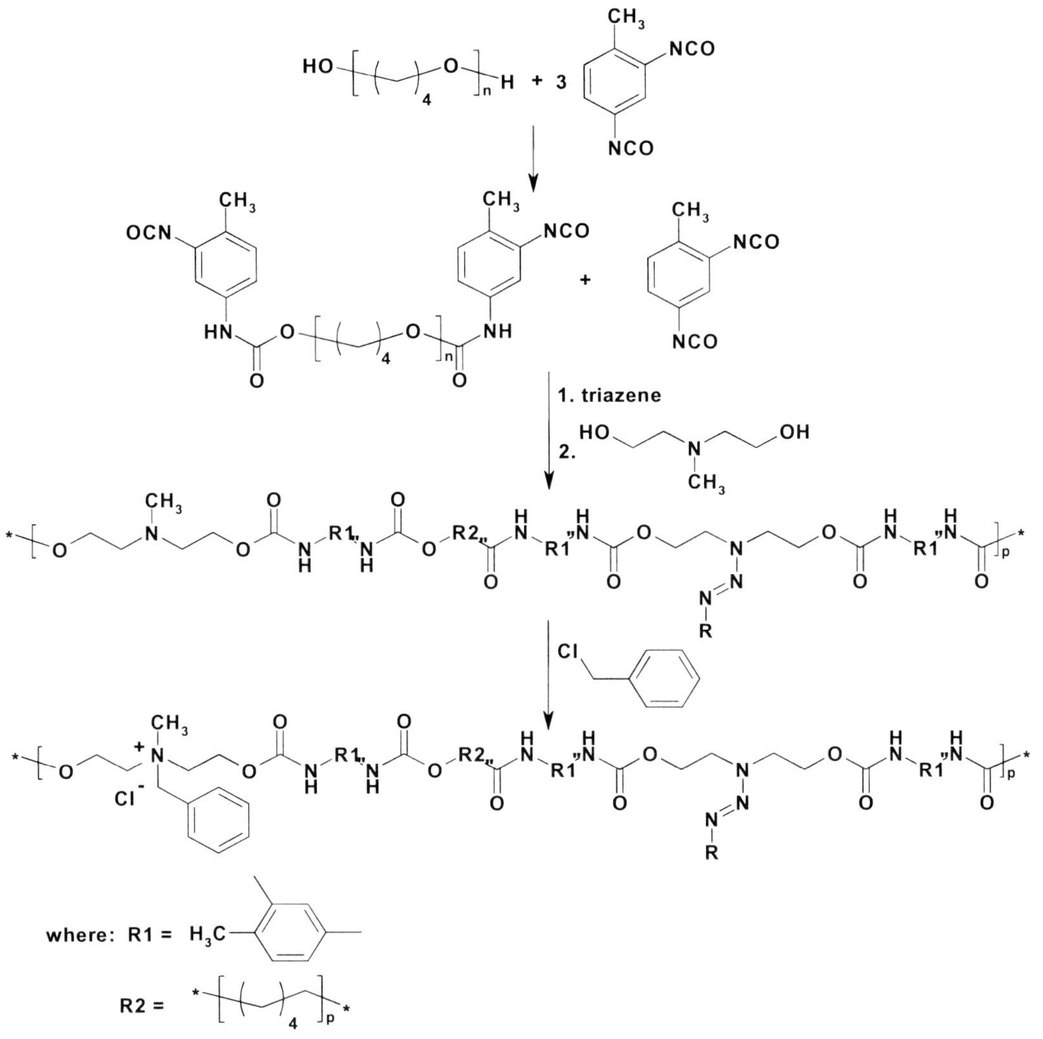

Scheme 5. Synthesis of triazene polyurethane cationomers PUC-T16, PUC-T17, PUC-T18 and PUC-T20.

To develop a predictive understanding of how the insertion of the ionic groups influences the polymers properties, the conversion of the nonionic polyurethanes into ionomeric structures through the quaternization of the aliphatic tertiary amino functionality from the aforementioned polymers with benzyl chloride were performed, the transformation degree being calculated form the elemental analysis data.

Due to the specific groups inserted on the polymeric backbone, all polymers have an intensive brown-red colour, indicative of the triazene chromophore and are soluble in common polar solvents like DMF, DMSO and DMAc. The peculiar property of the triazene polyurethanes is that by solutions casting may form coloured films with a typical microstructure to the urethane elastomers, a good adhesion on a variety of substrates and high optical quality.

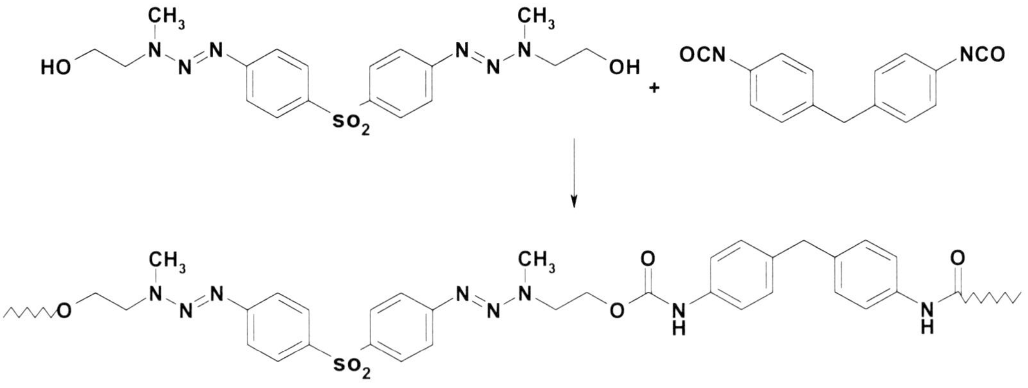

Scheme 6. Synthesis of hard triazene polyurethane PUH-T23

Taking into account the actual tendencies in microlithography, directed toward the obtaining of performing materials with remarkable properties, cheaper and easy to acquire, it is essential to find innovative modalities and techniques to achieve polymeric architectures able to generate microstructures by UV or laser irradiation. One of the methods consists in increasing the photosensitivity of polymers by the introduction of a higher content of photolabile units. Contextually, new hard photopolyurethanes resulted by the reaction between a diisocyanate (2,4 toluene diisocyanate or 4,4' methylene bis(phenylisocyanate)) and a triazene diol, were prepared and studied [118-120]. Scheme 6 depicts the synthetic pathway used for the synthesis of the hard triazene polyurethane (PUH-T23).

2.5.2. Photochemical Behaviour of Triazene Containing Materials

Evaluation of the triazene unit photobehavior represents a crucial element in directing polymer application toward microlithography and particularly, to dry etching ablation. The choice of this chromophoric structure as photosensitive moiety is mainly stimulated by the opportunity to correlate its absorption maximum in the UV spectrum with laser irradiation wavelength, through the including of appropriate substituents in the vicinity of the triazene group. Usually, the triazene chromophores present in their UV spectra a broad absorption maximum, attributed to π-π^* transitions, situated between 290 and 360 nm. Under UV irradiation, the triazene groups are decomposed with the release of small gaseous molecules (N_2, CO_2) – process reflected in their UV spectra by a gradual decreasing of the absorption maximum. The monitoring of this photobehavior with irradiation time allows an estimation of the rate and photodecomposition degree of the triazene chromophore. Photochemical behaviour of the triazene monomers was studied during irradiation of their solutions in methanol, using as UV irradiation source a high pressure mercury lamp. A typical profile of such photoprocess is represented in figure 10 for the triazene XVI. It may be noticed that with the increasing of the irradiation time, the amplitude of the UV absorption maximum is diminishing until a lowest limit, when is supposed that all triazene units from solution are decomposed, so that finally the UV absorption of the monomer is insignificant.

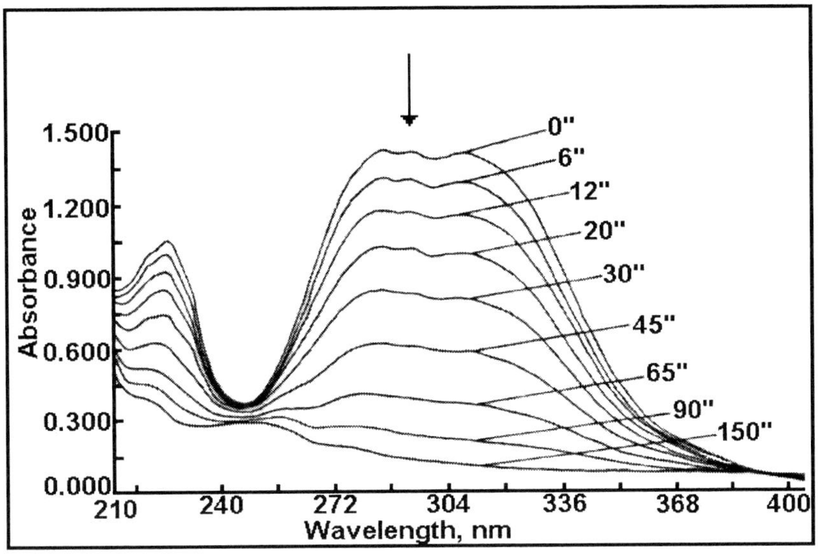

Figure 10. Changes in UV-vis absorption spectra of triazene monomer XVI in methanol solution upon UV light exposure. Reprinted with permission of [Ref. 117], Copyright (2004) – Wiley.

As in the case of azo derivatives, the photochemical ablation of triazene monomers is quantitatively described through a constant rate value, given by the expression:

$$k = \ln\frac{A_0 - A_\infty}{A_t - A_\infty} \times \frac{1}{t}$$

where A_0, A_t and A_∞ are the values of the absorbance at time 0, t and ∞, respectively. The kinetic representation obtained by plotting the logarithm of the reduced absorption against time indicates a first-order kinetic for all photoprocesses (Figure 11).

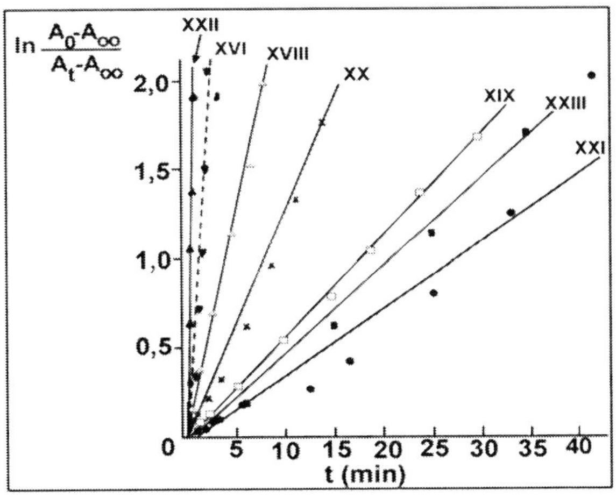

Figure 11. Kinetic evaluation of the triazene monomers photolysis at 298 K.

Moreover, as the kinetic slope is higher, the photodecomposition process is faster and as a consequence, the triazene compound presents a high photosensitivity. Analyzing the obtained data (table 4 and figure 11) it can be affirmed that the triazene monomers photodecomposition is principally dependent on the nature of the substituent attached to the aromatic ring, through the inductive effect exerted over triazene linkage stability.

The literature data [121] revealed that the presence of electron-donating groups at the aromatic ring has a major impact on the photolytic decay, favouring the triazene photodecomposition, whereas the electron-withdrawing groups increase the photochemical stability. Additionally, an important feature, observable in the case of bistriazene compounds, is referring to the great influence of the conjugation effects generated in the aromatic rings by the nature of the substituents in the chromophore neighbourhood.

To assess the influence of polymer architecture on the photolabile character of the triazene groups from polyurethanes, the photosensitivity of the chromophores both in polymer solutions and films was investigated. Therefore, the photochemical behaviour of the triazene polyurethanes in DMF solution was initially compared, in order to establish some correlations between the photosensitivity of monomers and polymers. Consequently, the triazene polyurethanes display a similar behaviour as the corresponding monomers, the only appeared differences being related to the lower photodecomposition rates than for monomers (Figure 12).

Table 4. Photochemical characteristics of triazene monomers and soft polymers

Sample	λ_{max} (nm)	k 10^{-3} (s^{-1})	Sample	λ_{max} (nm)	k 10^{-3} (s^{-1})	
	MeOH solution	MeOH solution		DMF solution	DMF solution	film
XVI	283, 309	16.9	PU-T16	284, 316	6,2	5,4
XVII	365	-	PU-T17	375	-	-
XVIII	287, 325	8.91	PU-T18	288, 327	2,67	1,85
XIX	285, 312	0.774	PU-T19	284, 315	0.35	0.31
XX	330, 343	1.06	PU-T20	346	0.41	0.29
XXI	360	0.53	PU-T21	335	0.26	0.12
XXII	296, 324	26.5	PU-T22	290, 330	3,09	20,3
XXIII	342	0.68	PU-T23	345	0,27	0,16

This dissimilarity is probably due to the possible steric hindrance appeared into the macromolecular backbone or to some recombination of the formed radicals in the "solvent cage" before deactivation, which may favour an enhancement of their lifetime. Considering the results achieved on the above triazene derivatives, a theoretical photolysis mechanism for the studied polyurethanes was proposed (Scheme 7).

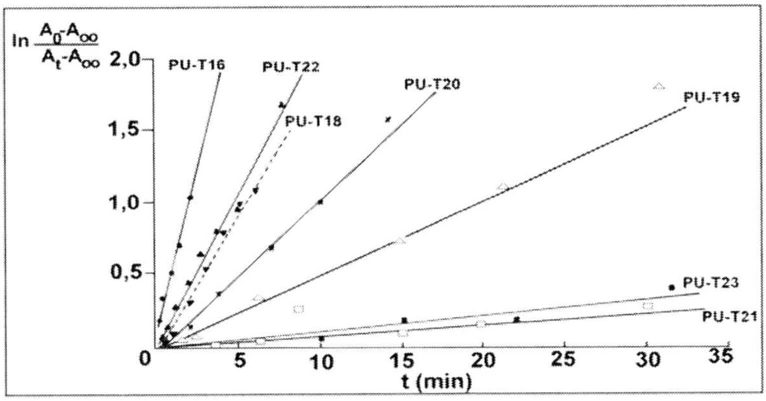

Figure 12. Kinetic evaluation of the triazene polyurethanes photodecomposition in DMF solution at 298 K.

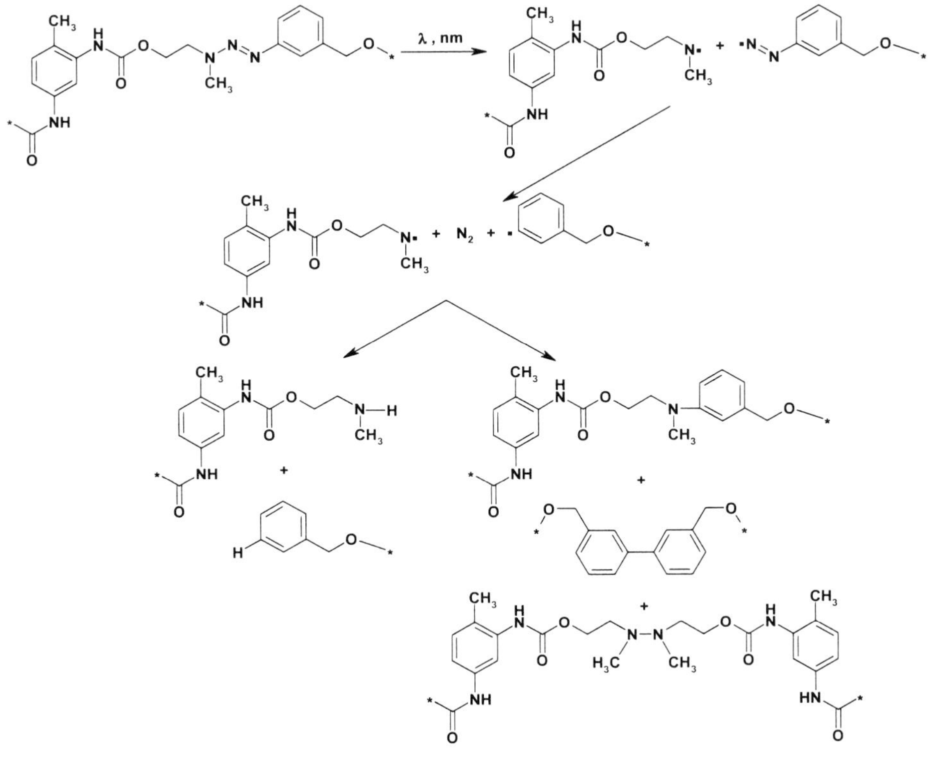

Scheme 7. Theoretical decomposition mechanism for triazene groups and the alternatives for radical stabilization.

It can assume that in the first step of the photoprocess, the *trans-cis* (E-Z) isomerization of the *N=N* double bond with the formation of more unstable thermodynamically *cis* isomer occurred. Once formed, this can be photochemically decomposed leading to aminyl and phenyldiazenyl macroradicals able to generate phenyl radical and molecular nitrogen by a β-scission of the last. As a result, the radical species formed would be stabilized by a pathway of hydrogen atom extraction or recombination, with different consequences on the polymeric chains length.

The photochemical behaviour of the triazene polyurethanes in film state represents a major feature required by dry etching applications. Therefore, the polymeric thin films coated on quartz were subjected to UV irradiation, to establish their ability to function as photolithographic materials. It has to be mentioned that all triazene polyurethanes form clear and homogeneous films by simple casting from DMF solution. Under UV lighting, they suffer a photodecomposition process similar with that found in solution, the obtained kinetic rate values being also lower than for the corresponding monomers. However, the achieved results are quite satisfactory for this type of materials, reason for that these can be recommended as promising photolithographic materials. Accordingly to the above presented mechanism, by an adequate selection of the polymeric films and UV irradiation through a mask, it can obtain negatively or positively working photoresists. Indeed, by subjecting to irradiation through a mask the polymeric films layered from polyurethanes in which the triazene units are incorporated in the main chain of the polymeric backbone, after this treatment, the irradiated area become more soluble and may be cleared away with chloroform, for example. This phenomenon appeared due to the diminishing of the polymer molar mass caused by triazene decomposition, the polyurethane acting as a positive photoresist. In the case of negative photoresists, the triazene groups are located in the side chain of the polymers and by UV irradiation they generate radicals which determine the crosslinking of the polymer and as a result, the remained film becomes insoluble.

The UV irradiation of the ionic polyetherurethanes with triazene linkages issued similarly as in the case of the nonionic forms, moreover, the kinetic constants presenting the lowest values from the class. The main advantage offered by the ionic forms consist in the possibility to utilize the ionomeric aqueous dispersions of triazene polyurethanes to cover different substrates, subsequently the triazene groups behaving under irradiation as in the case of the parent polymers.

The results attained on the above mentioned systems revealed that materials photosensitivity increased with an enhancement of the triazene content in the polymers [118], reason for that, new hard type triazene polyurethanes that include photolabile triazene units in every repeating segment were developed for some previously presented triazene monomers [119]. Table 5 includes several photochemical characteristics of this kind of novel polyurethanes.

Table 5. Photochemical properties of triazene hard polyurethanes

Sample	λ_{max} (nm)	k 10^{-3} (s^{-1})
	film	film
PUH-T16	292, 314	3,52
PUH-T17	370	-
PUH-T20	345	0.33
PUH-T21	352	1.13
PUH-T22	328	9.18
PUH-T23	342	0,57

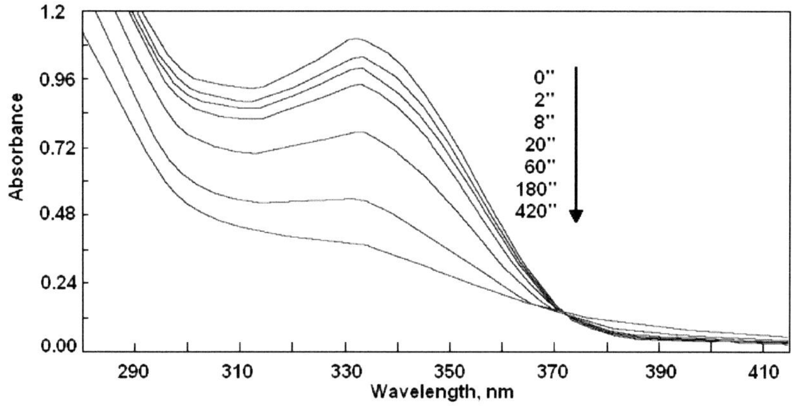

Figure 13. Modifications occurred in the UV-vis absorption spectra of hard type triazene polyurethane XVI in film during UV light exposure.

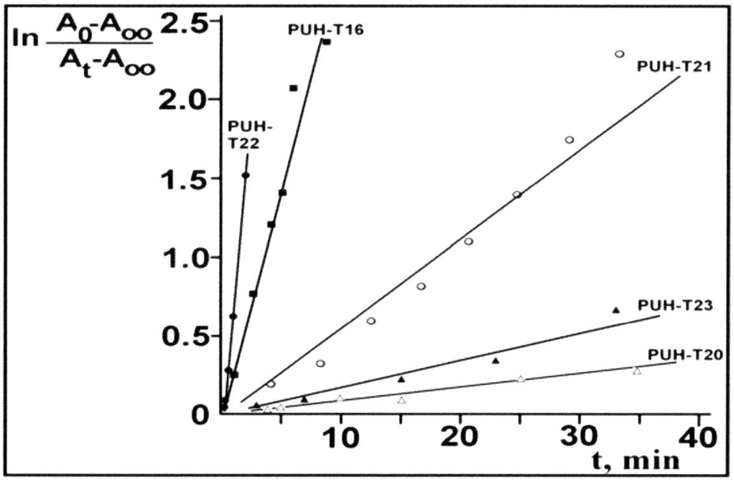

Figure 14. Kinetic evaluation of the hard triazene polyurethanes photolysis in film at 298 K.

It may be observed that generally, the polymers present a similar photolytic behaviour and by exposing the polymeric films to UV irradiation, a gradual decrease of the absorption maximum with the irradiation period was detected (e.g. in Figure 13 is represented the photolytic decay of triazene absorption in PUH-T22).

As in the case of soft polyurethanes, the hard type polymers exhibit an equivalent photoreactivity order, the photolytic behaviour being strongly influenced by the inductive or conjugation effects (Figure 14). The minor inversions produced in the polymers photosensitivity order can be explained by some physical or manipulation dissimilarities, and they not modify the general arrangement.

A specification must be done about the monomer and polymers with *p*-nitrophenyl triazene moiety (XVII, PU-17, and PUH-17, respectively) because in this case, even for prolonged irradiation times there are no modifications in the UV absorption spectrum.

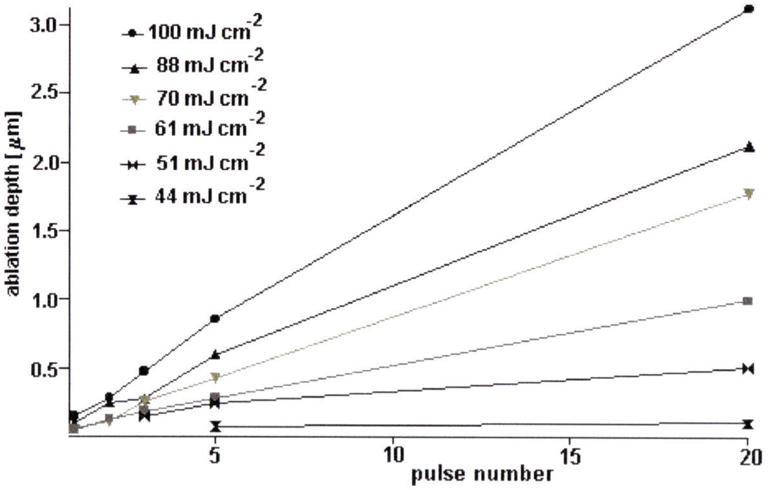

Figure 15. Plots of the etch depth versus pulse number for PUH-17 at different fluences. Reprinted with permission of [Ref. 119], Copyright (2005) – Elsevier.

We supposed that for these structures the electron-withdrawing effect of the substituent is very strong and the UV light is too weak to produce the triazene cleavage. This is the motivation why we firstly exposed a triazene-polyurethane film to laser irradiation. The ablation rate of the films based on *p*-nitrophenyl triazene polyurethane PUH-17 was obtained by determining the ablation depths as a function of pulse number (Figure 15).

The total ablated depth at a given fluence increases linearly with the number of pulses for all fluences, indicating that no incubation behaviour is present as expected for highly absorbing polymers ($\alpha_{lin} > 8000$ cm^{-1}). In the case of this polymer a threshold fluence of 52 ± 3 mJ cm^{-2} and an effective absorption coefficient for ablation, α_{eff} of 4.0 (± 0.2) x 10^{4} cm^{-1} was obtained.

Microscopic and profilometric analysis of the polymer film after irradiation at fluences above and below the ablation threshold revealed interesting features (Figure 16).

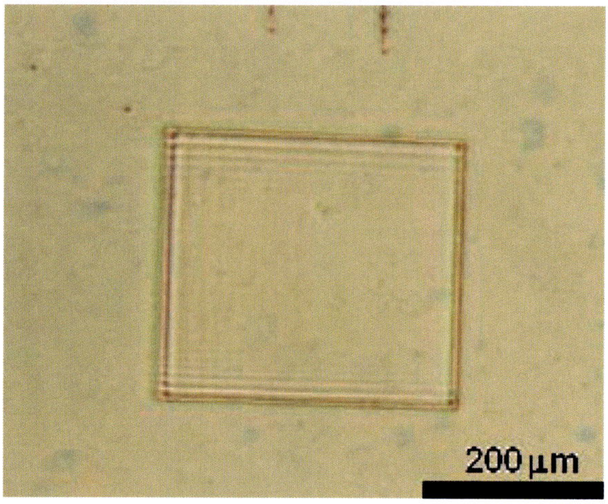

a

Figure 16. (Continued on next page.)

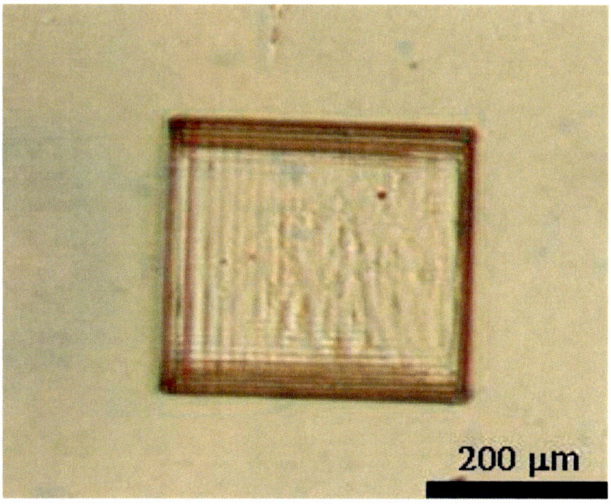

b

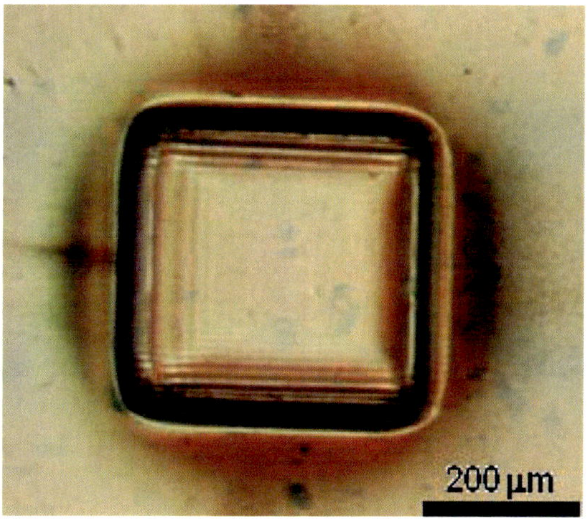

c

Figure 16. Optical micrographs of the triazene polyurethane PUH-17 irradiated at 33 mJ cm-2 (a), 51 mJ cm-2 (b) and 280 mJ cm-2 (c) after 20 pulses. Reprinted with permission of [Ref. 119], Copyright (2005) – Elsevier.

At fluences below the ablation threshold (e.g. at F = 33 mJ cm^{-2}), a surface swelling was detected, which may be attributed to the decomposition of the triazene group. This decomposition yields gaseous species that result in a volume increase and a surface swelling. At fluences around the threshold of ablation (F = 51 mJ cm^{-2}), the triazene polyurethane is already etched without pronounced redeposition of the ablation products (shown in the optical micrograph in Figure 16, b for 20 pulses). If much higher fluences and pulse numbers are applied, the polymer surface and the surrounding of the crater appear strongly modified. At a fluence of 280 mJ cm^{-2} for 20 pulses, the surface becomes black and carbonized (Figure 16, c).

A detailed analysis and comparison of the ablation parameter α_{eff}, F_{th}, d(100 mJ cm^{-2}) and material characteristics (λ_{max}, α_{lin} at 308 nm and at λ_{max}) for PUH-17 with other triazene polymers from literature [107] undoubtedly shows pronounced differences, i.e. that PUH-17 presents ablation relevant parameter inferior values. The marked difference between PUH-17 and other triazene-polymers with very similar ablation behaviour is not very clear, but may be due to the low absorption coefficient at the laser irradiation wavelength of 308 nm or the fact that the triazene group is placed in the side chain. This may cause in a release of the triazene fragment and the remaining of the main polymeric chain at least partially intact. The radical decomposition mechanism may even induce a cross-linking of the ablated polymer that is not any more photoactive, resulting in higher threshold fluence, lower ablation rates and surface modification.

CONCLUSIONS

Novel photopolyurethanes of cationomer type containing azobenzene chemically or electrostatically attached in the side chain were synthesized and characterized comparatively with some non-ionic polymers, whose isomerization indicated a diminished sensitivity to UV irradiation (365 nm) due to the presence of ammonium quaternary groups besides the chromophoric ones capable to control the photochemical properties specifically to the ionomeric matrix. The choice of the level of ionization, the content of chromophore and its location, as well as state of chromophore (solution, film) played an important role in determining the morphology and properties of these polymers.

Owing to the insertion of azobenzene unit in the soft segment of polyurethane backbone, the corresponding azo-cationomers exhibited in thin films a better photochromic activity explained by the increased molecular mobility of the azo chromophore from the flexible component but controlled of the electrostatic interactions between neighbouring chains. The effect exercised by the chromophoric groups as counterions of ammonium quaternary structure is also reflected by an improved photochromic response.

Among the synthesized triazeno-polyurethanes, the hard photopolymers bearing photolabile triazene moieties in the side chain showed a significant photosensitivity under UV irradiation due to the chromophore high content.

Laser ablation (308 nm) of the triazenopolymers in thin films achieved at intermediate fluences allowed a high resolution structuring of the polymeric surface. A comparison of the ablation parameters of these polymers with those that contain bistriazene in the main chain evidenced performing properties of the latter.

REFERENCES

[1] Handwerger, R. G.; Diamond, S. L. *Bioconjugate Chem.* 2007, 18, 717-723.

[2] Besson, E.; Gue, A.-M.; Sudor, J.; Korri-Youssoufi, H.; Jaffrezic, N.; Tardy, J. *Langmuir* 2006, 22, 8346-8352.

[3] Ito, Y. *Biotechnol. Prog.* 2006, 22, 924-932.

[4] Khire, V. S.; Harant, A. W.; Watkins, A. W.; Anseth, K. S.; Bowman, C. N. *Macromolecules* 2006, 39, 5081-5086.

[5] Yang, Y.; Wang, X.; Liu, L.; Xie, X.; Yang, Z.; Li, R. K. Y.; Mai, Y.-W. *J. Phys. Chem. C.* 2007.

[6] Li, Q.; Li, Z.; Zeng, F.; Gong, W.; Li, Z.; Zhu, Z.; Zeng, Q.; Yu, S.; Ye, C.; Qin, J. *J. Phys. Chem.* B. 2007, 111, 508-514.
[7] Crenshaw, B. R.; Weder, C. *Macromolecules* 2006, 39, 9581-9589.
[8] Dieterich, D.; Uhlig, K. "Polyurethanes" in Ullmann's Encyclopedia of Industrial Chemistry, 5th Ed., vol A21, 1992.
[9] Hepburn, C. Polyurethane elastomers, *Elsevier Appl. Sci.*, London, 1991.
[10] Bogart, J.; Gibson, P.; Cooper, S. L. J. Polym. Sci. Part B: *Polym. Phys.* 1983, 21, 65-95.
[11] Dieterich, D.; Keberle, W.; Witt, H. Angew. *Chem. Int. Ed.* 1970, 9, 40-50.
[12] Marx, C. L.; Caulfield, D. F.; Cooper, S. L. *Macromolecules* 1973, 6, 344-353.
[13] Hwang, K. S.; Yang, C.-Z.; Cooper, S. L. *Polym. Eng. Sci.* 1981, 21, 1027-1037.
[14] Nomula, S.; Cooper, S. L. *J. Phys. Chem.* B. 2000, 104, 6963-6972.
[15] Dieterich, D. in Adv. Urethane Ionomers, (H. X. Xiao, K. C. Frisch, Eds) Technomic, Lancaster, 1995, 1.
[16] Nomula S.; Cooper S. L. *J. Colloid Interface Sci.* 1998, 205, 331-339.
[17] Lee, J. C.; Kim, B. K. *J. Polym. Sci. Part A: Polym. Chem.* 1994, 32, 1983-1989.
[18] Matsunga, K.; Nakagawa, K.; Sawai, S.; Sonoda, O.; Tajima, M.; Yoshida, Y. *J. Appl. Polym. Sci.* 2005, 98, 2144-2148.
[19] Bräuer, M.; Oder, G. *Plaste Kautsch.* 1992, 39, 193-199.
[20] Yang, S.; Li, C.; Cooper, S. L. *J. Polym. Sci. Part B: Polym. Phys.* 1991, 29, 75-86.
[21] Radhakrishnan, G.; Ramesh S. *Europ. Polym. J.* 1996, 32, 993-998.
[22] Wallon, A.; Bergmann, K.; Haberle, K.; Maempel, L.; Rieger, J.; Schmiedberger, H.; Zosel, A. in Adv. Urethane Ionomers, (H. H. Xiao, K. C. Frisch, Eds), Technomic, Lancaster, 1995, 83.
[23] Ho, C. C.; Xiao, H. X.; Frisch, K. C.; McLean Y. in Adv. Urethane Ionomers, (H. X. Xiao, K. C. Frisch, Eds), Technomic, Lancaster, 1995, 111.
[24] Antony, P.; De, S. K. *Polym. Rev.* 2001, 41, 41–77.
[25] Reiff, H.; Dieterich, D. in Ionomers, (M. R. Tant et al., Eds), Blackie, London, 1997, 444.
[26] Leclerc, M. *Adv. Mater.* 1999, 11, 1491-1498.
[27] van den Berg, O.; Sengers, W. G. F.; Jager, W. F.; Picken, S. J.; Wubbenhorst, M. *Macromolecules* 2004, 37, 2460-2470.
[28] McQuade, D. T.; Pullen, A. E.; Swager, T. M. *Chem. Rev.* 2000, 100, 2537-2574.
[29] Natansohn, A.; Rochon, P. *Chem. Rev.* 2002, 102, 4139-4176.
[30] Argun, A. A.; Aubert, P.-H.; Thompson, B. C.; Schwendeman, I.; Gaupp, C. L.; Hwang, J.; Pinto, N. J.; Tanner, D. B.; MacDiarmid, A. G.; Reynolds, J. *R. Chem. Mater.* 2004, 16, 4401-4412.
[31] Samuel, I. D. W.; Turnbull G. A. *Chem. Rev.* 2007, 107, 1272-1295.
[32] Thomas, S. W. III; Joly, G. D.; Swager, T. M. *Chem. Rev.* 2007, 107, 1339-1386.
[33] Dugave, C.; Demange, L. *Chem. Rev.* 2003, 103, 2475-2532.
[34] Ostroverkhova, O.; Moerner, W. E. *Chem. Rev.* 2004, 104, 3267-3314.
[35] Wang, X.; Wang, L.; Li, H.; Tang, X.; Chang, F. C. *J. Appl. Polym. Sci.* 2000, 77, 184-188.
[36] Chen, H.; Chen, D.-Z.; Fan, Q.; Yu, X. *J. Appl. Polym. Sci.* 2000, 76, 2049-2056.
[37] Polizos, G.; Georgoussis, G.; Kyritsis, A.; Shilov, V.; Shevchenko, V.; Gomza, Y.; Nesin, S.; Klimenko, N.; Wartewig, S.; Pissis, P. *Polym. Int.* 2000, 49, 987-992.

[38] Kim, B.; Yang, J.; Yoo, S.; Lee, J. *Colloid Polym. Sci.* 2003, 281, 461-468.
[39] Wei, X.; Yu, X. *J. Polym. Sci., Part B: Polym. Phys.* 1997, 35, 225-232.
[40] Xiao, H. X.; Frisch, K. C.; Hsu, G. S. L.; Al-Salah, H. A. in Adv. Urethane Ionomers, (H. X. Xiao, K. C. Frisch, Eds) Technomic, Lancaster, 1995, 23.
[41] Al-Salah, H. A. *J. Macromol. Sci. Pure Appl. Chem.* 1998, A35, 763-780.
[42] Reinecke, H.; Mijangos, C. *Macromol. Chem. Phys.* 1998, 199, 2199-2204.
[43] Huang, S. L.; Chao, M. S.; Lai, J.; *J. Appl. Polym. Sci.* 1998, 67, 865-873.
[44] Yang, T.; Chin, W. Cherng, J.; Shau, M. *Biomacromolecules* 2004, 5, 1926-1932.
[45] Buruiana, E. C.; Olaru, M.; Strat, M.; Strat, G.; Simionescu, B. C. *Polym. Int.* 2005, 54, 1296-1303.
[46] Buruiana, T.; Diaconu, I.; Buruiana, E. C.; Han, X.; Guo, F. *Angew. Macromol. Chem.* 1997, 245, 139-147.
[47] Buruiana, T.; Buruiana, E. C.; Bestiuc, I. *Rev. Roum. Chim.* (Rom. Acad.) 1994, 39, 213-218.
[48] Li, X.; Fei, G.; Wang, H. *J. Appl. Polym. Sci.* 2006, 100, 40-46.
[49] Velankar, S.; Yang, C. Z.; Cooper, S. L. *Polym. Prepr.*(Am. Chem. Soc. Div. Polym. Chem.) 1996, 37, 400-401.
[50] Goddard, R. J.; Cooper, S. L. *J. Polym. Sci. Part B: Polym. Phys.* 1994, 32, 1557-1571.
[51] Ahsan, M. A.; Varma, S. C.; George, M. H.; Barrie, J. A. *Polym. Commun.* 1991, 32, 509-512.
[52] Buruiana, E. C.; Diaconu, I.; Buruiana, T.; Grigoriu, G.; Caraculacu, A. *Angew. Macromol. Chem.* 1991, 187, 51-59.
[53] Nurdin, N.; Helary, G.; Sauvet, G. *J. Appl. Polym. Sci.* 1993, 50, 663-670.
[54] Nurdin, N.; Helary, G.; Sauvet, G. *J. Appl. Polym. Sci.* 1993, 50, 671-678.
[55] Loveday, D.; Wilkes, G. L.; Bheda, M. C.; Shen, Y. X.; Gibson, H. W. *J. Macromol. Sci. Pure Appl. Chem.* 1995, 32, 1-27.
[56] Buruiana, E. C.; Buruiana, T.; Robila, G.; Airinei, A. *J. Polym. Mater.* 1999, 16, 7-12.
[57] Buruiana, T.; Caraculacu, A. *Rev. Roum. Chim.*(Rom. Acad.) 1991, 36, 1123-1127.
[58] Mahesh, G. N.; Sivaraman, A.; Tharanikkarasu, K. *J. Polym. Sci. Part A: Polym. Chem.* 1997, 35, 1237-1244.
[59] Mahesh, G. N.; Philip, T. G.; Radhakrishnan, G. *Polym. Bull.* (Berlin) 1996, 37, 737-742.
[60] Buruiana, T.; Buruiana, E. C. *J. Polym. Mater.* 2002, 19, 29-36.
[61] Su, F.-A.; Chen, S.-A.; Chan, W.-C. *J. Polym. Res.* 1994, 1, 123-130.
[62] Yang, C.-H.; Li, Y.-J.; Wen, T.-C. *Ind. Eng. Chem. Res.* 1997, 36, 1614-1621.
[63] Mohanty, S.; Krishnamurti, N. *J. Appl. Polym. Sci.* 1996, 62, 1993-2003.
[64] Nomula, S.; Cooper, S. L. *Macromolecules* 1997, 30, 1355-1362.
[65] Chan, W.-C.; Chen, S.-A. *Polymer* 1993, 34, 1265-1270.
[66] Sriran, V.; Aruna, P.; Tharanikkarasu, K.; Venkatis, W.; Rhadakrishnan, G. *J. Appl. Polym. Sci.* 2001, 81, 813-821.
[67] Velankar, S.; Cooper, S. L. *Macromolecules* 2000, 33, 382-394.
[68] Goddard, R. J.; Cooper, S. L. *Macromolecules* 1995, 28, 1390-1400.
[69] Goddard, R. J.; Cooper, S. L. *Macromolecules* 1995, 28, 1401-1406.
[70] Lee, J. C.; Kim, B. K. *J. Polym. Sci. Part A: Polym. Chem.* 1994, 32, 1983-1989.
[71] Xiao, H. X.; Frisch, K. C.; Hsu, G. S. L. *Prog. Rubber Plast. Technol.* 1991, 7, 211-242.

[72] Mohanty, S.; Krishnamurti, N. *J. Appl. Polym. Sci.* 1996, 62, 1993-2003.
[73] Wang, C. I.; Kuo, Y. M.; Chao, D. Y. *Polym. Adv. Tech.* 2000, 11, 127-135.
[74] Kuo, Y.-M.; Ku, K.-Y.; Lin, H.-C.; Wang, N.-H.; Chwang, C.-P.; Chao, D.-Y. *J. Appl. Polym. Sci.* 1998, 69, 2097-2105.
[75] Lee, J.-J.; Chi, Z.-K.; Kuo, Y.-M.; Chao, D.-Y. *J. Appl. Polym. Sci.* 1995, 57, 1005-1011.
[76] Al-Salah, H. A. *Acta Polym.* 1998, 49, 465-470.
[77] Buruiana, E. C.; Buruiana, T. *Polym. J.*(Tokio) 2001, 33, 42-48.
[78] Buruiana, E. C.; Buruiana T. *Polym. J.*(Tokyo) 2001, 33, 723-732.
[79] Buruiana, E. C.; Buruiana T. *Eur. Polym. J.* 2001, 37, 2505-2511.
[80] Buruiana, E. C.; Buruiana, T. *J. Photochem. Photobiol. A: Chem.* 2002, 151, 237-252.
[81] Buruiana, T.; Buruiana, E. C.; Airinei, A.; Grecu, I. *Eur. Polym.* J. 2001, 37, 343-348.
[82] Buruiana, T.; Airinei, A.; Buruiana, E. C.; Robila, G. *Eur. Polym. J.* 1997, 33, 877-880.
[83] Buruiana, T.; Airinei, A.; Buruiana, E. C.; Robila, G. *Angew. Makromol. Chem.* 1995, 226, 205-212.
[84] Buruiana, E .C.; Buruiana, T.; Airinei, A.; Robila, G.; Grecu, I. *Polym. J.* (Tokio) 1999, 31, 510-515.
[85] Kim, S. J.; Reneker, D. H. *Polym. Bull.* (Berlin) 1993, 31, 367-374.
[86] Brecl, M.; Malavasic, T. *J. Polym. Sci. Part A: Polym. Chem.* 1997, 35, 2871-2888.
[87] Tsutsumi, N.; Yoshizaki, S.; Sakai, W.; Kiyotsukuri, T. *Macromolecules* 1995, 28, 6437-6442.
[88] Nuyken, O.; Stebani, J.; Long, A. Angew. *Makromol. Chem.* 1995, 226, 143-160.
[89] Sung, C. S.; Lamarre, L.; Chung, K. H. *Macromolecules* 1979, 12, 666-669.
[90] Sung, C. S.; Lamarre, L.; Chung, K. H. *Polym. Prepr.(Am. Chem. Soc. Div. Polym. Chem*) 1981, 22, 277-278.
[91] Buruiana, T.; Airinei, A.; Buruiana, E.C.; Robila, G. *Angew. Makromol. Chem.* 1998, 258, 39-44.
[92] Buruiana, T.; Buruiana, E. C.; Airinei, A. *Rev. Roum. Chem.*(Rom. Acad.) 2000, 45, 697-701.
[93] Buruiana, E. C.; Buruiana, T.; Airinei, A.; Robila, G. *Angew. Makromol. Chem.* 1993, 206, 87-96.
[94] Kinugasa, S.; Hattori, S.; Hayashi, H. *Polym. Prepr. Jpn.* (Engl .Ed.) 1991, 40, 5-7.
[95] Buruiana, T.; Airinei, A.; Buruiana, E. C.; Robila, G. *J. Macromol. Sci. Pure Appl. Chem.* 1998, A32, 1397-1404.
[96] Ahlheim, M.; Lehr, F.; Kaatz, P. G.; Pretre, Ph.; Gunter, P. *Polym. Prepr.(Am. Chem. Soc. Div. Polym. Chem.)* 1994, 35, 250-251.
[97] Natansohn, A.; Rochon, P.; Ho, M.-S.; Barret, C. *Macromolecules* 1995, 28, 4179-4183.
[98] Watanabe, O.; Tsuchimori, M.; Okada, A. *J. Mater. Chem.* 1996, 6, 1487-1492.
[99] Eisenbach, C.D. *Polym. Prepr.*(Am. Chem. Soc. Div. Polym. Chem.) 1980, 21, 309-310.
[100] Mita, I.; Horie, K.; Hirao, K. *Macromolecules* 1989, 22, 558-603.
[101] Rau, H. In Photochemistry Process and Photochemical Reactions in Polymers, (J. F. Rabek, Ed.), CRC Press, Boca Raton, FL, 1990, 119.
[102] Buruiana, T.; Airinei, A.; Buruiana, E.C.; Robila, G. *J. Polym. Mater.* 1994, 11, 95-100.
[103] Lippert, T.; Wokaun, A. *Chimia* 2001, 55, 783-786.

[104] Kärkkäinen, A.; Tamkin, J.; Rogers, J.; Neal, D.; Homi, O.; Jabbour, G.; Rantala, J.; Descour, M. *Appl. Optics* 2002, 41, 3988-3998.
[105] Sato, H.; Nishio, S. *J. Photochem. Photobiol. C: Photochem. Reviews* 2001, 2, 139-152.
[106] Lippert, T.; Hauer, M.; Phipps, C. R.; Wokaun, A. *Appl. Phys. A* 2003, 77, 259–264.
[107] Wei, J.; Hoogen, N.; Lippert, T.; Nuyken, O.; Wokaun, A. *J. Phys. Chem. B* 2001, 105, 1267-1275.
[108] Lippert, T.; Dickinson, J. T.; Langford, S. C.; Furutani, H.; Fukumura, H.; Masuhada, H.; Kunz, T.; Wokaun, A. *Appl. Surf. Sci.* 1998, 117, 127-129.
[109] Lippert, T.; Nakamura, T.; Niino, T.; Yabe, A. *Macromolecules* 1996, 29, 6301-6309.
[110] Nuyken, O.; Dahn, U. J. Polym. Sci., Part A: *Polym. Chem. Ed.* 1997, 35, 3017-3025.
[111] Nuyken, O.; Dahn, U.; Wokaun, A.; Kunz, T.; Hahn, C.; Hessel, V.; Laudsiedel, *J. Acta Polym.* 1998, 49, 427-432.
[112] Stebani, J.; Nuyken, O.; Lippert, T.; Wokaun, A. *Makromol. Chem. Rapid Commun.* 1993, 14, 365-369.
[113] Stebani, J.; Nuyken, O.; Lippert, T.; Wokaun, A.; Stasko, A. *Macromol. Chem. Phys.* 1995, 196, 739-749.
[114] Furutani, H.; Fukumura, H.; Masuhara, H.; Lippert, T.; Yabe, A. *J. Phys. Chem. A* 1997, 101, 5742-5747.
[115] Nuyken, O.; Dahn, U.; Hoogen, N. *Polym. News* 1999, 24, 257-266.
[116] Buruiana, E. C.; Niculescu, V.; Buruiana, T. *J. Appl. Polym. Sci.* 2003, 88, 1203-1210.
[117] Buruiana, E. C.; Niculescu, V.; Buruiana, T. *J. Appl. Polym. Sci.* 2004, 92, 2599-2605.
[118] Buruiana, E. C.; Melinte, V.; Buruiana, T.; Simionescu, B. C. *J. Appl. Polym. Sci.* 2005, 96, 385-391.
[119] Buruiana, E. C.; Melinte, V.; Buruiana, T.; Lippert, T.; Yoshikawa, H.; Mashuhara, H. *J. Photochem. Photobiol. A: Chem.* 2005, 171, 261-267.
[120] Buruiana, E. C.; Melinte, V.; Buruiana, T.; Simionescu, B. C.; Lippert, T.; Urech, L. *J. Photochem. Photobiol. A: Chem.* 2007, 186, 270-277.
[121] Nuyken, O.; Scherer, C.; Baindl, A.; Brenner, A. R.; Dahn, U.; Gartner, R.; Kaiser-Rohrich, S.; Kollefrath, R.; Matusche, P.; Voit, B. *Prog. Polym. Sci.* 1997, 22, 93-183.

In: Advances in Functional Heterochain Polymers
Editor: Maria Cazacu
ISBN: 978-1-60456-598-0

Chapter 6

POLY(ETHER IMIDE)S FOR HIGH PERFORMANCE MATERIALS

Elena Hamciuc[1] and Corneliu Hamciuc
"Petru Poni" Institute of Macromolecular Chemistry, Aleea Gr. Ghica Voda 41A, Iasi, 700487, Romania

ABSTRACT

This chapter is a review concerning the progress on the synthesis, characterization and potential applications of poly(ether imide)s. It is focused on the recent development in this field. Different types of structurally modified poly(ether imide)s having flexible units, bulky groups, fluorine atoms, siloxane sequences, as well as hyperbranced compounds or poly(ether imide)s containing other heterocycles are presented. The properties of these polymers like solubility, film forming ability, thermal stability, glass transition temperature, mechanical and dielectric characteristics are discussed in correlation with their structures. Future perspectives of the polymers for use as advanced materials in high performance applications are also given.

1. INTRODUCTION

The development of new technologies has generated an increasing demand for the preparation of new polymeric materials with special properties. Among heterocyclic thermostable polymers, aromatic polyimides represent the most important class of heat resistant polymeric materials. They are well known to retain their excellent properties at high temperatures for long period of time. Some of these polymers have been widely used as high performance materials in the aerospace and electronic industries due to their excellent thermal stability, outstanding mechanical integrity, dimensional stability, good electrical properties and chemical resistance. However, polyimides are often insoluble and intractable in their fully imidized form, presenting serious processing difficulties. Conventional aromatic polyimides

[1] e-mail: ehamciuc@icmpp.ro

must be processed at their precursor stage, poly(amic-acid)s, which are unstable at room temperature. This process has some inherent limitations like the emission of volatile by-products during curing and storage instability of the poly(amic-acid) intermediate [1-3]. To overcome these problems, extensive research has been carried out in the past decades on the development of novel soluble and/or melt-processable aromatic polyimides that can be applied in the state of full imidation. Various efforts have been made on the synthesis of soluble and processable polyimides without much loss of their excellent thermostability. The introduction of kinks or flexible linkages, noncoplanar units or bulky lateral groups along the backbone was used to improve the solubility of these polymers. The main concept behind all these approaches is the reduction of several types of polymer chain-chain interactions, and the reduction of chain packing and charge transfer electronic polarization interactions [4, 5]. It has been generally recognized that aromatic ether linkages inserted in aromatic main chains provide them with a significantly lower energy of internal rotation. In general, such a structural modification leads to lower glass transition temperature (T_g) and crystalline melting temperatures, as well as significant improvement in solubility and other processing characteristics of the polymers, without greatly sacrificing other advantageous polymer properties [6, 7].

Poly(ether imide)s, having the general structure **1**, have received much attention because they can provide a compromise between good thermal stability and processability when compared to aromatic polyimides [8].

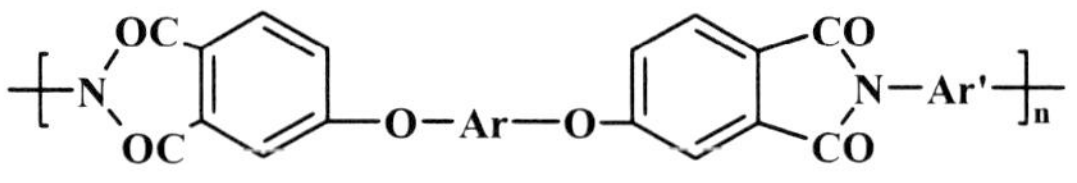

1

These polymers were developed as a result of the requirement for high-performance polymers that could be readily fabricated by standard plastic extrusion and injection molding processes. Poly(ether-imide)s prepared through phenoxy derivatives via nitro displacements, especially poly(ether imide) Ultem produced by General Electric, have achieved great commercial success [9, 10].

2. Synthesis of Poly(Ether Imide)s

The poly(ether imide)s are mainly prepared by cyclopolycondensation reactions of bis(ether anhydride)s with aromatic diamines, as well as by the aromatic nucleophilic substitution reactions of bishalogenoimides or dinitroimides with bisphenols.

2.1. Synthesis of Poly(Ether Imide)s by Cyclopolycondensation Reactions

Due to the large variety of the available monomers, bis(ether anhydride)s and aromatic diamines, different types of poly(ether imide)s have been synthesized to meet the requirements for new applications. The main used monomers and pathways to their synthesis are reviewed in following.

2.1.1. Synthesis of Bis(Ether Anhydride)s

Generally, the bis(ether anhydride) monomers were prepared by a sequence of reactions, as shown in scheme 1. The intermediate bis(ether dinitrile)s **2** were obtained from nucleophilic nitrodisplacement of 4-nitrophthalonitrile with the phenoxide ions of bisphenols in N,N-dimethylformamide (DMF). Then the resulting bis(ether dinitrile)s were hydrolysed to bis(ether diacid)s **3** by ethanolic potassium hydroxide solution, followed by the cyclodehydration with acetic anhydride to yield the final bis(ether anhydride)s **4** [11].

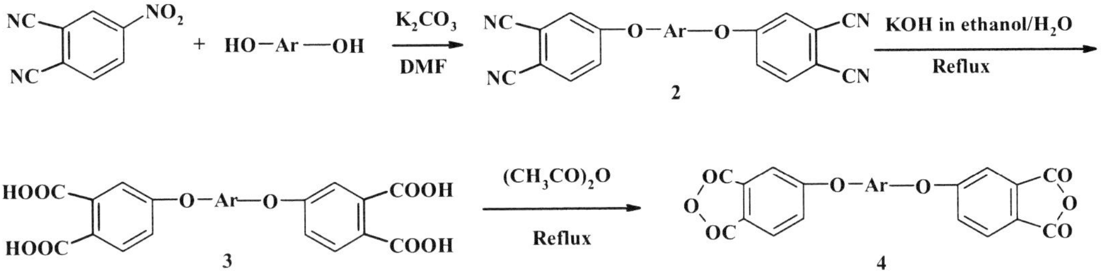

Scheme 1. Synthesis of bis(ether anhydride)s **4** starting from 4-nitrophthalonitrile and bisphenols.

Bis(ether anhydride)s can also be obtained by a process in three steps starting from N-phenyl- or N-alkyl-nitrophthalimides and bisphenols. The bis(N-phenylphthalimide ether)s **5** are hydrolyzed to bis(ether diacid)s **3,** and then cyclodehydrated to yield the corresponding bis(ether anhydride)s **4**, according to scheme 2 [12].

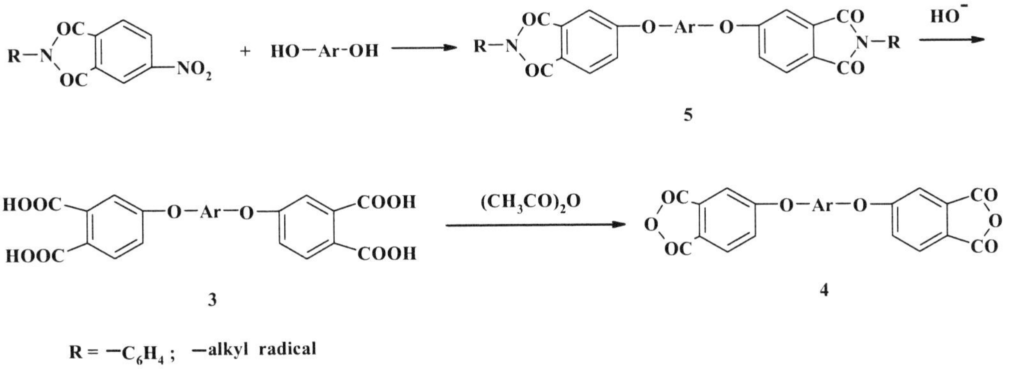

Scheme 2. Synthesis of bis(ether anhydride)s **4** starting from N-phenyl- or N-alkyl-4-nitrophthalimide and bisphenols.

Table 1 presents some of the reported bis(ether anhydride)s having various structures.

2.1.2. Synthesis of the Polymers

Poly(ether imide)s **7** were prepared by cyclopolycondensation reaction of bis(ether anhydride)s **4** with various aromatic or aliphatic diamines **6**, as shown in scheme 3. The preparation process takes place in two steps. A poly(amic acid) **7′** is synthesized in the first step by polycondensation reaction of a bis(ether anhydride) with a diamine, in a polar solvent such as N-methylpyrrolidone (NMP), DMF or N,N-dimethylacetamide (DMAc). In the second step the resulting poly(amic acid) **7′** is thermally or chemicaly cyclized to the corresponding poly(ether imide) structure **7**.

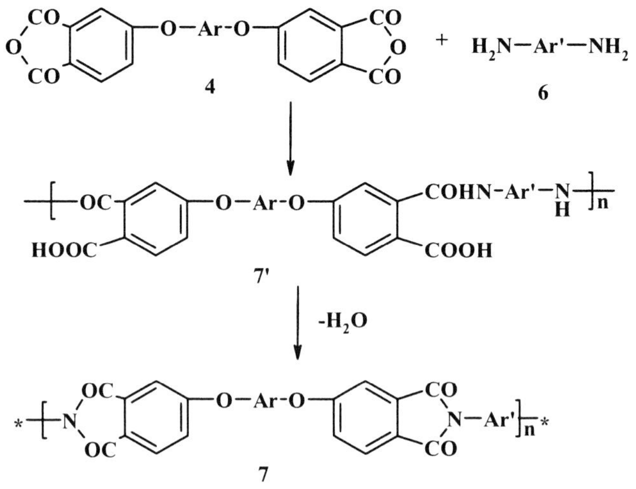

Ar = aromatic radical
Ar' = aromatic or aliphatic radical

Scheme 3. Synthesis of poly(ether imide)s **7**.

The thermal cyclization of poly(amic acid) intermediate to its imide structure, in the solid state, was performed by special thermal treatment heating the poly(amic acid) up to 250°C [8]. Also, the thermal cyclization can be carried out in solution, in the presence of xylene or *o*-dichlorobenzene to remove the water resulted from the cyclization reaction by azeotropic distillation [13]. The chemical cyclization was done in the presence of a mixture of acetic anhydride/pyridine [14].

Table 1. Some bis(ether anhydride)s used for the preparation of poly(ether imide)s

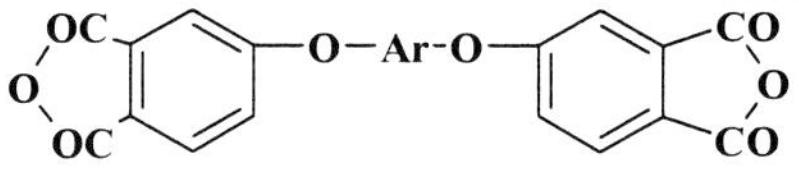

Crt. No.	Ar	Bis(ether anhydride)	Melting point (°C)	Ref.
1	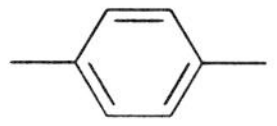	4,4′-(1,4-phenylenedioxy)-diphthalic anhydride	264-266	[11]
2	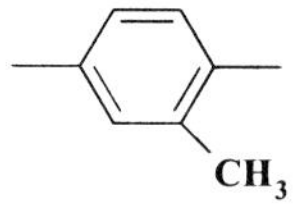	4,4′-(2,5-tolylenedioxy)-diphthalic anhydride	212-214	[11]
3	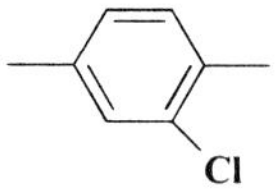	4,4′-(2-chloro-1,4-phenylenedioxy)diphthalic anhydride	197-200	[11]

Crt. No.	Ar	Bis(ether anhydride)	Melting point (°C)	Ref.
4	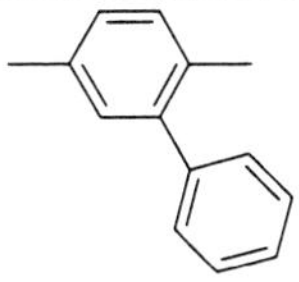	4,4′-(2,5-biphenylenedioxy)-diphthalic anhydride	224-226	[11]
5	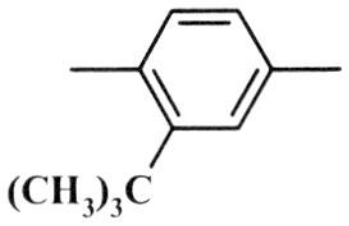	1,4-bis(3,4-dicarboxyphenoxy)-2-*tert*-butylbenzene dianhydride	211-213	[14]
6	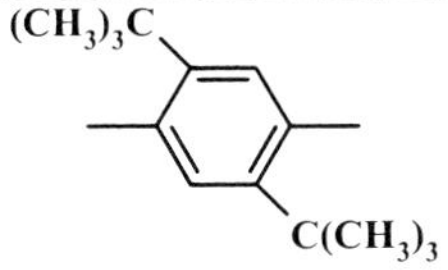	1,4-bis(3,4-dicarboxyphenoxy)-2,5-di-*tert*-butylbenzene dianhydride	253.7	[22]
7		1,2-bis(3,4-dicarboxyphenoxy)benzene dianhydride	187-188	[28]
8	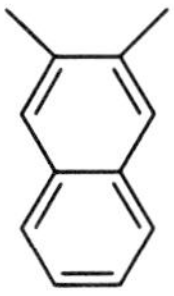	2,3-bis(3,4-dicarboxyphenoxy)-naphthalene dianhydride	264-265	[28]
9	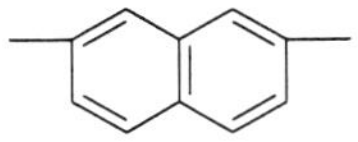	2,7-bis(3,4-dicarboxyphenoxy)-naphthalene dianhydride	158-160	[8]
10	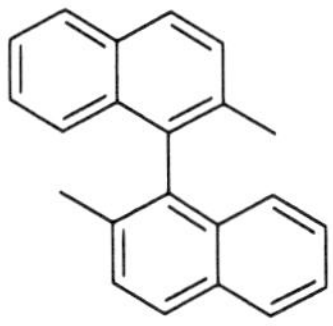	2,2′-bis(3,4-dicarboxyphenoxy)-1,1′-binaphthyl dianhydride	228-230	[30]
11	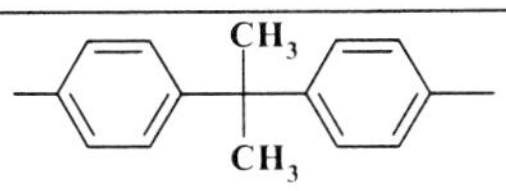	2,2-bis[4-(3,4-dicarboxyphenoxy)phenyl]-propane dianhydride	182-184	[12]
12	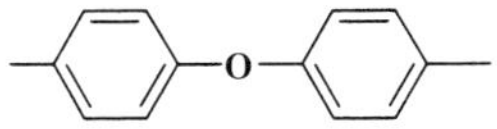	4,4′-bis(3,4-dicarboxyphenoxy)-diphenylether dianhydride	235	[12]
13	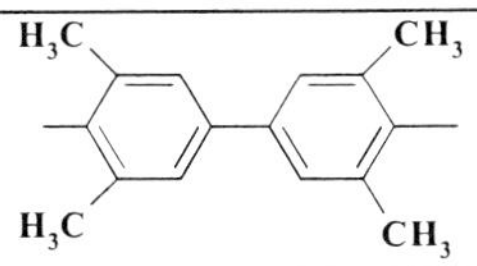	3,3′,5,5′-tetramethyl-4,4′-bis(3, 4-dicarboxyphenoxy)-biphenyl dianhydride	264-267	[12]
14	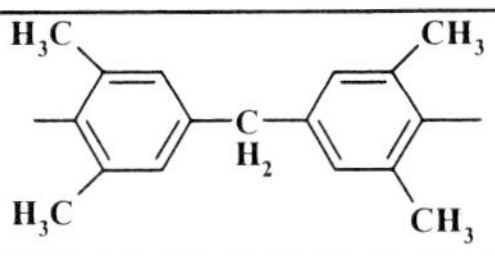	3,3′,5,5′-tetramethyl-4,4′-bis(3,4-dicarboxyphenoxy)-diphenylmethane dianhydride	223-225	[12]

Table 1. Continued.

Crt. No.	Ar	Bis(ether anhydride)	Melting point (°C)	Ref.
15	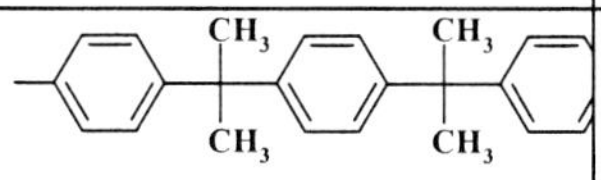	bis-[(3,4-dicarboxyphenoxy)-phenyl]-1,4-phenylene-diisopropylidene dianhydride	190-192	[12]
16	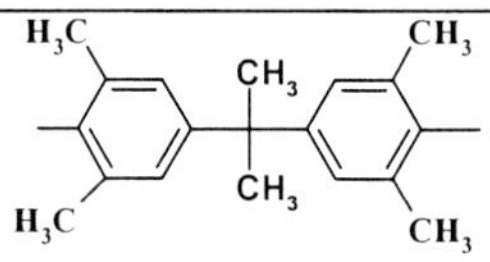	3,3′,5,5′-tetramethyl-2,2-bis[4-(3, 4-dicarboxyphenoxy)phenyl]-propane dianhydride	164-165	[36]
17	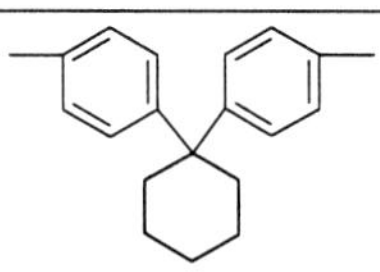	1,1-bis[4-(3,4-dicarboxyphenoxy)phenyl]-cyclohexane dianhydride	198-199	[42]
18	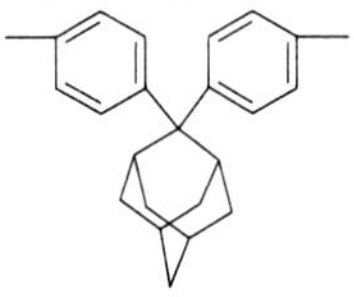	2,2-bis[4-(3,4-dicarboxyphenoxy)phenyl]-adamantane dianhydride	227	[43]
19	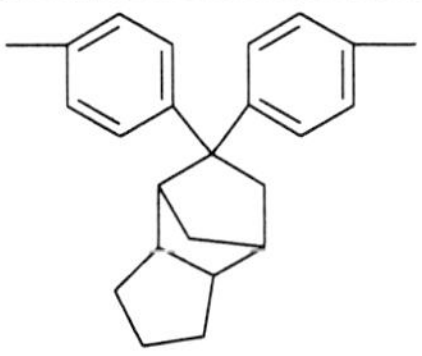	4,4′-[(octahydro-4,7-methano-5H-inden-5-ylidene)bis(1,4-phenylene)-dioxy]diphthalic dianhydride	162-165	[45]
20	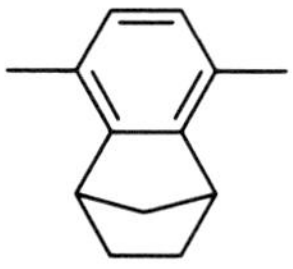	3,6-bis(3,4-dicarboxyphenoxy)-benzonorbornane dianhydride	212-215	[46]
21	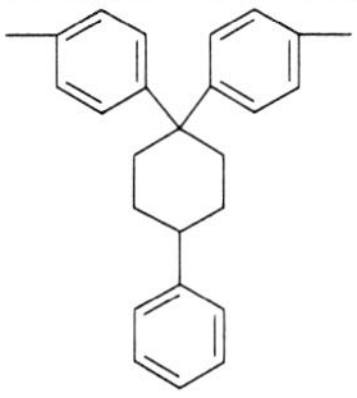	1,1-bis{4-[4-(3,4-dicarboxyphenoxy)phenyl]-4-phenylcyclohexane} dianhydride	217-218	[47]
22	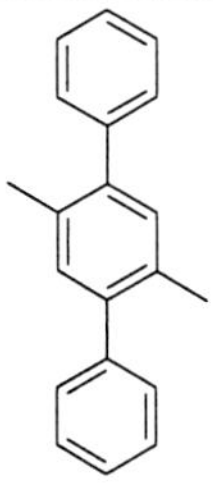	2′,5′-bis(3,4-dicarboxyphenoxy)-*p*-terphenyl dianhydride	305-308	[48]

Crt. No.	Ar	Bis(ether anhydride)	Melting point (°C)	Ref.
23	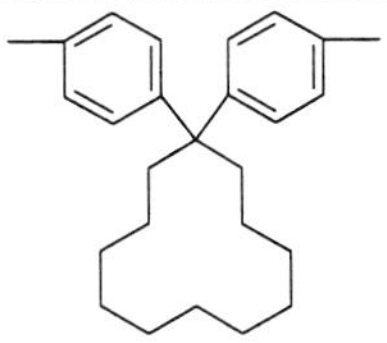	1,1-bis[4-(3,4-dicarboxyphenoxy)phenyl]-cyclododecane dianhydride	212-213	[49]
24	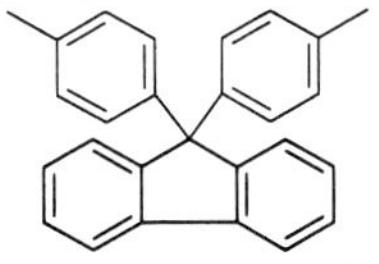	9,9-bis[4-(3,4-dicarboxyphenoxy)phenyl]-fluorene dianhydride	252-254	[51]
25	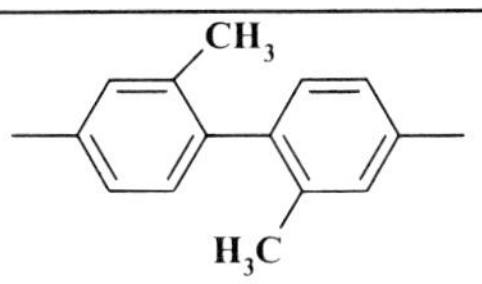	2,2′-dimethyl-4,4′-bis[4-(3,4-dicarboxyphenoxy)]-biphenyl dianhydride	217-218	[53]
26	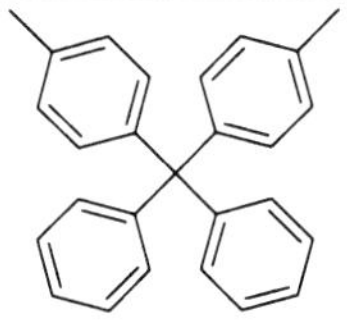	bis[4-(3,4-dicarboxyphenoxy)phenyl]-diphenylmethane dianhydride	262	[53]
27	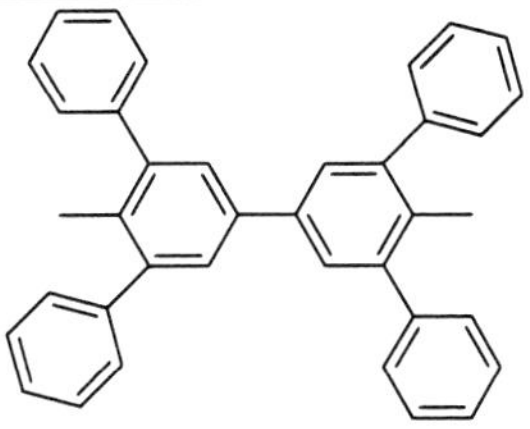	4,4′-bis(3,4-dicarboxyphenoxy)-3,3′,5,5′-tetraphenyl-1,1′-biphenyl dianhydride	276-279	[54]
28	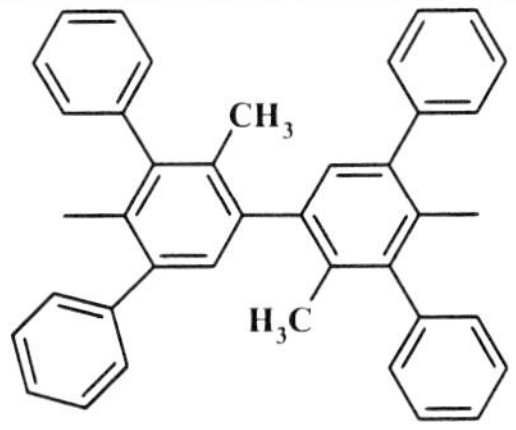	4,4′-bis(3,4-dicarboxyphenoxy)-2,2′-dimethyl-3,3′,5,5′-tetraphenyl-1,1′-biphenyl dianhydride	275-278	[54]
29	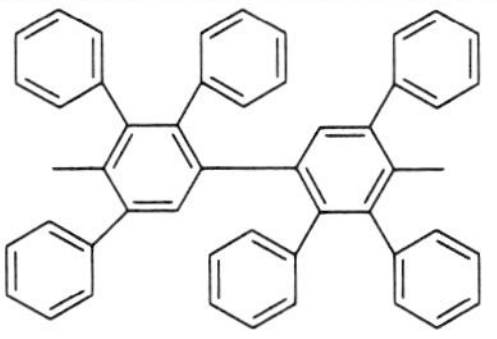	4,4′- bis(3,4-dicarboxyphenoxy)-2,2′,3,3′,5,5′-hexaphenyl - 1,1′-biphenyl dianhydride	305-307	[54]
30	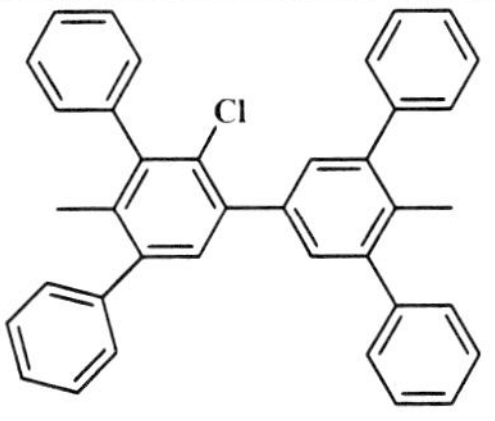	4,4′- bis(3,4-dicarboxyphenoxy)-2-chloro-3,3′,5,5′-tetraphenyl-1,1′-biphenyl dianhydride	232-236	[54]

Table 1. Continued.

Crt. No.	Ar	Bis(ether anhydride)	Melting point (°C)	Ref.
31	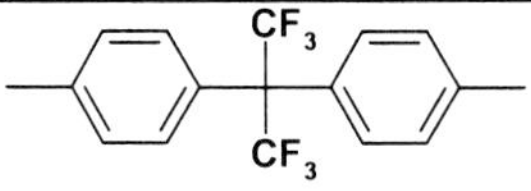	2,2-bis[4-(3,4-dicarboxyphenoxy)phenyl]-hexafluoropropane dianhydride	226-228	[60]
32	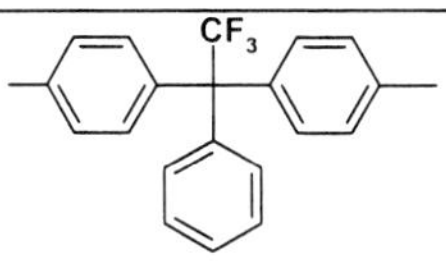	1,1′-bis[4-(3,4-dicarboxyphenoxy)phenyl]-1-phenyl-2,2,2-trifluoroethane dianhydride	98-107	[61]
33	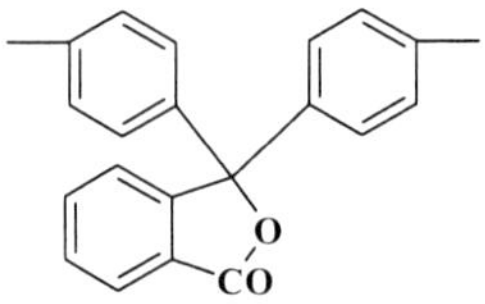	3,3-bis[4-(3,4-dicarboxyphenoxy)phenyl]-phthalide dianhydride	250-251	[62]
34	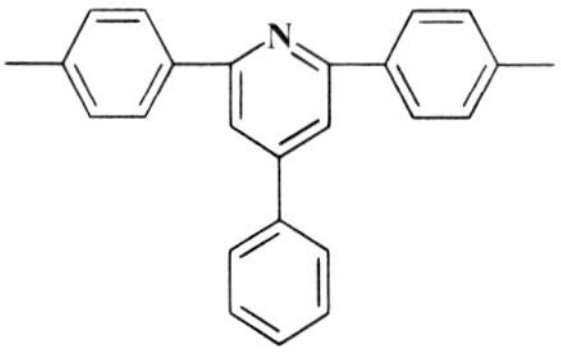	4-phenyl-2,6-bis[4-(3,4-dicarboxyphenoxy)phenyl]-pyridine dyanhydride	282	[97]

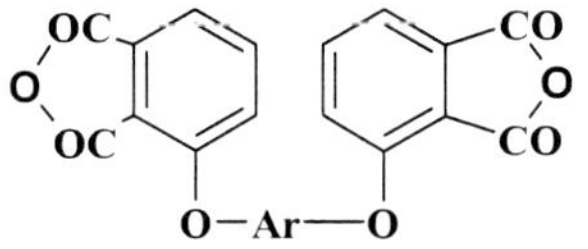

Crt. No.	Ar	Bis(ether anhydride)	Melting point (°C)	Ref.
35	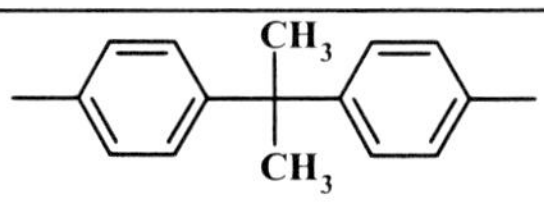	2,2-bis[4-(2,3-dicarboxyphenoxy)phenyl]-propane	187.8-189.2	[37]
36	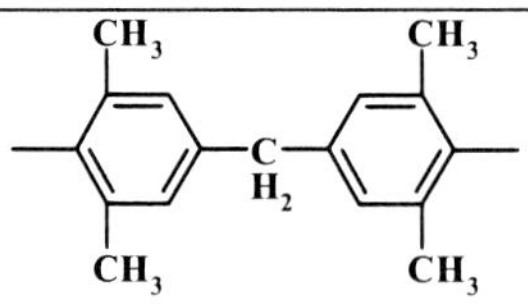	3,3′,5,5′-tetramethyl-4,4′-bis(2, 3-dicarboxyphenoxy)-diphenylmethane dianhydride	245.7-247.5	[37]
37	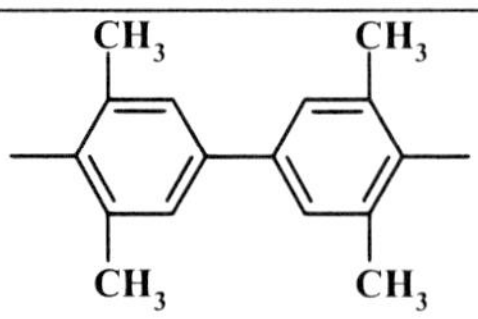	3,3′,5,5′-tetramethyl-4,4′-bis(2, 3-dicarboxyphenoxy)-biphenyl dianhydride	317.2-319.5	[37]
38	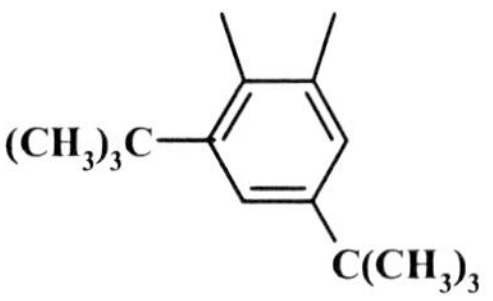	1,2-bis(2,3-dicarboxyphenoxy)-3,5-di-*tert*-butylbenzene dianhydride	208.7-215.0	[37]

Crt. No.	Ar	Bis(ether anhydride)	Melting point (°C)	Ref.
39	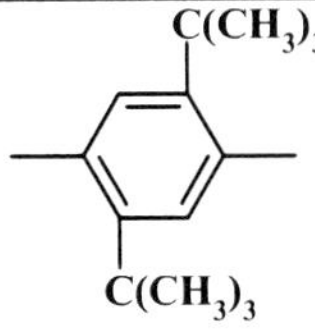	1,4-bis(2,3-dicarboxyphenoxy)-2,5-di-*tert*-butylbenzene dianhydride	322	[37]
40	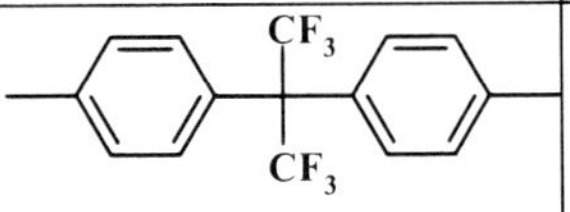	2,2-bis[4-(2,3-dicarboxyphenoxy)phenyl]-hexafluoropropane dianhydride	208-209	[37]

2.2. Synthesis of Poly(Ether Imide)s by Nucleophilic Aromatic Substitution Polymerization Technique

Poly(ether imide)s **9** were synthesized by nucleophilic aromatic substitution polymerization of bis(halogenoimide)s [15-17] or bis(nitroimide)s [18, 19] with bisphenol salts in polar aprotic solvents (Scheme 4).

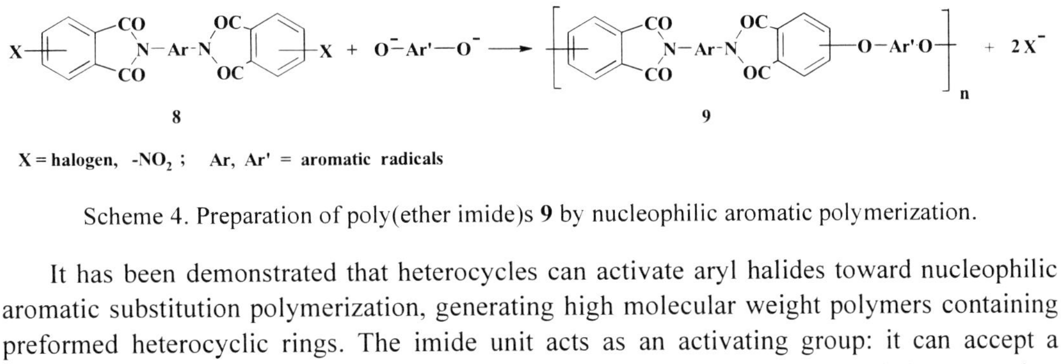

8 **9**

X = halogen, -NO_2 ; Ar, Ar' = aromatic radicals

Scheme 4. Preparation of poly(ether imide)s **9** by nucleophilic aromatic polymerization.

It has been demonstrated that heterocycles can activate aryl halides toward nucleophilic aromatic substitution polymerization, generating high molecular weight polymers containing preformed heterocyclic rings. The imide unit acts as an activating group: it can accept a negative charge and lowers the activating energy for the displacement of the halogen or nitro groups through an expected Meisenheimer complex. The nitroimides are of one order of magnitude more reactive than the corresponding fluoro derivatives and can react at room temperature in dipolar aprotic solvents [20]. Poly(ether imide)s derived from diphenols such as 4,4′-isopropylidenediphenol (bisphenol A) or 4,4′-dihydroxydiphenylether showed high molecular weight, while the polymers derived from mononuclear diphenols like hydroquinone or resorcinol exhibited low molecular weight.

Other poly(ether imide)s have been synthesized by polycondensation reaction of disodium bisphenol A with bis(chlorophthalimide)s in *o*-dichlorobenzene using a microwave oven. These reactions proceeded rapidly and were completed within 25 min by comparing with conventional heating polycondensation. The polymers having inherent viscosity of 0.55-0.92 dL/g were obtained [21].

3. Properties of Poly(Ether Imide)s

3.1. Poly(Ether Imide)s Based on Bis(Ether Anhydride)s Derived From Dihydroxybenzene and Dihydroxynaphthalene

Poly(ether imide)s containing methyl, chloro or phenyl substituents **10** were prepared by thermal imidization up to 270°C of the poly(amic acid)s resulting from the polycondensation reaction of different aromatic diamines with bis(ether anhydride)s, such as 4,4′-(1,4-phenylenedioxy)diphthalic anhydride, 4,4′-(2,5-tolylenedioxy)diphthalic anhydride, 4,4′-(2-chloro-1,4-phenylenedioxy)diphthalic anhydride and 4,4′-(2,5-biphenylenedioxy)diphthalic anhydride [11]. Aromatic bis(ether anhydride)s were synthesized in three steps starting from the reaction of 4-nitrophthalonitrile with potassium phenoxides of hydroquinone, methylhydroquinone, chlorohydroquinone and phenylhydroquinone, followed by alkaline hydrolysis and dehydration. Most of the polymers were soluble in polar aprotic solvents like NMP, dimethyl sulfoxide (DMSO) or *m*-cresol. The introduction of the chloro or phenyl substituents leads to a decreased crystallinity and an increase of the solubility of the polymers. The incorporation of methyl groups did not increase the solubility to a great extend and had low influence on the crystallinity. The polymers exhibited high thermal stability; thermogravimetric analysis showed that the temperature of 10% weight loss (T_{10}) of the polymers was above 500°C. The T_g of the polymers was in the range of 204-263°C. The methyl and chloro groups slightly reduced the thermostability of the polymers and increased the T_g, while the phenyl substituent did not significantly influence the thermal stability but decreased T_g.

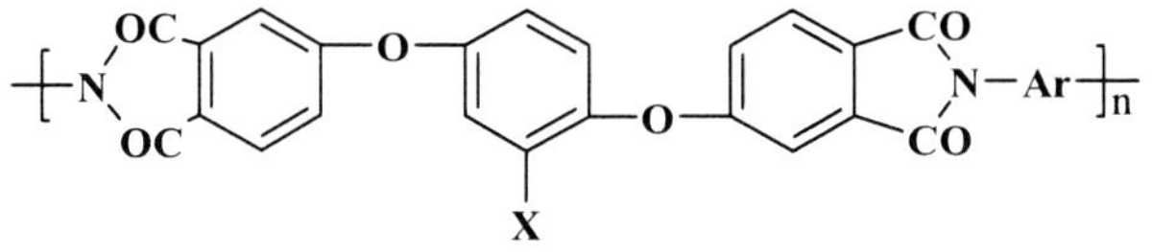

10

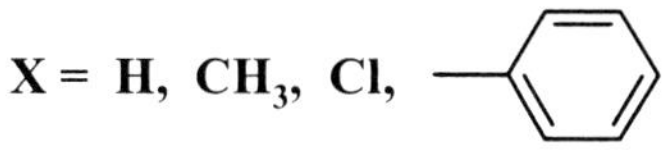

Poly(ether imide)s having pendent *tert*-butyl groups **11** were obtained by polycondensation reaction of 1,4-bis(3,4-dicarboxyphenoxy)-2-*tert*-butylbenzene dianhydride and various aromatic diamines [14]. The polymers were easily soluble in polar aprotic solvents, such as NMP, DMAc, DMF, or in phenolic solvents like *m*-cresol, and even in halogenated solvents like dichloromethane. The good solubility can be explained by the presence of *tert*-butyl group which decreases polymer inter- and intra-chain interactions and disturbs the molecular packing of the chains. All the polymers showed high thermal stability, with T_{10} in the range of 469-547°C in nitrogen and 462-541°C in air. The T_g was in the range of 217-278°C. The polymers keep high thermooxidative stability because the *tert*-butyl group on the aromatic ring is highly resistant to oxidation due to the lack of benzylic hydrogens. The polymer films exhibited high transparency and good mechanical properties. The light color of the polymers can be explained by the decreased intermolecular interactions due to low polarity of the *tert*-butyl groups. The bulky *tert*-butyl groups prevent the charge transfer complex formation between polymer chains through steric hindrance.

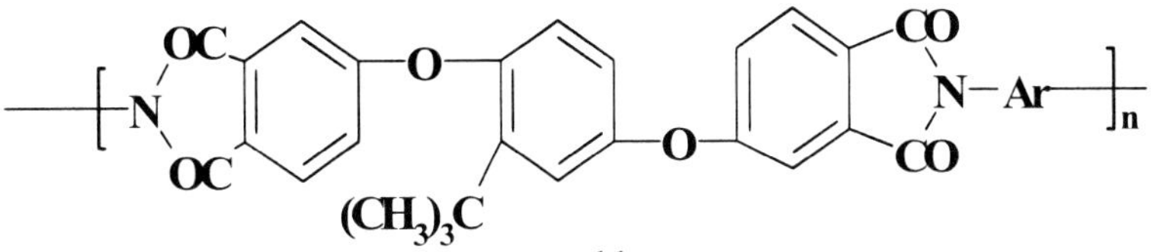

11

Poly(ether imide)s **12** exhibiting light color and good physical properties were obtained from 1,4-bis(3,4-dicarboxyphenoxy)-2,5-di-*tert*-butylbenzene dianhydride and certain aromatic diamines [22]. Most of the polymers showed excellent solubility in chlorinated solvents such as dichloromethane, chloroform and m-cresol. They were not easily soluble in amide-type polar solvents. The solubility of these polymers containing two *tert*-butyl groups in the unit structure was not better than that of poly(ether imide)s containing one *tert*-butyl group **11**, probably due to the molecular symmetry which increased close packing and thus reducing the solubility. The polymers could be cast into light-colored films, with good thermal stability and mechanical properties. The T_g of the polymers was in the range of 232-285°C and T_{10} occurred above 511°C in nitrogen and 487°C in air. The polymer films showed low dielectric constant of 2.71-3.54 at 1 MHz, and low moisture absorption of 0.18-0.66%.

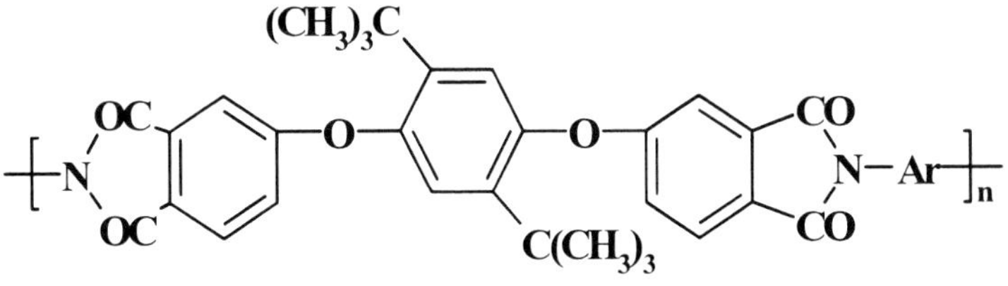

12

Two series of aromatic poly(ether imide)s **13** were synthesized from 1,4-bis(3,4-dicarboxyphenoxy)benzene dianhydride and various diamines, and their gas permeability coefficients to H_2, O_2 and N_2 were determined. In the case of polymers **13-I** based on *m*-phenylenediamine and methyl-substituted *m*-phenylenediamines, the increase of permeability was accompanied by a decrease of permselectivity [23]. Also, the poly(ether imide)s **13-I** prepared from 3,5-diaminobenzoic esters had lower permselectivity than the others, while the poly(ether imide) derived from 3,5-diaminobenzoic acid had much higher permselectivity due to cross-linking.

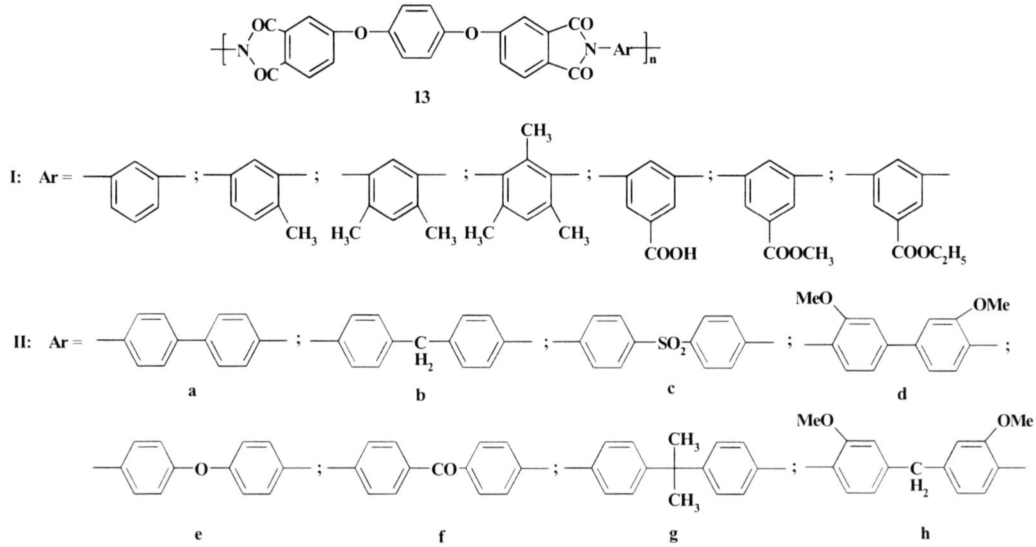

Polymer membranes having good gas separation properties were prepared from poly(ether imide)s **13-II** [24]. The gas permeability coefficients of these polymer membranes to H_2, CO_2, O_2, N_2 and CH_4 have been measured at pressure of 7 atm and in the temperature range of 30-150°C. The order of decrease of the permeability coefficients, in the case of polymers **13-II**, was the following:

$$\mathbf{g > c > b > e > f > a}$$

The permeability and permselectivity were mainly controlled by the fractional free volume of the polymers. The polymers obtained from diamines without side-groups on the phenyl rings showed an increase of the permeability and a decrease of permselectivity by increasing the fractional free volume. The polymer membranes **13-IId** and **13-IIh** containing methoxy side-groups on the aromatic rings exhibited high permeability coefficients to H_2 and high permselectivity coefficients for the gas pairs H_2/CH_4 and H_2/N_2. All the polymers had excellent thermal stability, chemical resistance and good mechanical properties.

Hydroxy-containing poly(ether imide)s **14** were prepared by solution polycondensation reactions of bis(ether anhydride)s, such as 1,2-bis(3,4-dicarboxyphenoxy)benzene dianhydride and 1,2-bis(3,4-dicarboxyphenoxy)-4-*tert*-butylbenzene dianhydride, with bis(*o*-aminophenol)s like 3,3′-dihydroxybenzidine, 3,3′-diamino-4,4′-dihydroxybiphenyl and 2,2-bis(3-amino-4-hydroxyphenyl)-hexafluoropropane [25]. The polymers were soluble in organic solvents, and showed good processability into transparent and flexible films by solution casting. They were amorphous and thermally stable up to 400°C, with T_g in the range of 241-277°C. Due to the presence of hydroxy groups in *ortho* position toward the imide nitrogens, these polymers exhibited thermal conversion to poly(ether benzoxazole)s by heating between 400°C and 500°C, under nitrogen. The pendent hydroxy groups along the backbones may be also useful as reactive sites for subsequent condensation reactions such as the covalent bonding of nonlinear optical chromophores [26].

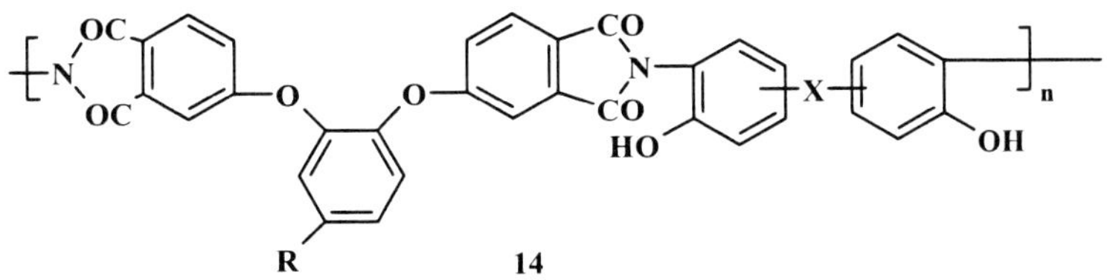

14

R = H, —C(CH$_3$)$_3$; X = — , 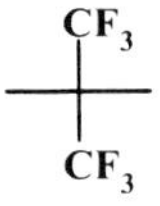

Aromatic poly(ether imide)s, homo- and co-polymers, were obtained by direct polycondensation reaction of 3- or/and 4-aminophenoxy phthalic acid in molten benzoic acid [27]. The homopolymer derived from 3-aminophenoxy phthalic acid was amorphous and soluble in organic solvents, while the homopolymer derived from 4-aminophenoxy phthalic acid was insoluble and intractable. The copolymers showed different properties, depending on their composition. Thus, at a concentration higher than 30% of the 3-aminophenoxy phthalic acid monomer, the copolymers were amorphous and soluble. At a concentration lower than 30% of 3-aminophenoxy phthalic acid, the copolymers were insoluble and intractable.

Amorphous poly(ether imide)s **15** based on aromatic diamines and bis(ether anhydride)s derived from catechol or 2,3-dihydroxynaphthalene exhibited good solubility and excellent thermal stability [28]. The presence of *ortho*-catenated aromatic units and ether linkages into the macromolecular chains of the polymers improved their solubility. The polymers were soluble in polar aprotic solvents, such as NMP, DMF, and in phenolic solvents like *m*-cresol. They showed good ability to form flexible thin films having good mechanical properties with tensile strength, elongation at break, and initial modulus of 47-108 MPa, 3-63%, and 1.3-1.9 GPa, respectively. The polymers had decomposition temperature above 500°C, and a T_g in the range of 193-224°C. The poly(ether imide)s containing naphthalene groups exhibited slightly higher T_g than their catechol-based analogues. This is due to the bulky 2,3-disubstituted naphthalene unit which decreased the chain flexibility.

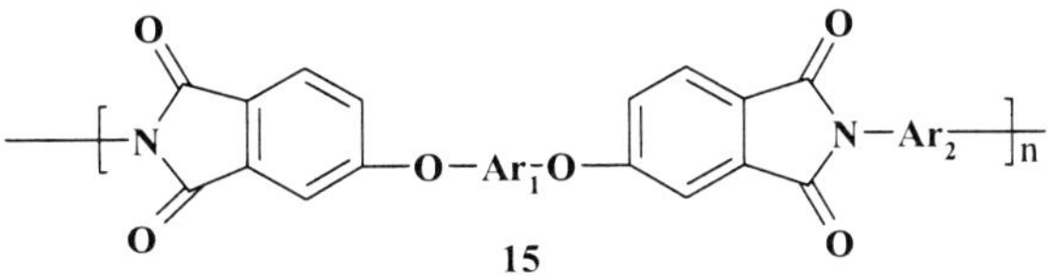

15

Ar$_1$ = 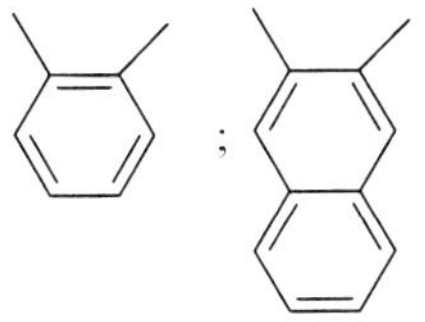 **Ar$_2$ = aromatic radical**

A poly(ether imide) with four ortho-catenated phenylene rings per unit structure has been synthesized starting from 1,2-bis(2′-aminophenoxy)benzene and 1,2-bis(3,4-dicarboxyphenoxy)benzene [29]. The polymer had a large proportion of oligomers. It was found that the oligomers contained macrocyclic compounds with 2 to 10 repeat units. The large sequence of *ortho*-catenated phenylene rings favors the macrocycle formation in competion with propagation.

Poly(ether imide)s **16** have been prepared by polycondensation reaction of 2,7-bis(3,4-dicarboxyphenoxy)naphthalene dianhydride with certain aromatic diamines [8]. The introduction of 2,7-diphenoxynaphthalene units into the polymer backbone did not lead to a significant increase of solubility. The polymers exhibited high thermal stability and moderate T_g, depending on the structure of the diamine monomer. They showed good film forming properties. The polymer films were amorphous and presented good mechanical properties with tensile strength of 89-136 MPa, elongation to break of 11-45% and initial modulus of 1.7-2.2 GPa.

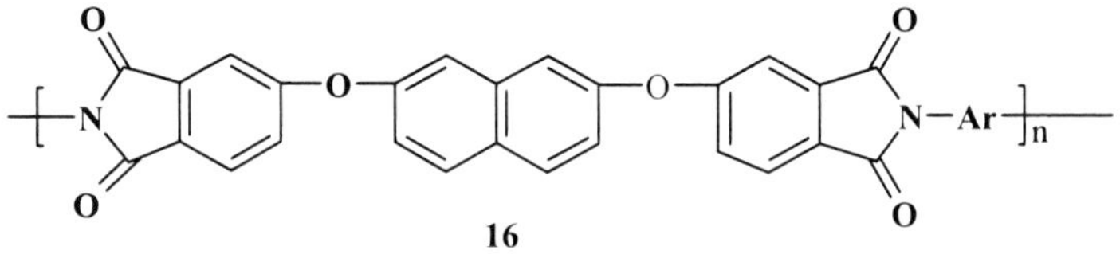

16

Poly(ether imide)s **17** were synthesized by the reaction of 2,2′-bis(3,4-dicarboxyphenoxy)-1,1′-binaphthyl dianhydride with various aromatic diamines [30]. Two synthetic routes were used for the synthesis of the bis(ether anhydride). In the first method N-phenyl-4-chlorophthalimide or N-methyl-4-nitrophthalimide reacted with 1,1′-bi-2-naphthol to obtain a bis(ether imide) which was then hydrolyzed and dehydrated to the corresponding bis(ether anhydride). 4-Nitrophthalonitrile was used instead of 4-substituted phthalimide in the second method. This second route proves to be superior to the first one because of the low reaction temperature and the high yield. The most of the polymers were soluble in organic solvents such as NMP, DMAc, pyridine or dichloromethane. The good solubility can be attributed to the presence of flexible ether linkages and non-coplanarity in 2,2′-substituted-1,1′-binaphthyl units. The compound derived from *p*-phenylenediamine was partial crystalline, while the other polymers were amorphous. The polymers exhibited high T_g values, and 5% weight loss temperatures in the range of 502-541°C in nitrogen and 473-537°C in air.

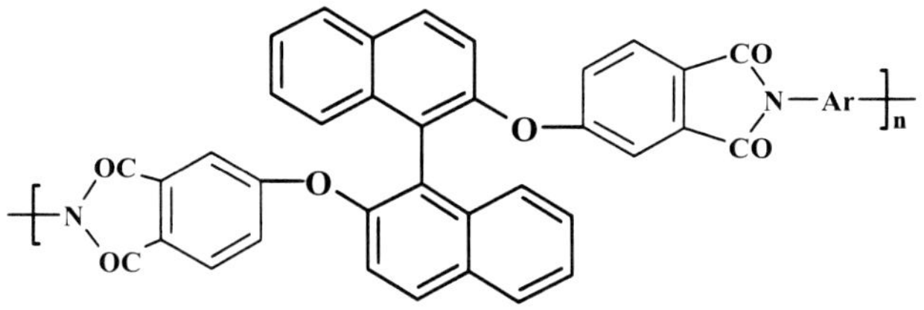

17

Aromatic poly(ether imide)s having high molecular weight were obtained from 2,6-bis(3,4-dicarboxyphenoxy)naphthalene dianhydride and different aromatic diamines [31]. The polymers were stable up to 500°C and showed T_g values in the range of 220-277°C. The films of the polymers derived from the rigid diamines, such as *p*-phenylenediamine and benzidine, crystallized during the thermal imidization process. The other polymers were amorphous.

3.2. Poly(Ether-Imide)s Containing Flexible Units

Poly(ether imide)s are an important class of high performance polymers which can be used in a number of diverse applications. The introduction of flexible groups, pendent bulky units, or both to polymer chains has been a general approach for changing the chemical structure of poly(ether imide)s. The incorporation of flexible linkages such as isopropylidene into the polymer backbone is expected to reduce the crystallinity and enhance the solubility and melt moldability of the poly(ether imide)s.

The most important member of this class of polymers is known as Ultem, a poly(ether imide) manufactured by the General Electric Company, derived from bisphenol A and *m*-phenylenediamine [32]. It is an amorphous polymer being soluble in DMAc and dichloromethane. This polymer is unharmed by hydrocarbons such as oil and gasoline, and shows high thermal stability, with a T_g of 215°C and a temperature of 5% weight loss of 528°C in air and 534°C in nitrogen atmosphere. At equilibrium, it absorbs 1.2 % of water, the yield stress of the wet polymer decreasing with 30% by comparing with the dry polymer at room temperature. A significant decrease in tensile strength takes place on heating, due to the lack of crystallinity and low T_g. Thus, from 25°C to 82°C, tensile strength reduced by half for dry polymer and by one-third for wet polymer.

A series of poly(ether imide)s **18** was synthesized from bis(ether anhydride)s containing flexible groups like ether, methylene, isopropylidene or rigid diphenyl unit, and aromatic diamines having flexible units [12]. The resulting poly(amic acid)s from the reaction of bis(ether anhydride)s with diamines were cyclodehydrated by chemical imidization in solution, using an excess of acetic anhydride mixed with an equimolar amount of pyridine. Most of the polymers were amorphous and exhibited high molecular weight. The T_g of the polymers was in the range of 200-300°C. They were soluble in organic solvents and showed good film forming ability. The permeability of the polymer membranes to CO_2 and CH_4 was investigated. Thus, due to structural units that hinder segmental rotation and reduce possibilities for conformational changes, the polymers exhibited high permeability which is associated with high T_g. This effect is more pronounced when the hindered moiety is rigid. Polymers having flexible units that do not hinder conformational changes showed relatively low T_g values and low permeabilities due to the more efficient chain packing of the flexible chains.

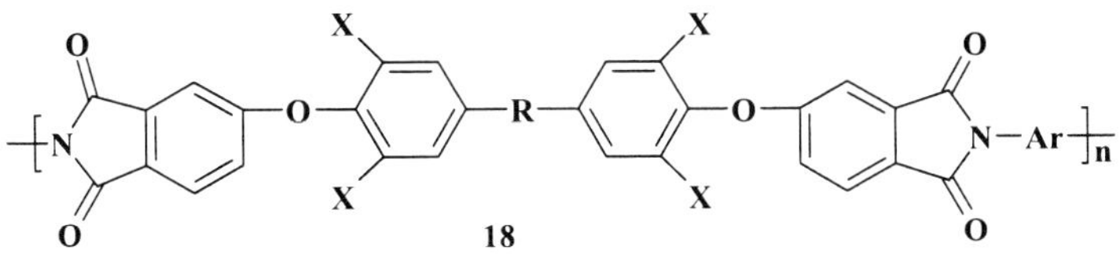

18

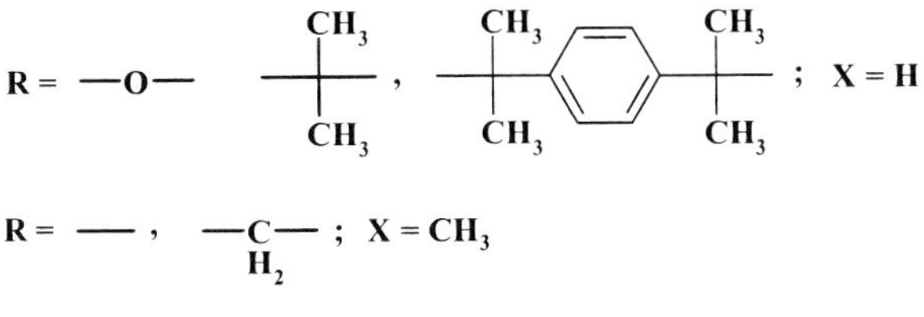

Poly(ether imide)s can be chemically modified to get special properties for specific applications. Among the chemical modification methods, atom transfer radical polymerization (ATRP) has been successfully applied. A poly(ether imide) derived from *m*-phenylene diamine and 2,2-bis[4-(3,4-dicarboxyphenoxy)phenyl]propane dianhydride was functionalized by chloromethylation and the resulting polymer was cast into microporous membrane by the phase inversion technique [33]. Amphiphilic poly(poly(ethylene glycol) methyl ether methacrylate)-block-poly(2,2,2-trifluoroethyl methacrylate) copolymer brushes were incorporated via surface initiated ATRP on the poly(ether imide) microporous membrane.

Phenylethynyl-terminated ether-imide oligomers were prepared by the reaction of 2,2′-bis[4-(3,4-dicarboxyphenoxy)phenyl]propane dianhydride with a stoichiometric excess of either *para, meta*, or isomeric mixtures of phenylenediamine and phenylethynylphthalic anhydride endcapper [34]. Thermal curing of the oligomers produced insoluble, solvent-resistant ductile films with desirable T_g values and excellent thermooxidative stability.

Aromatic poly(ether imide)s were obtained starting from the reaction of 3,3′,5,5′-tetrabromo-2,2-bis[4-(3,4-dicarboxyphenoxy)phenyl]propane dianhydride with various diamines [35]. The polymers were amorphous and soluble in polar organic solvents like NMP, DMAc, tetrahydrofuran (THF) and chloroform. They exhibited high T_g in the temperature range of 256-303°C, and high thermal stability having T_{10} above 470°C in nitrogen.

Highly soluble poly(ether imide)s with tetramethyl and isopropylidene groups **19** were synthesized by polycondensation reaction of 3,3′,5,5′-tetramethyl-2,2-bis[4-(3,4-dicarboxyphenoxy)-phenyl]propane dianhydride with different diamines [36]. The polymers were amorphous and showed excellent solubility in organic solvents. The tetramethyl substituents on the phenylene units hindered the dense packing of the polymer chains and reduced the intermolecular forces between the polymer chains causing a decrease in crystallinity. The polymers exhibited good thermal properties having T_g in the range of 244-282°C and T_{10} above 463°C, in nitrogen. Flexible and tough films with good mechanical properties were obtained through solution casting from DMAc.

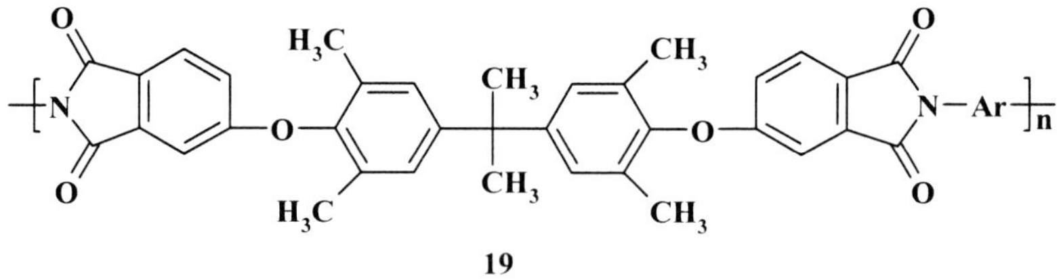

19

Partially aliphatic high molecular weight poly(ether imide)s **20**, containing aliphatic segments or dicyclohexylmethane groups coming from the diamine monomer, have been prepared using an ester-acid high-temperature solution imidization method [13]. In this method, the aliphatic or cycloaliphatic diamine was added to the ester-acid intermediate resulting from the reaction of bis[4-(3,4-dicarboxyphenoxy)phenyl]propane dianhydride with ethanol in the presence of a tertiary amine catalyst. This synthetic procedure eliminated the need for anhydrous solvents and overcame the problem of salt formation observed in the case of the traditional two-step poly(amic acid) route. By increasing the aliphatic content into the macromolecular chains of aromatic poly(ether imide)s the properties like T_g, thermal stability, refractive index and dielectric constant decreased, while the coefficient of thermal expansion increased.

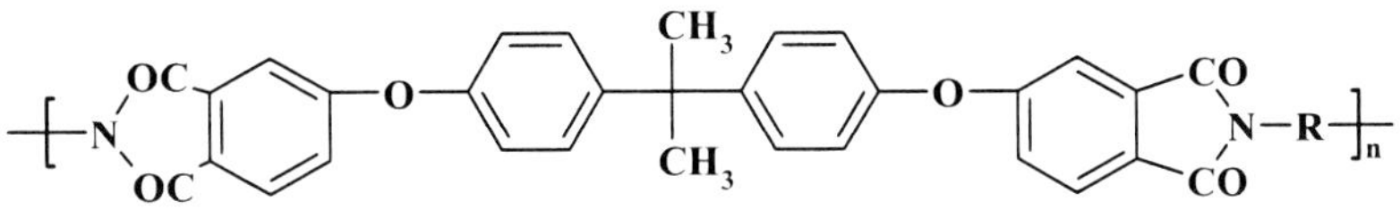

20

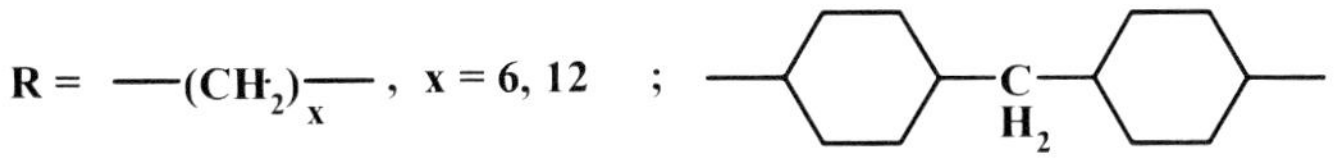

Poly(ether-imide)s incorporating 3- or 4-phthalimide groups **21** have been obtained by solution polycondensation of bis(ether anhydride)s having 3- or 4-phthalic anhydride units with a variety of aromatic diamines, followed by chemical imidization of the poly(amic acid)s intermediates [37]. Generally, the bis(ether anhydride)s containing 4-phthalic anhydride units gave high-molecular-weight polymers, while the bis(ether anhydride)s with 3-phthalic anhydride units led to low-molecular-weight products. The investigation of some low-molecular-weight products by electrospray-ionization-mass spectrometry concluded that the products consisted of small cyclic or linear oligomers depending on the structure of the diamine.

It is known that alkyl side chains attached to rigid polymers such as aromatic polyesters and polyimides decrease their melting points and T_g and increase their solubility. They often induced liquid-crystalline behavior. The attachment of alkyl side chains to aromatic polyimides was used in the development of alignment layers for liquid crystal displays. Poly(ether imide)s based on bis[4-(3,4-dicarboxyphenoxy)phenyl]propane dianhydride and various diamines containing multiple alkyl side chains were also reported [38]. The effect of alkyl side chains on the thermal, mechanical and optical properties of aromatic poly(ether imide)s was studied. The polymers had good solubility in chlorinated solvents and could be solution cast into transparent tough films. Compounds having the side chains with at least 16 methylene units could be crystallized. These polymers with multiple alkyl side chains showed higher elongation but lower tensile strength and module by comparing with the commercial Ultem poly(ether imide). The optical properties were investigated under stress by stretching a polymer under force above its T_g at 85°C, at different drawing ratios. The film's in-plane index parallel to the drawing direction increased and in-plane refractive index perpendicular to the drawing direction decreased during drawing due to the orientation of the main chains in the drawing direction. The film's in-plane birefringence increased with the drawing ratios. The potential applications of these polymers include stretchable, birefringence optical films.

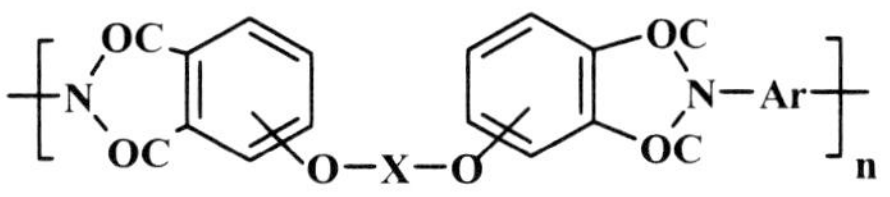

21

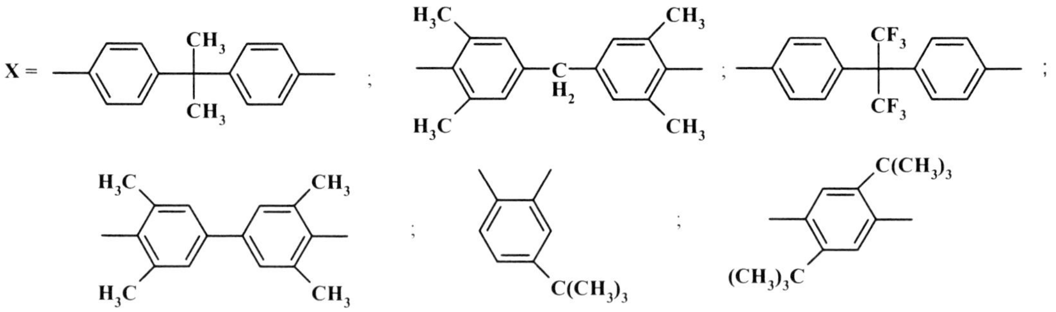

3.3. Poly(Ether Imide)s Containing Bulky Groups

The presence of pendent bulky groups along the polymer backbone has been shown to impart solubility as well as better mechanical and thermal properties to the resulting polymers, that is of particular importance in aromatic heterocyclic polymers with rigid chains. Polymers containing pendent bulky groups are potential candidates for gas separation membranes. They can be processed into membranes having improved selectivity without loss of permeability [39-41].

Poly(ether imide)s **22** having cyclohexylidene cardo groups were prepared from 1,1-bis[4-(3,4-dicarboxyphenoxy)phenyl]cyclohexane dianhydride and different aromatic diamines via a conventional two-stage polycondensation process [42]. The polymers were soluble in organic solvents and were easily processed into flexible films having good tensile properties. They showed high decomposition temperature above 490°C and a T_g of 200-234°C. These polymers exhibited relatively higher glass transition temperatures, better solubility in organic solvents and better mechanical properties by comparing with Ultem poly(ether imide).

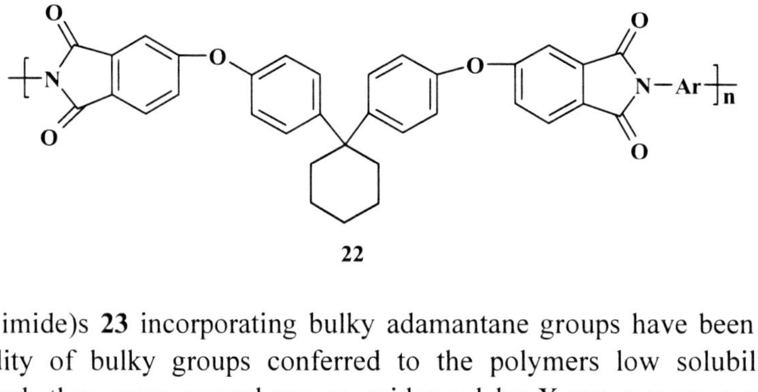

22

Poly(ether imide)s **23** incorporating bulky adamantane groups have been reported [43]. The high rigidity of bulky groups conferred to the polymers low solubility in organic solvents, although they were amorphous as evidenced by X-ray measurements. The T_g of these polymers was very high, in the 242-317°C range, due to the restricted mobility imposed by the presence of bulky adamantane units. Also, the polymers showed high thermal stability having the T_{10} in the range of 514-538°C in nitrogen and 511-527°C in air.

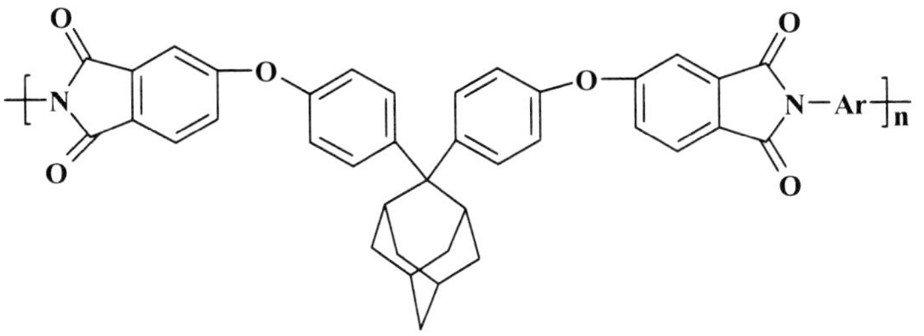

23

Poly(ether imide)s **24** have been synthesized by solution polycondensation of bis(ether anhydride)s containing pendent adamantyl group with aromatic diamines [44]. The intermediate poly(amic acid)s were chemically imidized with acetic anhydride/pyridine (1:1). The bis(ether anhydride)s have been prepared by the nitro displacement reactions between 4-nitrophthalonitrile and aromatic diols, such as 4-(adamantyl-1)-1,2-dihydroxybenzene, 4-(adamantyl-1)-1,3-dihydroxybenzene and 2-(adamantyl-1)-1,4-dihydroxybenzene. The T_g increased in going from 1,2 to 1,3 to 1,4-catenation of the adamantyl-phenylene units, and was in the range of 250-300°C. It was higher than those of similar polymers having methyl or *tert*-butyl substituents and than poly(ether imide)s containing adamantane units in the main chain. The polymers were soluble in aprotic solvents and many of them were soluble even in chloroform.

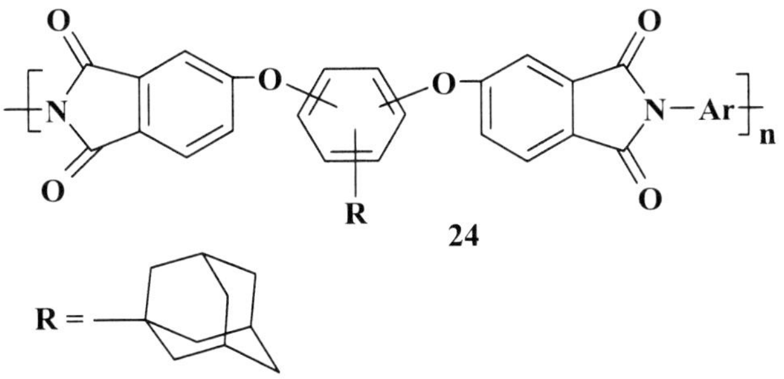

24

A bis(ether anhydride) which contains cardo group was synthesized by the reaction of 4,4′-(octahydro-4,7-methano-5H-inden-5-ylidene)-bisphenol with 4-nitrophthalonitrile [45]. Poly(ether imide)s **25** based on this dianhydride and certain aromatic diamines have been obtained and their physical properties were investigated. The introduction of bulky alicyclic pendant groups into the structure of the polymers increased the solubility without much decrease in the thermal stability. The polymers showed high solubility at room temperature in polar aprotic solvents such as DMAc or NMP, and even in less polar solvents like dichloromethane or THF. The transparent, flexible and tough films were obtained by casting the polymer solutions.

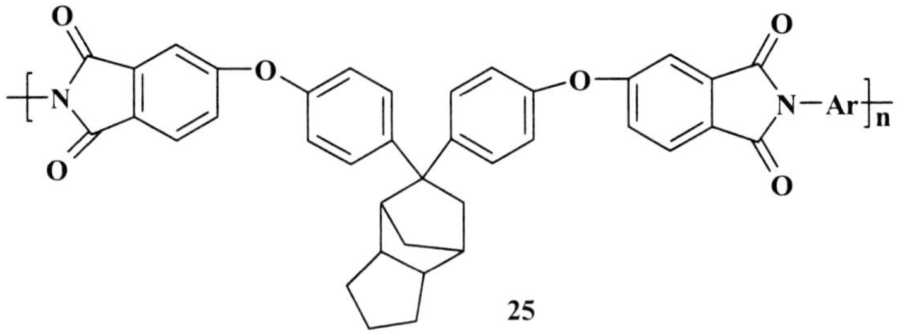

25

Poly(ether imide)s **26** were prepared starting from the reaction of 3,6-bis(3,4-dicarboxyphenoxy)benzonorbornane dianhydride with various diamines [46]. The polymers were completely amorphous but their solubility in organic solvents was limited. They were thermally stable up to 500°C, and exhibited good mechanical properties with initial modulus in the range of 1.97-3.08 GPa.

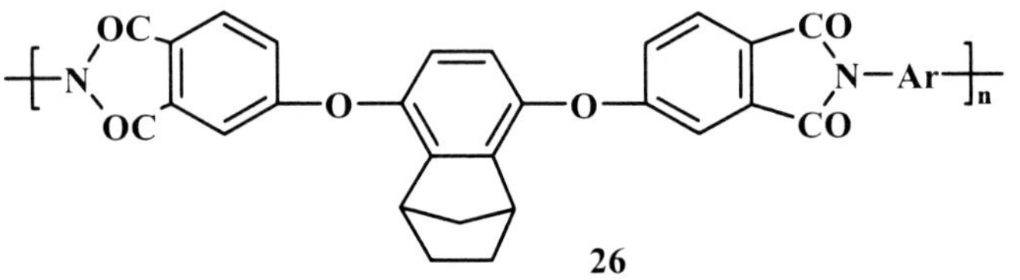

26

Poly(ether imide)s **27** containing 4-phenylcyclohexane pendent groups have been synthesized by a two-step polycondensation process of 1,1-bis[4-(3,4-dicarboxyphenoxy)phenyl]-4-phenylcyclohexane dianhydride and different diamines [47]. The incorporation of pendent phenylcyclohexyl groups into the polymer chain disturbs the close packing of the macromolecules improving the solubility. As a result the polymers were easily soluble in polar solvents such as NMP, DMF, and pyridine, as well as in less polar solvents like cyclohexanone, THF and chloroform. Most of them dissolved in chloroform at a concentration higher than 30% wt. The films made from these polymers had high tensile modulus around 2.0 GPa indicating that they are strong materials. The polymers exhibited high thermal stability showing T_{10} in the range of 458-497°C, and a T_g in the 226-255°C interval. By comparing the properties of poly(ether imide)s containing 4-phenylcyclohexane groups with those of polymers having cyclohexyl groups [42] it was found that the former showed higher T_g values, better solubility, and higher tensile strengths than the latter.

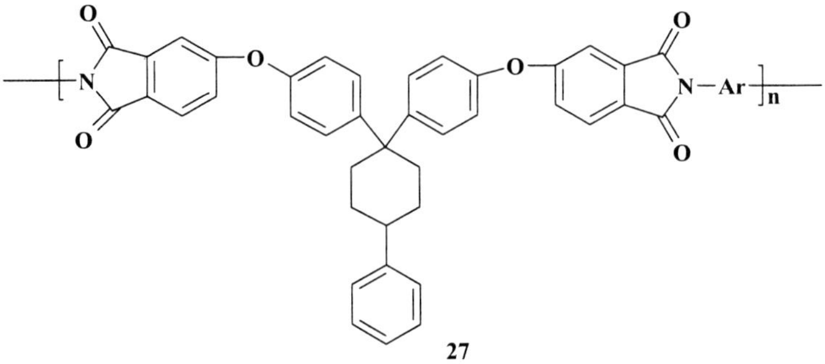

27

Poly(ether imide)s **28** based on 2′,5′-bis(3,4-dicarboxyphenoxy)-*p*-terphenyl dianhydride displayed high solubility in common organic solvents when compared with conventional aromatic polyimides [48]. Good quality, free-standing films could be prepared by casting the polymer solutions. The investigation of mechanical properties concluded that the films are strong and tough polymeric materials with tensile strength of 71-129 MPa, elongation to break of 10-35%, and initial modulus of 1.71-2.17 GPa. The polymers showed high thermal and thermooxidative stability, with decomposition temperature over 500°C and T_g in the 214-276°C range.

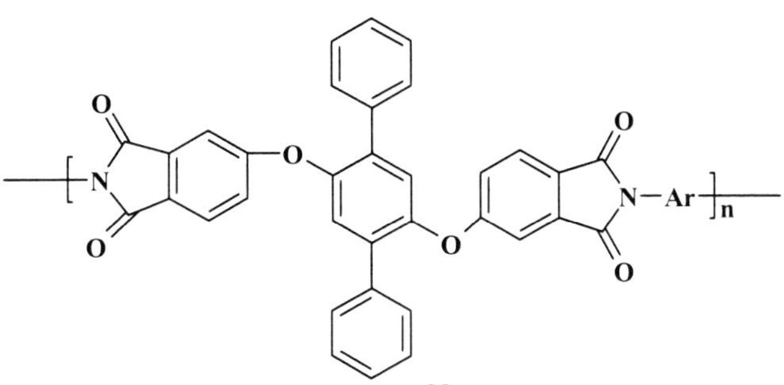

28

Poly(ether imide)s **29** derived from 1,1-bis[4-(3,4-dicarboxyphenoxy)phenyl]-cyclododecane dianhydride and various aromatic diamines were reported [49].

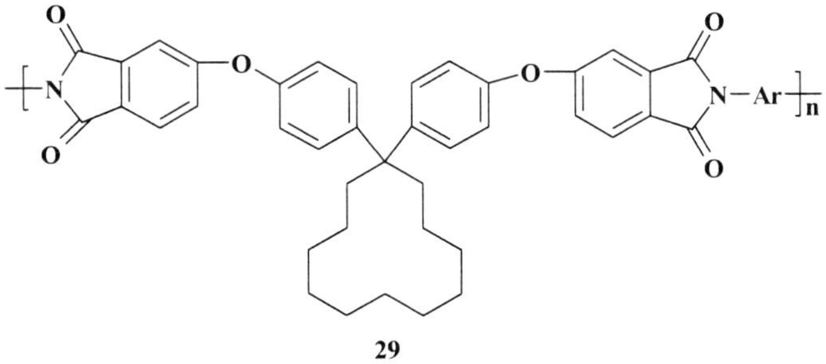

29

The polymers showed excellent solubility in organic solvents like NMP, DMAc, DMF, pyridine and THF. They were amorphous due to the presence of bulky pendent cyclododecane groups that increased the disorder of polymer chains. Transparent and flexible polymer films, having good mechanical properties, were obtained by solution casting from DMAc. The polymers exhibited good thermal properties, the T_g was in the range of 230-262°C and T_{10} above 434°C in nitrogen and 407°C in air. By comparing these poly(ether imide)s with the Ultem poly(ether imide) it was observed that the polymers containing cyclododecane groups have shown better properties such as higher T_g and higher T_{10}.

Cardo poly(ether imide)s having good optical transparency were synthesized from 2,2′-bis(3,4-dicarboxyphenoxy)-9,9′-spirobifluorene dianhydride and certain aromatic diamines [50]. These polymers exhibited remarkable solubility in common organic solvents at room temperature, and showed high thermal stability. The T_g of the polymers was of 240-293°C and 5% weight-loss temperatures were higher than 500°C.

Poly(ether imide)s **30** were prepared from 9,9-bis[4-(3,4-dicarboxyphenoxy)phenyl]fluorene dianhydride and various aromatic diamines via a conventional two-stage process [51, 52]. The polymers were soluble in organic solvents and could be processed into flexible and tough films, except those derived from *p*-phenylenediamine, *m*-phenylenediamine, and benzidine. They exhibited high thermal stability and excellent optical clarity at wavelength greater than 400 nm being suitable for optical applications. The polymers showed relatively high refractive index, probably due to the high concentration of π electrons in their repeating units. It was found that the refractive index values could be controlled by varying the dianhydride and diamine composition, being in the range of 1.60-1.80. Polymers with high refraction indexes have large applications in optics and photonics because of their ability to reduce reflection losses at interfaces. One advantage of using poly(ether imide)s consists in the fact that these polymers can be obtained from solution in completely imidized form, thus avoiding the use of poly(amic acid)s more sensitive to moisture and temperature variations. Also, during the thermal imidization of poly(amic acid)s could appear voids that contribute to light scattering.

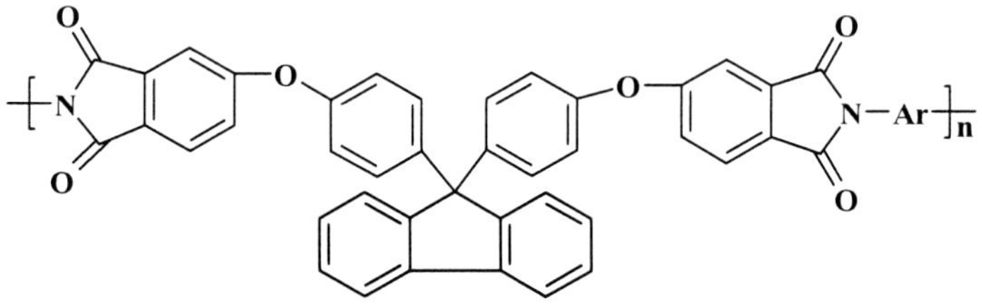

30

Poly(ether imide)s **31** have been obtained from 2,2′-dimethyl-4,4′-bis[4-(3,4-dicarboxyphenoxy)biphenyl dianhydride or bis[4-(3,4-dicarboxyphenoxy)phenyl]-diphenylmethane dianhydride with aromatic diamines [53]. All the polymers were amorphous. The presence of noncoplanar conformation of 2,2′-dimethylsubstituted biphenylene units or bulky phenyl groups on the diphenylmethylene units decreases the molecular packing thus lowering the crystallinity. These polymers were soluble in organic solvents such as NMP, DMAc, pyridine, THF and even in chloroform. They exhibited high thermal stability, having T_{10} above 489°C in nitrogen and a T_g in the range of 224-256°C. The isothermal gravimetric analysis revealed that the polymers showed 7.0-10.5% weight loss after heating at 350°C for 20 h in static air. The polymers with 2,2′-dimethyl-4,4′-biphenylene units possess slightly lower solubility but higher thermal stability and mechanical properties than those with diphenylmethylene linkages.

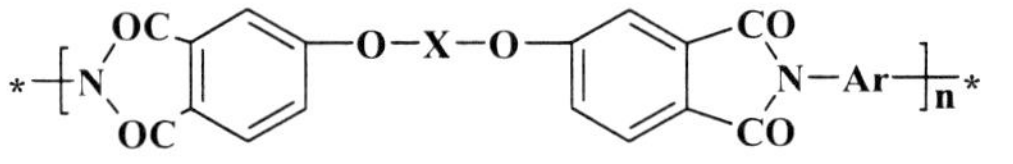

31

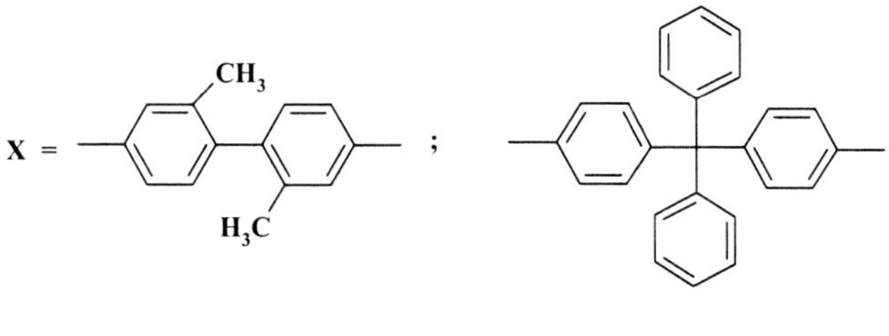

Poly(ether imide)s **32** having rigid phenyl-substituted biphenyl units were synthesized by the reaction of *p*- or *m*-phenylenediamine with several bis(ether anhydride)s, such as 4,4′-bis(3,4-dicarboxyphenoxy)-2,2′-dimethyl-3,3′,5,5′-tetraphenyl-1,1′-biphenyl dianhydride, 4,4′-bis(3,4-dicarboxyphenoxy)-3,3′,5,5′-tetraphenyl-1,1′-biphenyl dianhydride, 4,4′-bis(3,4-dicarboxyphenoxy)-2,2′,3,3′,5,5′-hexaphenyl-1,1′-biphenyl dianhydride and 4,4′-bis(3,4-dicarboxyphenoxy)-2-chloro-1,1′-biphenyl dianhydride [54]. The polymers were soluble in organic solvents, and transparent films could be cast from their chloroform solutions. The bulky phenyl substituents in the 2,2′-position force noncoplanarity between the phenylene rings of the biphenylene moiety to provide a rigid group which increases the free volume of the polymers and disturb the molecular packing thus increasing the solubility. All polymers showed high T_g values in the range of 240-317°C and remarkable thermooxidative stability, the temperature of 5% weight loss being above 500°C in air and nitrogen.

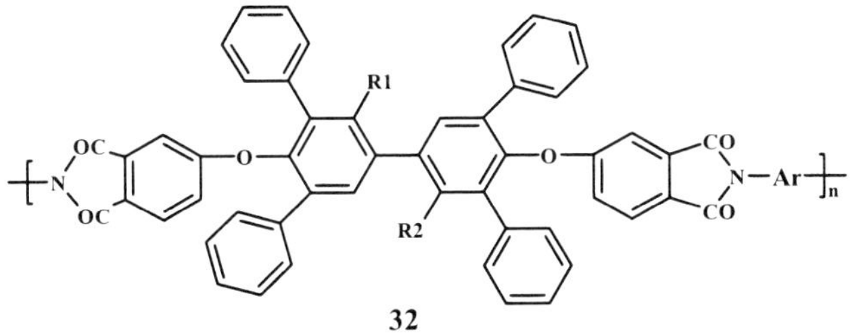

32

R1, R2 = H, CH_3, -C_6H_5, Cl
Ar = *o*-C_6H_4 or *p*-C_6H_4

Recently, asymmetric poly(ether imide)s **33** derived from 2-(3,4-dicarboxyphen-4-yl)-4-(3′,4′-dicarboxyphenoxy-4-phenyl)-2,3-phthalazin-1-one dianhydride and various aromatic diamines were reported [55]. The polycondensation reaction took place in one step, in *m*-cresol. The introduction of asymmetric and noncoplanar phthalazinone units into polymer backbones improved their solubility with the retention of thermostability. All the polymers were soluble in polar organic solvents like *m*-cresol and NMP, and those containing additionally flexible methylene or ether linkages were soluble even in chloroform. Most of them were able to be cast as clear, flexible, and tough films that showed good mechanical properties. The polymers exhibited high thermal stability, with T_g in the range of 289-326°C, and the temperature of 5% weight loss higher than 482°C.

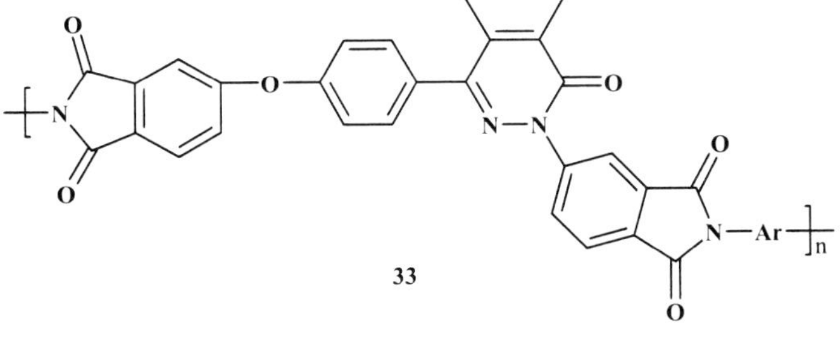

33

3.4. Fluorinated Poly(Ether Imide)s

Fluorinated polyimides are known to have low dielectric constant, low water absorption, high transparency, and high solubility compared to fluorine-free polyimides [56]. The incorporation of hexafluoroisopropylidene groups into polymer backbones enhances the polymer solubility without sacrificing thermal stability. The retention of high thermal stability is attributed to the strong C–F bonds. Other effects of the hexafluoroisopropylidene groups are the increased glass transition temperature and flame resistance with a concomitant decreased crystallinity and water absorption. The bulky hexafluoroisopropylidene groups also serve to increase the free volume of the polymers, thus improving their electrical insulating characteristics [57-59]. Fluorinated poly(ether imide)s **34** derived from 2,2-bis[4-(3,4-dicarboxyphenoxy)phenyl]hexafluoropropane dianhydride and various aromatic diamines were reported [60]. These polymers were soluble in aprotic polar solvents such as NMP, DMF, and DMAc, and even in less polar solvents like pyridine and dichloromethane, and showed good film forming properties. The good solubility can be attributed to the combined favorable effects of hexafluoroisopropylidene groups and aromatic ether linkages in the macromolecular structure. The polymer films exhibited good mechanical and electrical properties. They showed a dielectric constant in the range of 2.8-3.2 at 1 kHz, and a refractive index of 1.54-1.59. A decrease of the dielectric constant, refractive index, and water absorption was noticed by increasing the fluorine content. The polymers were thermally stable, they did not decompose below 560°C in nitrogen.

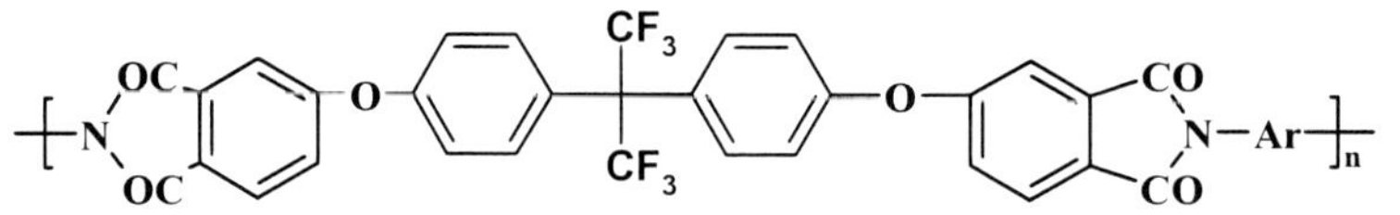

34

Poly(ether imide)s **35**, containing one trifluoromethyl group per structural unit, were prepared by two step polycondensation reaction of 1,1′-bis[4-(3,4-dicarboxyphenoxy)phenyl]-1-phenyl-2,2,2-trifluoroethane dianhydride with various diamines [61]. The polymers showed high molecular weight and inherent viscosities of 0.22-1.29 dL/g. They were soluble in polar solvents such as NMP, DMF, DMAc and dichloromethane. The good solubility of these polymers can be explained by the presence of voluminous phenyl and trifluoromethyl groups coming from the dianhydride segment. The polymers were easily processed into flexible films having good mechanical properties, with tensile modulus of 1.6-2.5 GPa. T_g was in the range of 207-264°C and T_{10} was over 520°C.

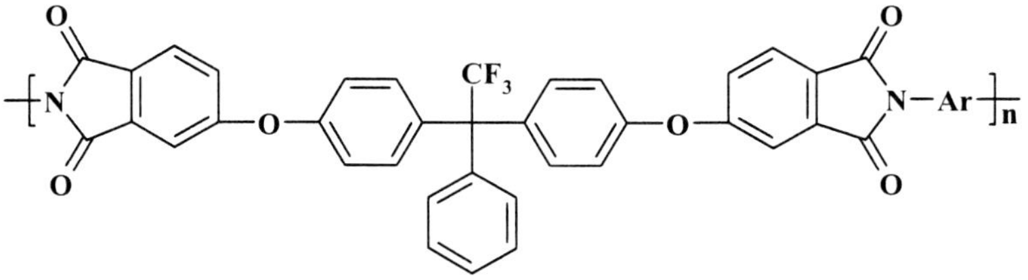

35

Colorless poly(ether imide)s **36** with trifluoromethyl groups coming from the diamine segment were obtained by polycondensation reaction of 3,3-bis[4-(3,4-dicarboxyphenoxy)phenyl]phthalide dianhydride with different fluorinated bis(ether amine)s [62]. The polymers were soluble in polar organic solvents such as DMF and in less polar solvents like dioxane, THF and chloroform. Their solubility could be improved concomitant with the increase of the thermal stability by the introduction of voluminous phthalide groups into their macromolecular chains. Also, the bulk phthalide groups disturb the chains packing reducing the interchain interactions to enhance solubility. On the other hand, it is known that a diamine with trifluoromethyl groups can contribute to the optical transparence of polyimide films, enhances the solubility and thermal stability and reduces the dielectric constant and moisture absorption [14, 63]. Generally, most polyimide films show coloration from light yellow to brown because of intermolecular charge-transfer complex formation or highly conjugated aromatic structures. In addition, the coloration of polyimide films is affected by the imidization method. Thus, a polyimide film prepared by thermal imidization could be colored because of heating and oxidation in air. Chemical imidization can avoid oxidation and coloration [64]. The yellowness index of polymer films **36** ranged from 3.1 to 9.5. These polymers were characterized by high T_g, in the 237-297°C range, and decomposition temperature over 490°C. The char yields were more than 54 wt% at 800°C in nitrogen. The polymer films exhibited good mechanical properties and dielectric constant in the range of 3.05-3.56 at 1 MHz, which were lower than those of the similar nonfluorinated polymers.

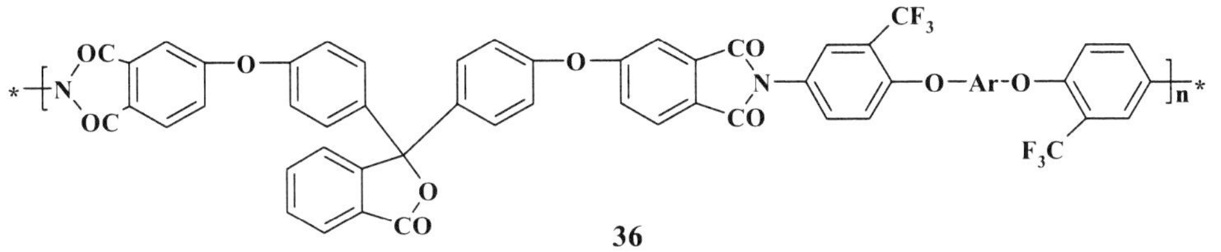

36

Fluorinated poly(ether imide)s **37** containing *ortho*-catenated phenylene rings have been synthesized by the chemical imidization of the intermediate poly(amic acid)s resulting from the polycondensation reaction of 1,2-bis(3,4-dicarboxyphenoxy)benzene dianhydride with certain trifluoromethyl-substituted bis(ether amine)s [65]. The polymers were easily soluble in organic solvents and showed good film forming ability. The polymer films were colorless having low yellowness index of 4.34-6.35. The polymers exhibited T_g in the range of 185-270°C, softening temperatures in the range of 184-275°C and T_{10} higher than 466°C. They showed low moisture absorption of maximum 0.7% and low dielectric constant of 2.78-3.26 at 10 kHz. The strong electronegativity of fluorine atoms results in very low polarizability of C-F bonds, thus decreasing the dielectric constant. Also, the trifluoromethyl groups disturb the chain packing thus increasing the free volume, a factor that produces a decrease of the dielectric constant. The low water uptake values ensure that the polymers have stable dielectric performance. Due to the good combination of properties these polymers can be potential materials for optoelectronic applications.

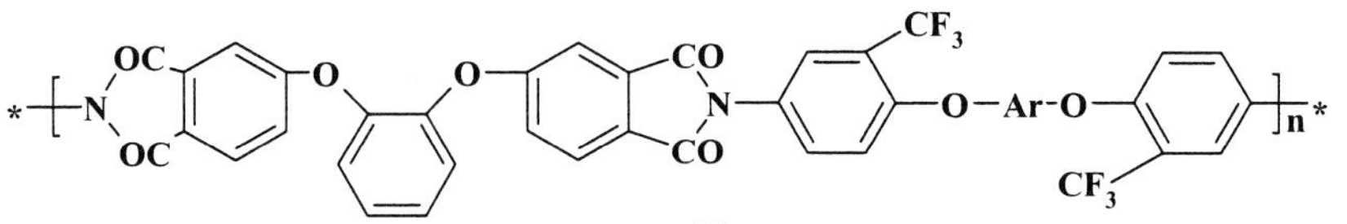

37

Poly(ether imide)s **38** having pendant perfluoroalkyl chains were prepared by the reaction of bis[4-(3,4-dicarboxyphenoxy)phenyl]propane dianhydride with a mixture of *m*-phenylenediamine and 2,3-bis(perfluoroalkylmethyl)butane-1,4-diamine [66]. The perfluorinated pendent chains should bring interesting hydrophobic properties which, combined with hydrophilic properties of the imide groups, lead to interesting polymeric materials. The aliphatic units coming from the segments of the fluorinated diamine are flexible and allow considerable changes in the chains conformation. The polymers showed a T_g in the range of 110-217°C, and their decomposition temperature was higher than 370°C in inert atmosphere and 325°C in air. The good combination of properties and processability may provide an easy preparation of dense non-porous films, either by solution casting or melt processing.

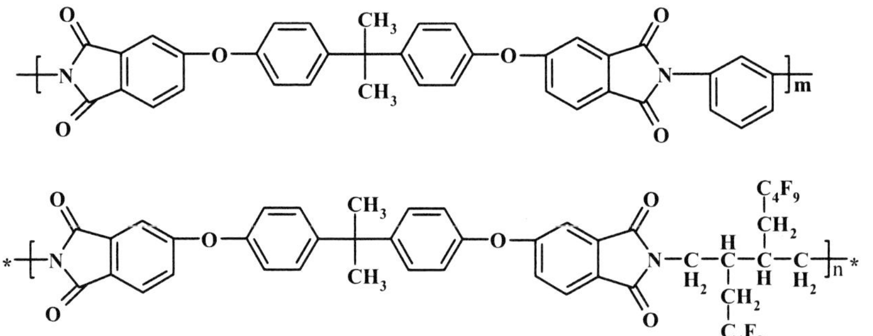

38

Compatible polymer blends with various compositions based on a fluorinated poly(ether imide) synthesized from the bis[4-(3,4-dicarboxyphenoxy)phenyl]propane dianhydride and 2,3-bis(perfluoroalkylmethyl)butane-1,4-diamine and a poly(ether imide) prepared from the same bis(ether anhydride) and *m*-phenylenediamine, were obtained by codissolution in a common solvent [67]. A good correlation was found between T_g of the polymer blends and the content of fluorinated monomer. Thus, the T_g decreased from 217°C to 100°C as the concentration of the segments coming from the fluorinated diamine increased. Irrespective of the concentration of the fluorinated aliphatic segments only one T_g was observed from DSC curves for polymer blends, which is a good indication of the good compatibility and, as a result total miscibility of the two phases. The polymer blend films were obtained by casting the solution of the mixed polymers in NMP or THF. Cast-evaporated films from fluorinated copolyetherimides and blends having a molar concentration of fluorinated monomer units lower than 0.15 gave hydrophobic non-porous membranes that can have potential applications to eliminate organic polluants from water.

3.5. Siloxane-Containing Poly(Ether Imide)s

In general, polysiloxanes have many important properties: oxidative stability, gas permeability, and dielectric behavior [68]. However, their low strength and modulus and low T_g have partly limited their application as high performance materials. On the other hand polyimides are a class of polymers widely used in advanced microelectronics, aerospace, and printed circuit industries, due to their excellent thermal stability, mechanical and electrical properties. But these polymers are processed with great difficulty because many of them are insoluble and infusible. Various efforts have been made on the synthesis of soluble and processable polyimides without much sacrifice of their excellent thermostability. The incorporation of polysiloxane segments into a polyimide macromolecular chain introduces large changes in its properties. Polyimide-polysiloxane copolymers show some useful properties such as high adhesion, ease of processability, good thermooxidative stability, lower surface tension, enhanced chemical and radiation resistance, excellent dielectric properties, etc. They are high performance materials for microelectronic applications, adhesion and gas separation membranes [69, 70]. Generally, these copolymers are synthesized by the polycondensation reaction of a dianhydride with an aromatic diamine and an amine telechelic polysiloxane, by transimidization between an aminopyridine terminated oligoimide and an amine N-capped oligosiloxane or via hydrosilylation reaction between a hydrosilane telechelic H-functionalized polysiloxane and an allyl telechelic polyimide [71, 72]. Several kinds of polysiloxane-polyimide copolymers have been prepared and studied as potential materials for advanced applications [73-77].

A poly(ether-imide)-polydimethylsiloxane copolymer was prepared by polycondensation reaction of a bis(ether anhydride), namely bis-[(3,4-dicarboxyphenoxy)phenyl]-1,4-phenylene-diisopropylidene dianhydride, with an equimolar quantity of 4,4′-(1,4-phenylenediisopropylidene)bisaniline and a bis(aminopropyl)polydimethylsiloxane oligomer of controlled molecular weight [78]. A reference poly(ether imide) without siloxane sequences was prepared by thermal imidization of poly(amic-acid) intermediate resulting from the solution polycondensation reaction of the same bis(ether anhydride) with the aromatic diamine. The polymers were easily soluble in polar organic solvents and showed good thermal stability, decomposition temperature being above 400°C. The introduction of polydimethylsiloxane segments resulted in materials having a dielectric constant of 3.00 at 1 MHz and room temperature, which make them potential candidates for future high performance applications in microelectronics or in related fields.

The development of new sensor materials with improved sensitivity and selectivity is very important for practical applications. It was observed that materials for sensors with high sensitivity toward chlorinated organic vapors included hybrid polymers containing both hard and soft blocks, and among them silicone block polyimides have this property. Thus, poly(ether imide)-polydimethylsiloxane copolymers **39** have been obtained from the reaction of the 2,2-bis[4-(3,4-dicarboxyphenoxy)phenyl]propane dianhydride with *m*-phenylenediamine and *α,ω*-(3-aminopropyl)oligodimethylsiloxane of controlled molecular weight [79]. These copolymers exhibited ultrasensitive sensor properties detecting parts-per-billion concentrations of chlorinated organic solvent vapors. Coatings of different thickness were used to confirm the gravimetric nature of sensor response and to assess its reproductibility. The detection limits for some organic solvent vapors, such as percloroethylene, trichloroethylene, cis-1,2-dichloroethylene, trans-1,2-dichloroethylene, 1,1-

dichloroethylene and vinyl chloride were 0.6, 3, 6, 6, 11 and 13 ppb, respectively. The high sensitivity and long-term stability of these sensor materials make them attractive as potential candidates to detect toxic industrial organic contaminants at trace levels in groundwater.

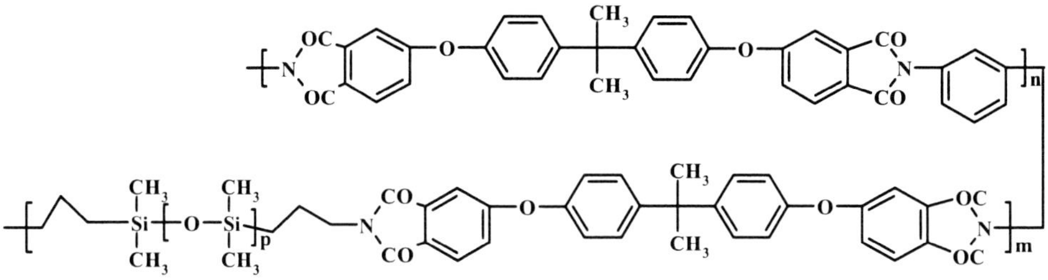

39

3.6. Hyperbranched Poly(Ether Imide)s

Hyperbranched polymers have received considerable attention due to their attractive properties such as higher solubility, lower viscosity, and multiple end groups offering the possibility for further modification and special applications compared to their linear analogues [80-82]. They were first introduced by Flory in 1952 [83]. Thirty years later, Kricheldorf synthesized aromatic polyester copolymers from AB and AB_2 monomers [84]. Kim and Webster reported hyperbranched polyphenylenes by one-step polycondensation of AB_2 monomers in 1990 [85]. Since then, the interest in hyperbranched polymers has rapidly grown.

Hyperbranched polymers are highly branched macromolecules that are generally obtained through a one-step polymerization process of AB_x type monomers. Owing to multi-functionality physical properties, such as solubility in organic solvents and T_g, can be controlled by the chemical modification of the end functional groups. Thus, they may be considered practical alternatives to dendrites for use in applications where the perfectly regular structure of highly branched macromolecules is less important.

Dendrites are built of AB_x monomers coupled in several layers around a central core molecule where each layer is called a generation. Dendrites can be synthesized by two methods: the divergent growth which starts from a polyfunctional core molecule and proceeds radially outward, and the convergent growth which begins at chain ends and involves the formation of dendritic wedges, the last step being the coupling of the dendrons to the core. Hyperbranched polymers are generally composed of three parts, dendritic, linear, and terminal units, and may be considered to be irregular analogues of dendritic macromolecules. They still maintain many of the properties found in their more perfectly defined dendrimer counterparts. The *degree of branching* (*DB*) has been defined as a factor to explain the structure of hyperbranched polymers [86]:

$$DB = (D+T)/(D+T+L)$$

where D is the number of dendritic units, T is the number of terminal units and L is the number of linear units.

One step synthesis allows hyperbranched polymers to be more readily available for potential applications. This significant feature has lead to the development of novel synthetic routes for the preparation of such polymers. To prepare hyperbranched poly(ether imide)s, it is important to use a method that involves stable AB_2 monomers that are easily synthesized and purified. Two typical synthetic methods have been employed for their preparation. The first involves the condensation of dianhydride and diamine monomers, thus generating a poly(amic acid), followed by cyclodehydration to form the imide ring. The second method includes aromatic nucleophilic substitution reactions of phenoxide nucleophiles with nitro- or fluoro-substituted phthalimides to form the ether linkage. Because of the high reactivity of amine and anhydride groups, no stable AB_2-type monomer that contains one amino group and two anhydride groups (or one anhydride group and two amine groups) appears to be available for the preparation of hyperbranched poly(ether imide)s [81].

A hyperbranched poly(ether imide) **42** having terminal phenolic groups has been synthesized by aromatic nucleophilic substitution polymerization of a monomer **41** which contains an activated aryl fluoride group and two phenolic units, namely N-{4-[1,1-di(4-hydroxyphenyl)ethyl]phenyl}-4-fluorphthalimide (Scheme 5) [87].

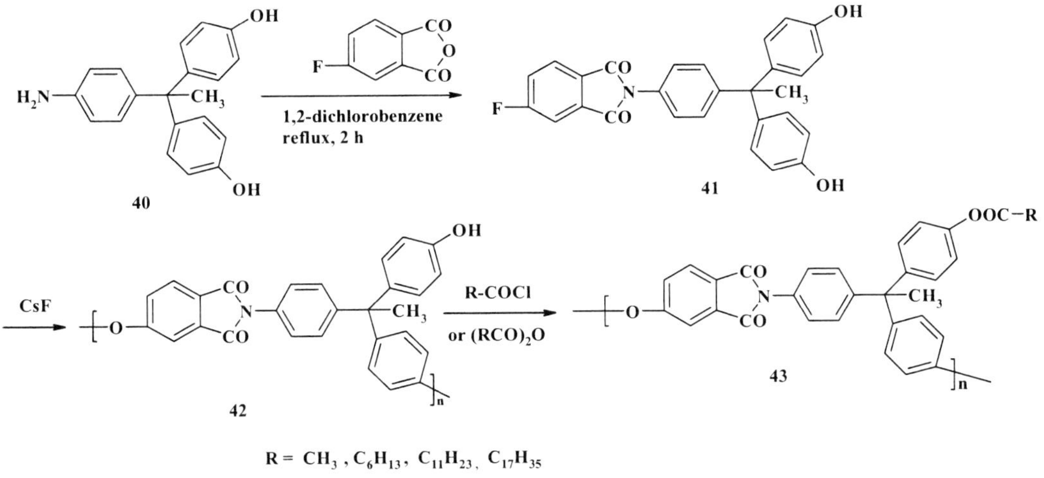

Scheme 5. Synthesis of hyperbranched poly(ether imide)s **42** and **43**.

The monomer was prepared by the reaction of 1-(4-aminophenyl)-1,1-bis(4-hydroxyphenyl)ethane, **40**, with 4-fluorophthalic anhydride. The polymerization reaction was performed in diphenylsulfone at 235°C in the presence of CsF. Scheme 5 only shows the overall compositional repeat unit, even though the hyperbranched polymer contained a combination of dendritic, linear and terminal units. The degree of branching, estimated by ^1H NMR spectroscopy using model compounds, was 55%. The polymer was soluble in organic solvents like DMSO, DMF and THF. Its terminal phenolic groups were modified by reaction with different acid chlorides or anhydrides to give the corresponding ester derivatives **43**. The nature of the chain-end groups influences some properties of the polymer such as T_g and solubility. The transition from the polar hydroxyl function to non-polar aliphatic end groups results in a decrease in T_g due to the reduction of the intermolecular interactions. The modified polymers were also soluble in organic solvents such as toluene, dichloromethane, chloroform, THF and DMF.

An aromatic hyperbranched poly(ether imide) **46** was prepared starting from an AB_2 monomer **45** [88]. The synthesis of the monomer was performed in two steps. First, 3,5-dihydroxyaniline reacted with 4-fluorophthalic anhydride in refluxing acetic acid to give 3,5-dihydroxyphenyl-4-fluorophthalimide, **44**. In the second step the phenolic groups were protected by reaction with *tert*-butyldimethylsilyl chloride to give the monomer **45**. The polycondensation of the monomer was performed in dimethylsulfone, at 240°C, in the presence of CsF as catalyst, to yield hyperbranched poly(ether imide) **46** (Scheme 6). The degree of branching of polymer has been determined by ^{1}H NMR spectroscopy studies of model compounds as being 55%. The number-average molecular weight and weight-average molecular weight of the polymer were 19200 g/mol and 47500 g/mol, respectively. The polymer exhibited high thermal stability in air with T_{10} occurring at 530°C. The DSC measurements showed no T_g before decomposition. However, optical microscopy revealed a visual softening temperature near 300°C.

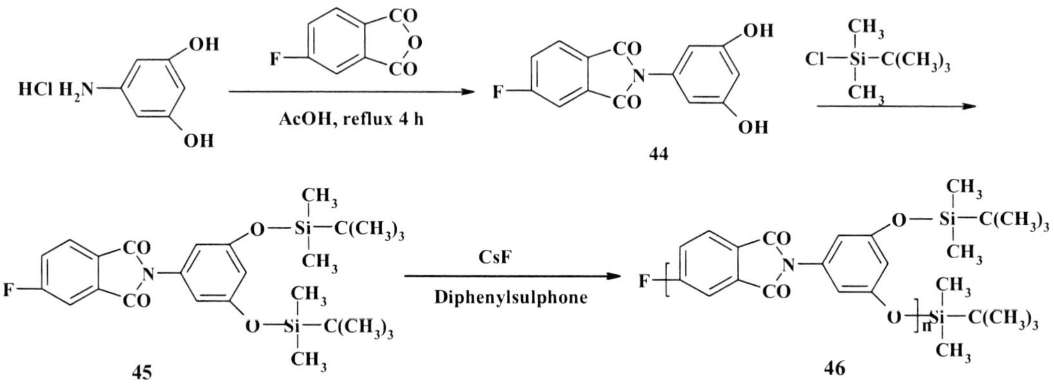

Scheme 6. Synthesis of hyperbranched poly(ether imide) **46**.

Poly(ether imide)s copolymers **47** having different degrees of branching were synthesized under identical reaction conditions with varying AB/AB_2 (mol/mol) compositions, as it can be seen in scheme 7 [89]. The degree of branching was up to 0.66. The research aim was to observe the effects that degree of branching and the number of end groups per molecule have on physical properties of the polymer. T_g was in the range of 182-212°C. An increasing of T_g and a decreasing of the solubility was observed by increasing the AB content of the copolymers. A similar effect was observed for the thermal stability. The T_{10} was 515°C for x=0, and 582°C for x=1. Also, the inherent viscosity increased gradually with increasing AB content up to 75% beyond which a sharp rise was observed, suggesting an architectural change from a globular hyperbranched to an extended star-type molecule. The copolymers having a high degree of branching were soluble in organic solvents such as DMAc, DMF, THF, ethyl acetate. Tough free-standing films were prepared from copolymers with AB content ≥85%.

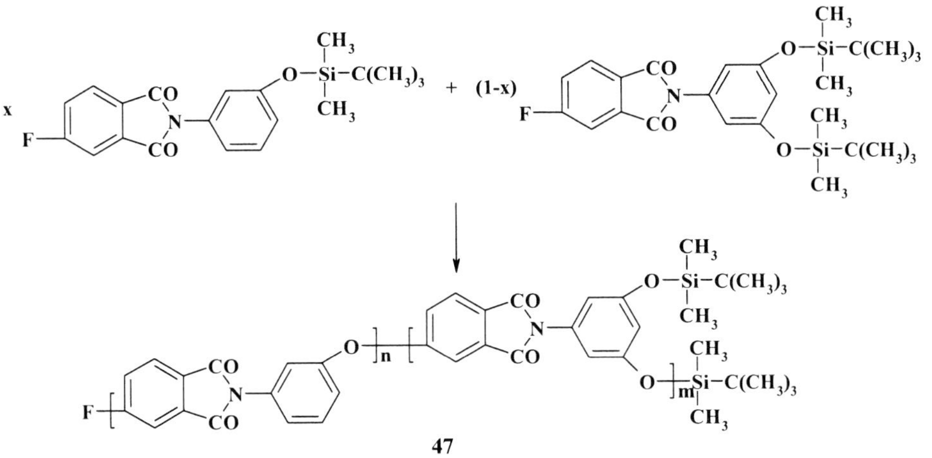

47

x = 0.00, 0.25, 0.5, 0.75, 0.85, 0.90, 0.925, 0.95, 1.00

Scheme 7. Synthesis of hyperbranched poly(ether imide)s **47**.

Hyperbranched poly(ether imide)s having hydroxy end groups were obtained from AB_2 monomers containing a chloro-substituted phthalimide group [90]. The monomers used in the synthesis of such polymers were N-[3- or 4-bis(4-hydroxyphenyl)toluoyl]-4-chlorophthalimide and N-{3- or 4-[1,1-bis(4-hydroxyphenyl]ethylphenyl}-4-chlorophthalimide. Hyperbranched poly(ether imide)s **49** were obtained by self-condensation of the corresponding monomers **48**, at high temperature in the presence of potassium carbonate, as shown in scheme 8. The degree of branching of the polymers, determined by ^{1}H NMR spectroscopy, was 0.6. The inherent viscosity of the polymers was in the range of 0.12-0.16 dL/g. They were soluble in NMP, DMAc, and THF. Their T_g was in 177-230°C range, and the temperature of 5% weight loss varied in the interval of 240-281°C. The low decomposition temperature can be explained by the presence of hydroxyl groups which are not thermally stable. To improve the thermal stability of the polymers, the hydroxyl end groups were modified by acylation with acetic anhydride or benzoyl chloride. The maximum conversion of acetylation and benzoylation of phenolic hydroxyl end groups was greater than 85% and 90%, respectively. The modified hyperbranched poly(ether imide)s **50** exhibited much higher thermal stability than the parent polymers, and lower T_g probably due to loss of H-bonding and the plasticization of the more flexible end groups. The hyperbranched poly(ether imide)s can also be modified by nucleophilic substitution reaction. By using N-phenyl-4-chlorophthalimide as electrophile agent of reaction, more than 98% conversion was obtained. The modified polymers with imide end groups showed the highest inherent viscosity and the best thermal stability.

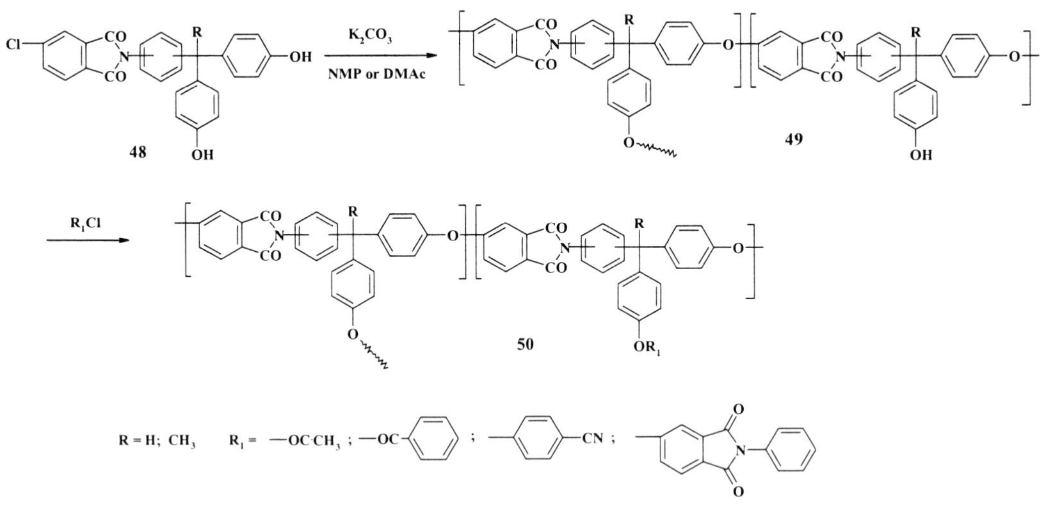

Scheme 8. Synthesis of hyperbranched poly(ether imide)s **49** and **50**.

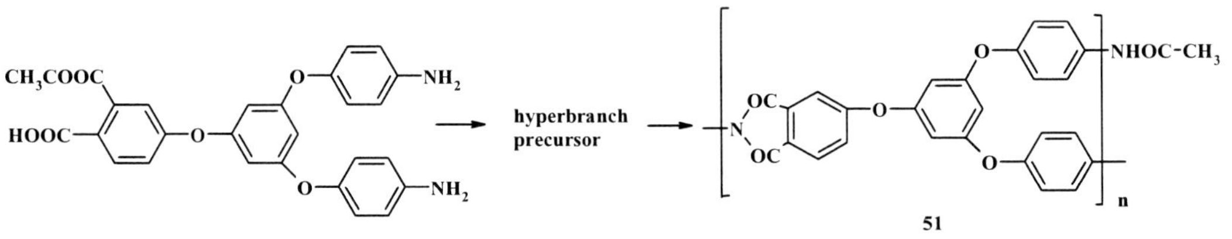

Scheme 9. Synthesis of hyperbranched poly(ether imide) **51**.

Hyperbranched polymer **51** was prepared from an AB_2 type monomer in two steps (Scheme 9) [91]. In the first step the direct self-polycondensation of the monomer was carried out in the presence of (2,3-dihydro-2-thioxo-3-benzoxazolyl)phosphonic acid diphenyl ester (DBOP) as condensing agent, in NMP, at room temperature, to give a hyperbranched precursor without heterocyclic structure. In the second step the precursor imidization was performed thermally (2h at 300°C) or chemically by using acetic anhydride and pyridine in DMSO. The cyclized polymer **51** obtained by using chemical imidization was soluble in NMP and DMSO. The degree of branching was 0.48. The T_g was of 193°C and T_{10} was of 470°C, in nitrogen.

Hyperbranched poly(ether imide)s **52** have been synthesized by A_2+B_3 polymerization following two methods: method I involves the polymerization of tri(phthalic anhydride) with 1,4-phenylenediamine, and method II refers to the polymerization of the isomeric tri(phthalic acid methyl ester) with 1,4-phenylenediamine (Scheme 10) [92]. In the case of the polycondensation reaction between two monomers in which one of them is a bifunctional compound and the other one is a trifunctional compound, a major problem is how to avoid the gelation. The polymerization by method I was difficult to control and often lead to gelation due to high reactivity between anhydride and amine functional groups. The polymerization by method II was performed under mild conditions by using DBOP as condensing agent. Gelation could be avoided when the polycondensation reaction was conducted at a concentration lower than of 0.1 g/mL. The polymers **52** were soluble in polar solvents such as

NMP, DMF, DMAc, and could be processed into films with good mechanical properties. The inherent viscosity was in the range of 0.17-0.97 dL/g, and the weight-average molecular weight was in the range of 33600-125000 g/mol. The degree of branching, estimated by ^1H NMR spectroscopy was of 0.52-0.56. The polymers exhibited high thermal stability having a T_g in the range of 212-235°C, and the temperature of 5% and T_{10} varied in the limit of 480-510°C and 520-560°C, respectively.

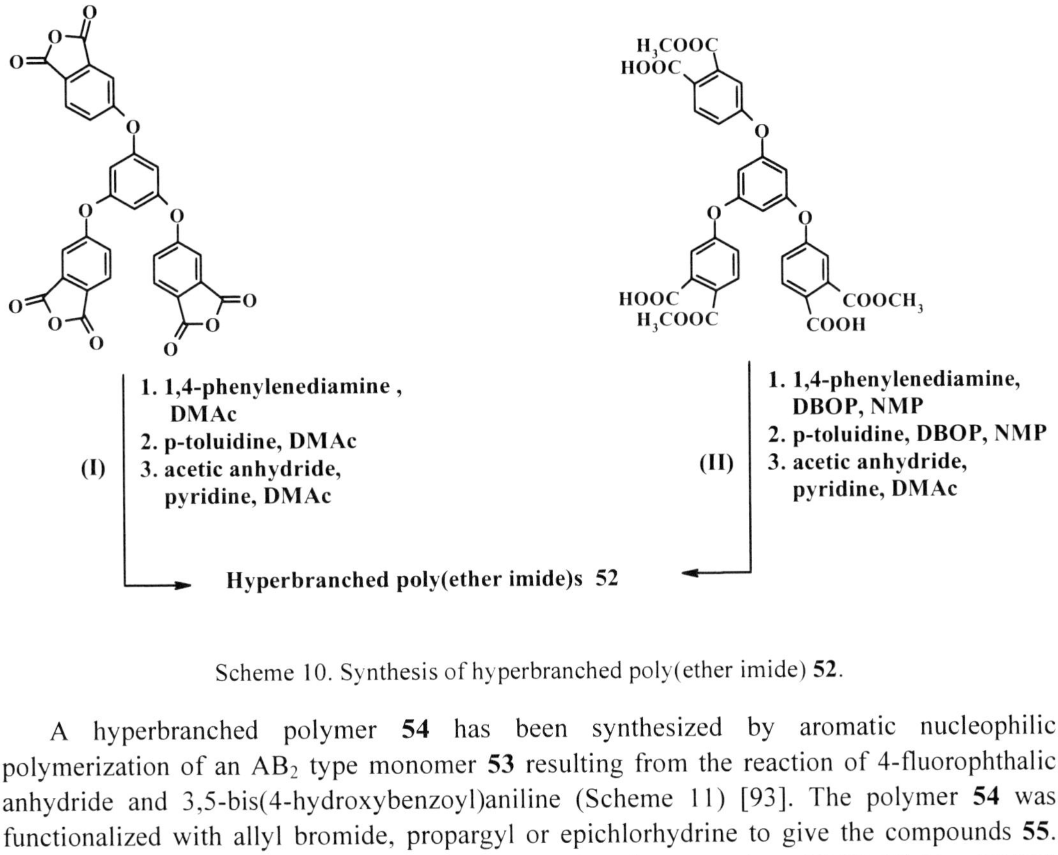

Scheme 10. Synthesis of hyperbranched poly(ether imide) **52**.

A hyperbranched polymer **54** has been synthesized by aromatic nucleophilic polymerization of an AB_2 type monomer **53** resulting from the reaction of 4-fluorophthalic anhydride and 3,5-bis(4-hydroxybenzoyl)aniline (Scheme 11) [93]. The polymer **54** was functionalized with allyl bromide, propargyl or epichlorhydrine to give the compounds **55**. The modified polymers **55** were soluble in organic solvents such as DMF, DMAc, DMSO and THF. The T_g of the polymers was in the domain of 122-225°C. The temperature of 5% weight loss varied in the interval of 318-416°C in air and 330-431°C in nitrogen.

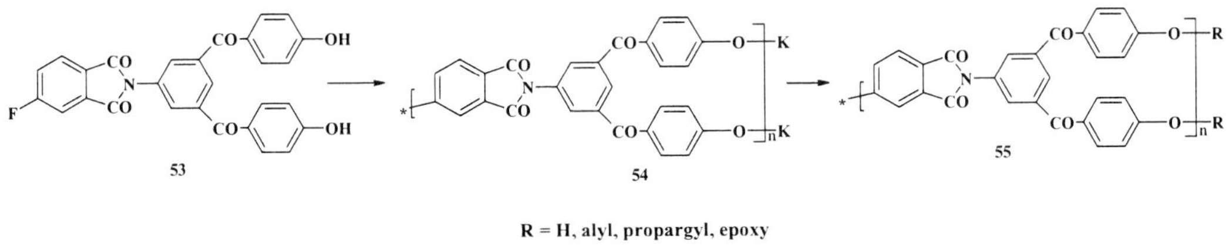

Scheme 11. Synthesis of hyperbranched poly(ether imide)s **54** and **55**.

A hyperbranched poly(ether imide) **56** that doesn't contain linear units was reported [94]. The degree of branching was determined by ^1H NMR spectroscopy as being 1. The polymer was soluble in organic solvents like NMP, DMF, THF, and showed a T_g of 170°C. The temperature at which the polymer loss 5% weight was of 365°C.

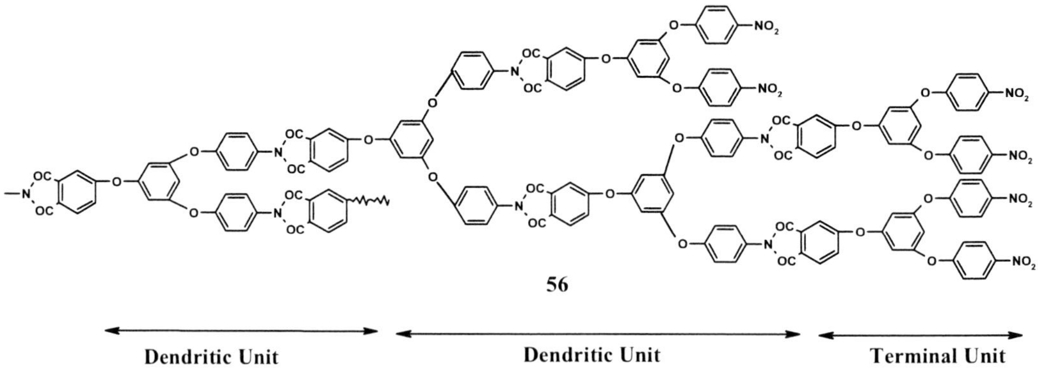

Dendrimers poly(ether imide)s have been also reported [95, 96]. The synthesis of a dendritic unit **60** is illustrated in scheme 12.

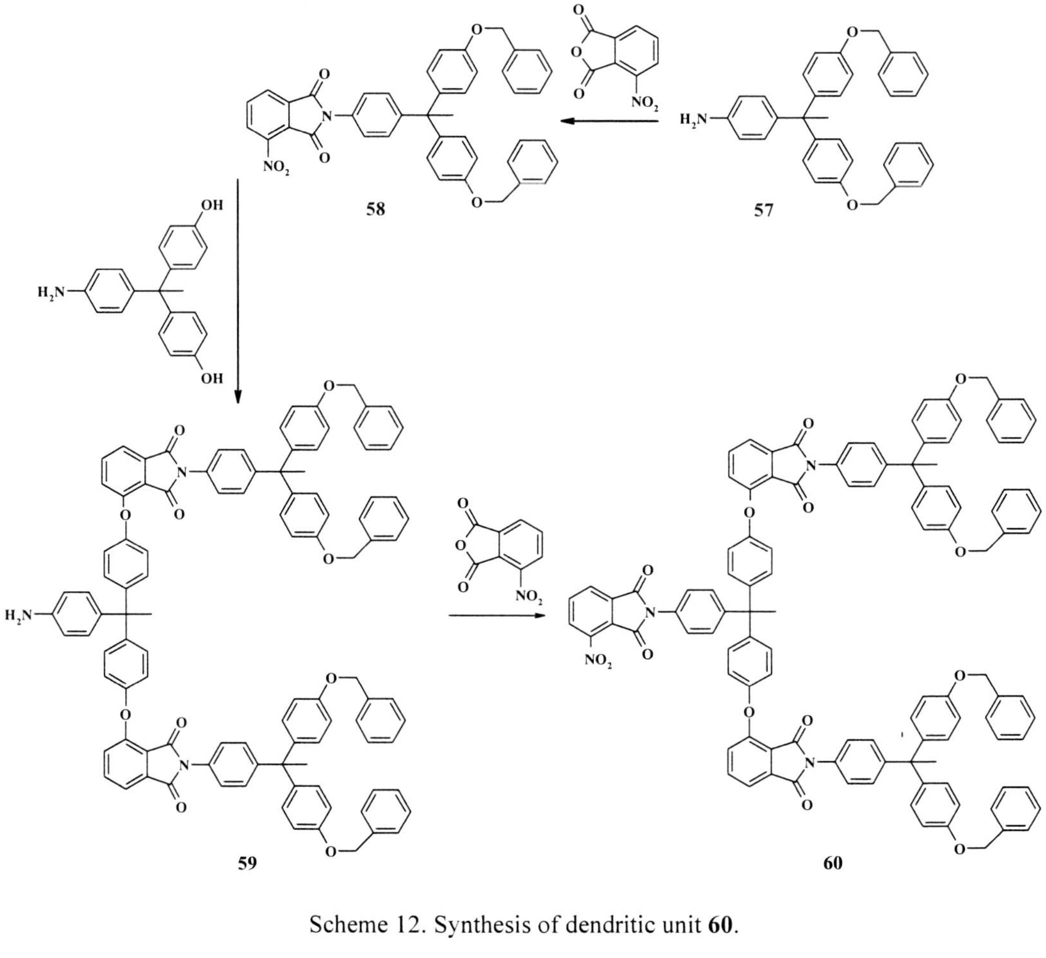

Scheme 12. Synthesis of dendritic unit **60**.

First, the compound **57** containing benzyl-ether units, namely 1,1-bis(4-benzyloxy)phenyl-1-(4-aminophenyl)ethane, was obtained by the reduction of the corresponding nitro-derivative resulting from the reaction between 1,1-bis(4-hydroxyphenyl)-1-(4-nitrophenyl)ethane and benzylbromide. The dendritic unit **60** was synthesized by the reaction of 3-nitrophthalic anhydride with compound **59**, resulting from the interaction of 1-(4-aminophenyl)-1,1-bis(4-hydroxyphenyl)-ethane with the compound **58**. Finally, the dendritic units having nitro group activated by the imide ring are attached to the polyfunctional compound, which is 1,1,1,-tris(4-hydroxyphenyl)ethane, to give the dendrimer poly(ether imide). The dendritic macromolecules are pure and show a narrow distribution of molecular weight. The selective removing of benzylether end groups allows preparing dendrimers containing phenolic end groups, which may be subsequently modified by etherification reactions. The nature of end groups has a high influence on the thermal stability of dendrimers and on their solution behavior.

3.7. Poly(Ether Imide)s Containing Other Heterocycles

The introduction of heteroaromatic rings into the macromolecular chains of poly(ether imide)s could improve their properties, especially the thermal stability and the solubility in organic solvents. Aromatic poly(ether imide)s containing pyridine rings **61** were obtained by a two steps conventional polycondensation process of 4-phenyl-2,6-bis[4-(3,4-dicarboxyphenoxy)phenyl]pyridine dyanhydride with different diamines [97]. The polymers are amorphous and have good solubility in organic solvents such as NMP, DMF, DMSO and *m*-cresol, due to the presence of flexible ether linkages and polarizability of nitrogen atom in pyridine ring. They are thermally stable having the temperature of 5% weight loss higher than 523°C and a T_g of 223-256°C. The films prepared from these polymers were flexible and tough, having the tensile strength of 70.7-97.6 MPa, tensile modulus of 1.34-2.82 GPa, and elongation at break of 7.9-9.7%.

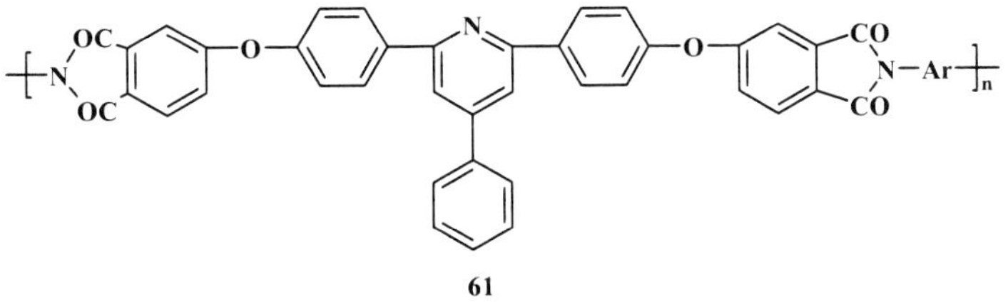

61

Aromatic polymers having 1,3,4-oxadiazole rings in the main chain are also well known for their thermal resistance in oxidative atmosphere, good hydrolytic stability, low dielectric constant and good mechanical properties [98]. Poly(1,3,4-oxadiazole-ether-imide)s with flexible isopropylidene groups were prepared by thermal imidization in solution of poly(amic-acid) intermediates resulting from the reaction of bis-[(3,4-dicarboxyphenoxy)phenyl]-1,4-phenylenediisopropylidene dianhydride, with different aromatic diamines containing preformed 1,3,4-oxadiazole ring, such as 2,5-bis(*p*-aminophenyl)-1,3,4-oxadiazole, 2,5-bis[*p*-(4-aminophenoxy)phenyl]-1,3,4-oxadiazole, and 2-(4-dimethylaminophenyl)-5-(3,5-diaminophenyl)-1,3,4-oxadiazole. Poly(1,3,4-oxadiazole-ether-imide)-polydimethylsiloxane copolymers were also obtained by polycondensation reaction of the same bis(ether-anhydride) with equimolar quantities of an aromatic diamine having 1,3,4-oxadiazole ring and a bis(aminopropyl)polydimethylsiloxane oligomer of controlled molecular weight [99]. All the polymers were easily soluble in polar organic solvents such as NMP and DMAc. They showed good thermal stability, the decomposition temperature being above 400°C. Solutions of the polymers in NMP exhibited fluorescence, having maximum emission wavelength in the range of 370-412 nm.

Fluorinated poly(1,3,4-oxadiazol-ether-imide)s **62** were prepared by solution polycondensation reaction of 2,2-bis[4-(3,4-dicarboxyphenoxy)phenyl]hexafluoropropane dianhydride with different aromatic diamines having preformed 1,3,4-oxadiazole ring [100]. The polymers had good solubility in organic solvents such as NMP, DMF, pyridine and chloroform. They were thermally stable up to 410°C and exhibited a T_g in the temperature range of 183-217°C. Solutions of the polymers derived from 2,5-bis(*p*-aminophenyl)-1,3,4-oxadiazole and 2-(4-dimethylaminophenyl)-5-(3,5-diaminophenyl)-1,3,4-oxadiazole exhibited fluorescence in the blue region of the visible spectrum, at 411 nm and 424 nm, respectively.

Poly(phenylquinoxaline)s are known to possess many desirable properties including excellent thermal stability, low dielectric constant, high T_g and good mechanical properties. These polymers showed an excellent potential for high-temperature applications along with very good storageability at ambient temperature [101]. Aromatic poly(phenylquinoxaline ether imide)s **63** were synthesized by solution polycondensation of aromatic diamines containing preformed phenylquinoxaline groups with bis(ether anhydride)s having isopropylidene, hexafluoroisopropylidene, cyclohexylidene or fluorene units [102, 103]. The polymers were soluble in different organic solvents such as NMP, DMF, DMAc and chloroform. They could be processed from NMP solutions into thin and ultrathin pinhole-free films, as was observed by atomic force microscopy measurements. They showed high thermal stability, having the decomposition temperature higher than 470°C, in air, and a T_g in the range of 210-238°C.

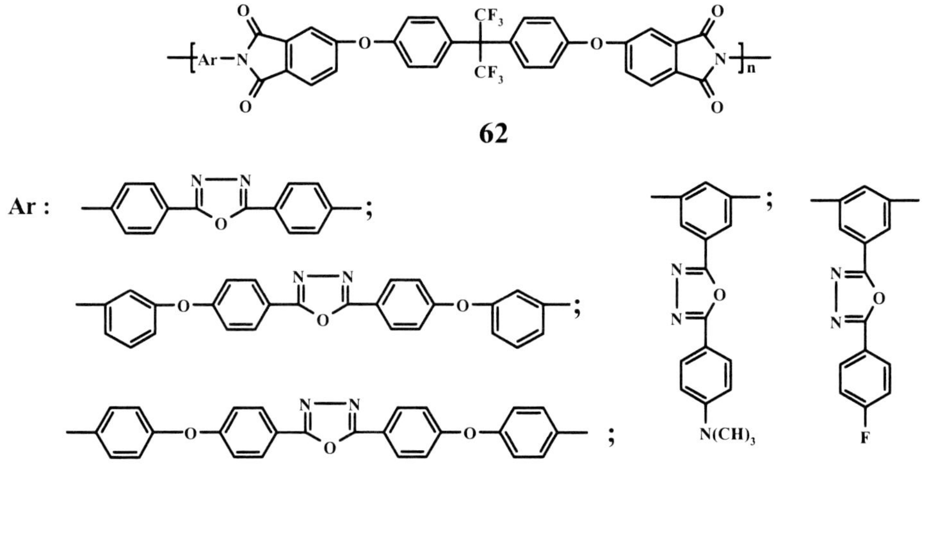

62

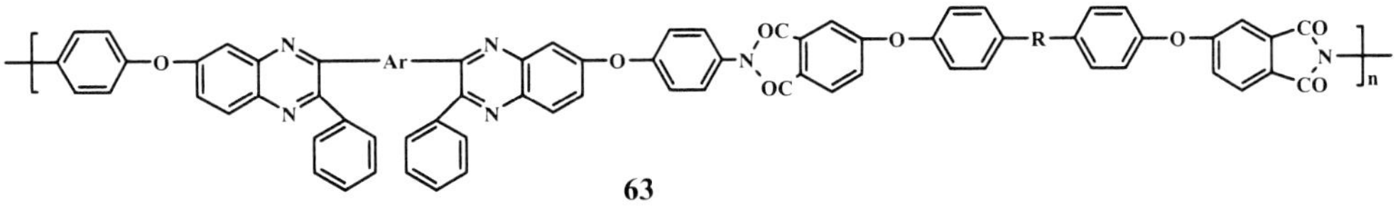

63

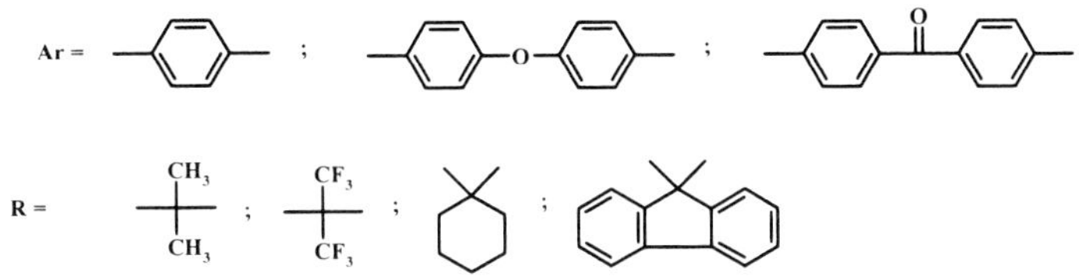

4. Applications of Poly(Ether Imide)s

Aromatic poly(ether imide)s are injection moldable polymers with a large area of applications in many fields including microelectronics, the aerospace industry, and for engineering parts employed in severe conditions. The most important member of this class of polymers is Ultem, a poly(ether imide) manufactured by the General Electric Company. The Ultem poly(ether imide) has important applications in aerospace both as the neat polymer and as the reinforced composite, especially with glass-fibre or graphite reinforcement, in printed wiring boards, under-the-hood automotive uses [32].

Poly(ether imide)s have very attractive features for gas separation applications, such as excellent thermal stability, a good film-forming capacity and a high gas separation performance. Some of the potential applications of poly(ether imide)s membranes in gas separation are: the recovery of H_2 from chemical industrial gases containing CH_4, N_2 or CO, the separation of acidic gases (CO_2, H_2S) from natural gases, and the enrichment of O_2 and N_2 from air for industrial and instrumentation uses [104-108]. Poly(ether imide)s with pendent *tert*-butyl groups may have potential applications such as flexible solar radiation protectors, orientated films in liquid crystal display devices, optical waveguides for communications, interconnectors, and optical half-waveplates for planar lightwave circuits, because they show high transparency of polymer films which is very important for such uses [14]. The potential applications of related polymers incorporating multiple alkyl side chains refer to stretchable, birefringence optical films [38]. Nonlinear optical functionalized poly(ether imide)s are promising candidates for applications such as frequency modulators, linear electrooptic devices and optical storage [109, 110].

The poly(ether imide)s with poly(dimethylsiloxane) sequences can be useful as sensor materials for chlorinated organic solvents [79]. A practical application should be the determination of toxic industrial organic contaminants at trace levels in groundwater. The unique properties of the imide-siloxane copolymeric materials make them especially attractive for applications in microeloectronics and as structural adhesives.

Hyperbranced poly(ether imide)s can be used in blends with linear polymers and as specialty additives for other high-performance polymer materials.

Due to the good balance of properties, poly(ether imide)s are good candidates for composite matrices and polymer blends [111-114]. Organic-inorganic hybrids of poly(ether imide)s and silica were prepared. A highly dispersed silica phase in poly(ether imide) matrix was obtained by in situ hydrolysis and condensation of tetraethoxysilane in the organic polymer solution in the presence of small amounts of aminosilanes. Membranes with different morphology were prepared. The membranes with higher inorganic content had a better stability under pressure [111].

5. Conclusions

Poly(ether imide)s are high performance engineering thermoplastics that combine the excellent thermal and oxidative stability of aromatic polyimides with easy processability of polyethers. Generally, they are injection moldable polymers mainly prepared by two methods. One is the polycondensation of aromatic diamines with bis(ether anhydride)s of aromatic tetracarboxylic acids, followed by thermal or chemical cyclodehydration. The other method is the aromatic nucleophilic substitution reaction of bishalogenoimides or dinitroimides with bisphenols.

Poly(ether imide)s can be structurally modified with the aim to meet the requirements for new applications. Chemical modification is performed by incorporation of certain units into the main chain of poly(ether imide)s using appropriate monomers, in principal aromatic diamines and bis(ether anhydride)s containing various preformed groups, to give polymers having desired properties. Thus different polymers with flexible units, bulky groups, fluorine atoms, siloxane sequences, as well as hyperbranced compounds or poly(ether imide)s containing other heterocycles were prepared, and their properties were studied.

Poly(ether imide) are of special interest owing to a convenient balance of properties such as high excellent thermal stability, remarkable solubility, strengths and moduli, good ductility, flame resistance and electroinsulating properties. The most of them are non-crystalline and possess moderate glass transition temperatures. Also, they are suitable candidates for composite matrices. Due to their good film-forming capacity by casting from solution, poly(ether imide)s are attractive materials for gas separation membranes, biomedical applications, and others. The polymer films or coatings can be used in advanced microelectronic and optoelectronic applications.

Acknowledgements:

Part of this research has been performed in the frame of the Projects CEEX-MATNANTECH 10/2005 and 97/2006 for which the authors acknowledge financing authority.

References

[1] Harris, F. W. In: Polyimides, Wilson, D.; Stenzenberger, H. D.; Hergenrother, P. M. Eds, Chapmanand Hall: New York, 1990, 1-37.

[2] Sato, M. In: Handbook of Thermoplastics, Olabisi O., Ed, New York, Marcel Dekker, 1997, 665- 699.

[3] Ree, M.; Shin, T. J.; Lee, S. W. Korea *Polym. J.* 2001, 9, 1-19.
[4] Huang, S. J.; Hoyt, A. E. *Trends Polym. Sci.* 1995, 3, 262-271.
[5] de Abajo, J.; de la Campa, J. G. *Adv. Polym. Sci.* 1999, 140, 23-59.
[6] Yang, C. P.; Hsiao, F. Z. *J. Polym. Research* 2003, 10, 181-193.
[7] Yang, C. P.; Hsiao, S. H.; Yang, C. C. *J. Polym. Research* 2004, 11, 23-29.
[8] Hsiao, S. H.; Chu, K. Y. *J. Polym. Sci.: Part A: Polym. Chem.* 1997, 35, 2281-2287.
[9] Wirth, J. G.; Heath, D. R.; *US Pat.* 3730946, 1973
[10] Takekoshi, T.; Kochanowski, J. E.; Manello, J. S.; Webber M. J. *J. Polym. Sci.: Polym. Symp.*1986, 74, 93-108.
[11] Hsiao, S. H.; Dai L. R. *J. Polym. Sci: Part A: Polym. Chem.* 1999, 37, 665-675.
[12] Eastmond, G. C.; Paprotny, J.; Webster, I. *Polymer* 1993, 34, 2865-2874.
[13] Eichstadt, A. E.; Ward, T. C.; Bagwell, M. D.; Farr, I. V.; Dunson, D. L.; McGrath, J. E. *Macromolecules* 2002, 35, 7561-7568.
[14] Yang, C. P.; Hsiao, S. H.; Yang, H. W. Macromol. *Chem. Phys.* 2000, 201, 409-418.
[15] Wirth, J. G.; Heath, D. R. *US Pat.* 3787364, 1974.
[16] Williams, F. J. *US Pat.* 3847869, 1974.
[17] Davies, M.; Hay, J. N.; Woodfine, B. *High Perform. Polym.* 1993, 5, 37-45.
[18] Wirth, J. G.; Heath, D. R.; *US Pat.* 3838097, 1974.
[19] Takekoshi, T.; Wirth, J. G.; Heath, D. R.; Kochanowski, J. E.; Manello, J. S.; Webber, M. J. *J. Polym. Sci: Part A: Polym. Chem.* 1980, 18, 3069-3082.
[20] Williams, F. J.; Donahue P. E. *J. Org. Chem.* 1977, 42, 3414-3419.
[21] Gao, C.; Zhang, S.; Gao, L.; Ding, M. *J. Applied. Polym. Sci.* 2004, 92, 2415-2419.
[22] Yang, C. P.; Hsu, M. Y. *J. Polym. Sci.: Part A: Polym. Chem.* 2005, 43, 1270-1284.
[23] Li, Y.; Ding, M.; Xu, J. *Eur. Polym. J.* 1996, 32, 1313-1317.
[24] Li, Y.; Ding, M., Xu, J. *Polym. Int.* 1996, 40, 57-62.
[25] Hsiao, S. H.; Chen, Y. J. *High Perform. Polym.* 2000, 12, 515-524.
[26] Chen T. A.; Jen A. K. Y.; Cai,Y.; *J. Am. Chem. Soc.* 1995, 117, 7295-1296.
[27] Buzin, P. V.; Yablokova, M. Y.; Kuznetsov, A. A.; Smirnov, A. V.; Abramov, I. G. *High Perform. Polym.* 2004, 16, 505-514.
[28] Hsiao, S. H.; Yang, C. P.; Chu, K. Y. *Macromolecules* 1997, 30, 165-170.
[29] Eastmond, G. C.; Paprotny, J. *Polymer*, 2004, 45, 1073-1078.
[30] Mi, Q.; Gao, L.; Ding, M. *Polymer,* 1997, 38, 3663-3668.
[31] Hsiao, S. H.; Liou, G. S.; Chen, S. H. *J. Polym. Sci.: Part A: Polym. Chem.* 2000, 36, 1657-1665.
[32] Sroog C. E., in Polyimides, Wilson, D.; Stenzenberger, H. D.; Hergenrother, P. M. Eds, Chapmanand Hall: New York, 1990, 252-284.
[33] Cheng, Z.; Zhu, X.; Kang, E. T.; Neoh, K. G. *Macromolecules* 2006, 39, 1660-1663.
[34] Tan, B.; Vasudevan, V.; Lee, Y. J.; Gardner, S.; Davis, R. M.; Bullions, T.; Loos, A. C. ; Parvatareddy, H.; Dillard, D. A.; McGrath, J. E.; Cella, J. *J. Polym. Sci.: Part A: Polym. Chem.* 1997, 35, 2943-2954.
[35] Liou G. S.; Wang, J. S. B.; Tseng, S. T.; Tsiang, R. C. C. *J. Polym. Sci.: Part A: Polym. Chem.* 1999, 37, 1673-1680.
[36] Liaw, D. J.; Chen, I. W.; Chen, W. H.; Lin, S. L. *J. Polym. Sci: Part A: Polym. Chem.* 2002, 40, 2556-2563.
[37] Eastmond, C. C.; Paprotny J.; Pethrick, R. A.; Santamaria-Mendia, F. *Macromolecules* 2006, 39, 7534-7548.

[38] Wang, D. H.; Shen, Z.; Cheng, S. Z. D.; Harris, F. W. *Polymer* 2007, 48, 2572-2584.

[39] McHattie, J. S.; Koros, W. J.; Paul, D. R. *Polymer,* 1992, 33, 1701-1711.

[40] Koros, W.J; Fleming, G. K. *J. Membr. Sci.* 1993, 83, 1-80.

[41] Pixton, M. R; Paul, D. R. *Polymer* 1995, 36, 3165-3172.

[42] Yang, C. P.; Yu, C. W. *J. Polym. Sci.: Part A: Polym. Chem.* 2001, 39, 788-799.

[43] Hsiao, S. H.; Lee, C. T.; Chern, Y. T. *J. Polym. Sci: Part A: Polym. Chem.* 1999, 37, 1619-1628.

[44] Eastmond, G. C.; Gibas, M.; Paprotny, J. *Eur. Polym. J.* 1999, 35, 2097-2106.

[45] Yang, C. P.; Chen, J. A. *J. Appl. Polym. Sci.* 1999, 73, 987-996.

[46] Hsiao, S. H.; Huang, T. *J. Polym. Sci.: Part A: Polym. Chem.* 2002, 40, 1712-1725.

[47] Liaw, D. J.; Hsu, C. Y.; Hsu, P. N.; Lin, S. L. *J. Polym. Sci.: Part A: Polym. Chem.* 2002, 40, 2066- 2074.

[48] Hsiao, S. H.; Chung, C. L.; Lee, M. L. *J. Polym. Sci: Part A: Polym. Chem.* 2004, 42, 1008-1017.

[49] Liaw, D. J.; Chen, I. W.; Yang, M. C. *Macromol. Chem. Phys.* 2002, 203, 2170-2176.

[50] Reddy, D. S.; Shu, C. F.; Wu, F. I. *J. Polym. Sci.: Part A: Polym. Chem.* 2002, 40, 262-268.

[51] Hsiao, S. H.; Li, C. T. *J. Polym. Sci.: Part A: Polym. Chem.* 1999, 37, 1403-1412.

[52] Mercado, R.; Wang, Z.; Flaim, T.; DiMenna, W.; Senapati, U. Proceeding of SPIE: Organic Photonic Materials and Devices VI, 5351, J. G. Groge, T Kaino, ed, 2004.

[53] Liaw, D. J.; Liaw, B. Y.; Hsu, P. N.; Hwang, C. Y. *Chem. Mater.* 2001, 13, 1811-1816.

[54] Kim, W. G.; Hay, A. S. *Macromolecules* 1993, 26, 5275-5280.

[55] Wang, J. Y.; Liao, G. X.; Liu, C.; Jian, X. G. *J. Polym. Sci: Part A: Polym. Chem* 2004, 42, 6089-6097.

[56] Mercer, F. W.; Mckenzie, M. T. *High Perform. Polym.* 1993, 5, 97-106.

[57] Bruma, M.; Fitch, J. W.; Cassidy, P. E. *J. Macromol. Sci.-Rev. Macromol. Chem. Phys.* 1996, C36, 119-159.

[58] Hamciuc, C.; Hamciuc, E.; Bruma, M.; Klapper, M.; Pakula, T. *Polym. Bull.* 2001, 47, 1-8.

[59] Yang C. P.; Chen, R. S.; Chen, K. H. *J. Polym. Sci.: Part A: Polym. Chem.* 2003, 41, 922-938.

[60] Lee, D. H.; Koo, S. Y.; Kim, D. Y.; Choi, H. J. *J. Appl. Polym. Sci.* 2000, 76, 249-257.

[61] Koo, S. Y.; Lee, D. H.; Choi, H. J.; Choi, K. Y. *J. Appl. Polym. Sci.* 1996, 61, 1197-1204.

[62] Yang, C. P.; Su, Y. Y. *J. Polym. Sci.: Part A: Polym. Chem.* 2006, 44, 3140-3152.

[63] Yang, S. Y. ; Ge, Z. Y.; Yin, D.X.; Liu, J. G. ; Li, Y. F., Fan, L. *J. Polym. Sci.: Part A: Polym. Chem.* 2004, 42, 4143-4152.

[64] Yang, C. P.; Su, Y. Y. *Polymer* 2003, 44, 6311-6322.

[65] Chung, C. L.; Yang, C. P.; Hsiao, S. H. *J. Polym. Sci.: Part A: Polym. Chem.* 2006, 44, 3092-3102.

[66] Larhrafi, M.; Mas, A.; Laurent, P.; Blancou, H.; Schue, F. *Polym. Int.* 1999, 48, 931-933.

[67] Larhrafi, M.; Mas, A.; Toreis, N.; Blancou, H.; Schue, F. *Polym. Int.* 2003, 52, 1795-1798

[68] Cazacu, M.; Racles, C. Materiale siliconice, obtinere, proprietati si aplicatii, Ed. Performantica, 2007, Iasi.

[69] Fitzgerald, J. J.; Tunney, S. E.; Landry, M. R. *Polymer* 1993, 34, 1823-1832.

[70] Furukawa, N., Yamada, Y.; Kymura, Y. *High Perform. Polym.* 1997, 9, 17-31.

[71] Yamada, Y.; Furukawa, N. *Polym. J.* 1997, 29, 923-930.

[72] Andre, S.; Guida-Pietrasanta, F.; Rousseau, A.; Boutevin, B. *J. Polym. Sci: Part A: Polym. Chem.* 2001, 39, 2414-2425.

[73] Furukawa, N.; Yuasa, M.; Kimura Y. *J. Polym. Sci.: Part A: Polym. Chem.* 1998, 36, 2237-2245.

[74] Mahoney, C. M.; Gardella, J. A. Jr.; Rosenfeld, J. C. *Macromolecules* 2002, 35, 5256-5266.

[75] Krea, M.; Roizard, D.; Moulai-Mostefa, N.; Sacco, D. *J. Memb. Sci.* 2004, 241, 55-64.

[76] Arnold, C. A.; Summers, J. D.; Chen, Y. P.; Bott, R. H.; Chen, D.; McGrath, J. E. *Polymer* 1989, 39, 986-995.

[77] Hamciuc, E.; Hamciuc, C.; Cazacu, M.; Lisa, G.; Okrasa, L. J. *Macromol. Sci. Part A: Pure and Appl. Chem.* 2007, 44, 1069–1078.

[78] Hamciuc, C.; Hamciuc, E.; Cazacu, M.; Okrasa, L. *Polym. Bull.* 2008, 59, 825-832.

[79] Potyrailo, R. A.; Sivavec, R. M. *Anal. Chem.* 2004, 76, 7023-7027.

[80] Malmström, E.; Hult, A. *J. Macromol. Sci.-Rev. Chem. Phys.* 1997, C37, 555-579.

[81] Jikei, M.; Kakimoto, M. *Prog. Polym. Sci.* 2001, 26, 1233-1285.

[82] Voit, B. *J. Polym. Sci: Part A: Polym. Chem.* 2000, 38, 2505-2525.

[83] Flory, P. J. *J. Am. Chem. Soc.* 1952, 74, 2718-2723.

[84] Kricheldorf, H. R.; Zang, Q. Z.; Schwarz, G. *Polymer*, 1982, 23, 1821-1829

[85] Kim, Y. H.; Webster, O. W. *J. Am. Chem. Soc.* 1990, 112, 4592-4593.

[86] Hawker, C. J.; Lee, R.; Fréchet, J. M. *J. Am. Chem. Soc.* 1991, 113, 4583-4588.

[87] Wu, F. I.; Shu, C. F. *J. Polym. Sci: Part A: Polym. Chem.* 2001, 39, 2536-2546.

[88] Thompson, D. S.; Markoski, L. J.; Moore, J. S. *Macromolecules* 1999, 32, 4764-4768.

[89] Markoski, L. J.; Thompson, J. L.; Moore, J. S. *Macromolecules* 2000, 33, 5315-5317.

[90] Li, X.; Li, Y.; Tong, Y.; Shi, L.; Liu, X. *Macromolecules* 2003, 36, 5537-5544.

[91] Yamanaka, K.; Jikei, M.; Kakimoto, M. *Macromolecules* 2000, 33, 1111-1114.

[92] Hao, J.; Jikei, M.; Kakimoto, M. *Macromolecules* 2002, 35, 5372-5381.

[93] Baek, J. B.; Qin, H.; Mather, P. T.; Tan, L. S. *Macromolecules* 2002, 35, 4951-4959.

[94] Yamanaka, K.; Jikei, M.; Kakimoto, M.; *Macromolecules* 2001, 34, 3910-3915.

[95] Leu, C. M.; Chang, Y. T.; Shu, C. F.; Teng, C. F.; Shiea, J. *Macromolecules*, 2000, 33, 2855-2861.

[96] Leu, C. M.; Shu, C. F.; Teng, C. F.; Shiea, J. *Polymer* 2001, 42, 2339-2348.

[97] Wang, X.; Li, Y.; Zhang, S.; Ma, T.; Shao, Y.; Zhao, X. *Eur. Polym. J.* 2006, 42, 1229-1239.

[98] Schulz, B.; Bruma, M.; Brehmer, L. *Adv. Mater.* 1997, 9, 601-613.

[99] Hamciuc, E.; Hamciuc, C.; Cazacu, M. *Eur. Polym. J.* 2007, 43, 4739-4749.

[100] Hamciuc, C.; Hamciuc, E.; Bruma, M. *Polymer*, 2005, 46, 5851-5859.

[101] Bruma, M. In: Handbook of Thermoplastics, Olabisi, O., Ed, New York, Marcel Dekker, 1997, 771-798.

[102] Hamciuc, E.; Hamciuc, C.; Bruma, M.; Belomoina, N. M. Russ. *Chem. Bull. Int. Ed.* 2004, 53, 2035-2040.

[103] Lungu, R.; Hamciuc, E.; Bruma, M.; Szesztay. M.; Muller, P.; Belomoina, N. M. *Rev. Roum. Chim.* 2005, 50, 579-587.

[104] Kneifel, K.; Peinemann K. V. *J. Membr. Sci.* 1992, 65, 295-307.

[105] Peinemann, K. V.; Maggioni, J. F.; Nunes, S. P. *Polymer* 1998, 39, 3411-3416.
[106] Bruma, M.; Hamciuc, E.; Sava, I.; Robison, J.; Alentiev, A. Yu.; Yampolskii, Yu. P. In: Proceedings of the 7th European Technical Symposium on Polyimides and High Performance Functional Polymers, Abadie, M. J. M. and Sillion, B., Eds, Montpellier, France, 2005, 263-273.
[107] Wang, D.; Li, K.; Teo, W. K. *J. Membr. Sci.* 1998, 138, 193-201
[108] Shieh, J. J.; Chung, T. S. *Ind. Eng. Chem. Res.* 1999, 38, 2650-2658.
[109] Lee, Y. B.; Woo, H. Y.; Yoon, C. B.; Shim, H. K. *J. Mater. Chem.* 1999, 9, 2345-2350.
[110] Song, N.,; Men, L.; Gao, J. P.; Bai, Y.; Beaudin, A. M. R.; Yu, G.; Wang, Z. Y. *Chem. Mater.* 2004, 16, 3708-3713.
[111] Nunes, S. P.; Peinemann, K. V.; Ohlogge, K.; Alpers, A.; Keller, M.; Pires, A.T.N. *J. Memb. Sci.* 1999, 157, 219-226.
[112] Woo, E. M.; Yau, S. N.; *Macromolecules* 1997, 30, 3626-3631.
[113] Zhao, L.; Li, L., Tian, J.; Zhuang, J.; Li, S. *Composites: Part A* 2004, 35, 1217-1224.
[114] Shen, L. Q.; Xu, Z. K.; Yang, Q.; Sun, H. L.; Wang, S. Y.; Xu, Y. Y. *J. Appl. Polym. Sci.* 2004, 92, 1709-1715.

In: Advances in Functional Heterochain Polymers
Editor: Maria Cazacu
ISBN: 978-1-60456-598-0

Chapter 7

BISMALEIMIDES-BASED ADDITION POLYIMIDES

Constantin Gaina and Viorica Gaina[1]
"Petru Poni" Institute of Macromolecular Chemistry, Iasi, 700487, Romania

ABSTRACT

Polyimides consisting of aromatic rings coupled by imide linkages can be obtained either by condensation or polyaddition. While the former procedure consists in a step-growth polymerization and results in linear structure, the latter is a heat-activated addition polymerization of diimides and results in a network structure. The latter category will be approached in this chapter: addition polyimides obtained by polymerisation of the bismaleimides, polyaddition of nucleophiles (diamines, bisphenols) to bismaleimides or by Diels-Alder cycloaddition.

INTRODUCTION

Polyimides continue to play an important role in the advancement of technologies as diverse as aerospace materials to medicine to microelectronics. High-temperature stability remains the most important among many desirable properties. This feature coupled with low-moisture absorption makes aromatic polyimides a leading candidate for present and future aerospace composite materials applications. Typical aromatic polyimides are obtained by a condensation reaction between desired diamine and dianhydride monomers in an aprotic polar solvent to yield high molecular weight polyamic acids. Ring closure is accomplished by either thermal or chemical treatments to give polyimides, which exhibit high thermal, chemical, and oxidative stability.

Addition polyimides are best defined as low molecular weight, at least difunctional monomers or prepolymers or mixtures thereof that carry functional reactive terminations and imide functions on their backbone. The reactive endgroups can undergo homo- and/or copolymerizations by thermal or catalytical means.

[1] e-mail: vgaina@icmpp.ro.

The maleimide group can undergo a variety of chemical reactions. The reactivity of the double bond is a consequence of the electron withdrawing nature of the two adjacent carbonyl groups, which create a very electron-deficient double bond, and therefore is susceptible to homo- and copolymerizations. Such polymerizations may be induced by free radicals or anions. Nucleophiles such as primary and secondary amines, phenates, thiophenates, carboxylates, etc. may react via the classical Michael addition mechanism. The maleimide group furthermore is a very reactive dienophile and can therefore be employed in a variety of Diels Alder reactions. An important chemical reaction of the maleimide group is the "ene" reaction with allylphenyl compounds.

Within this section the most important chemical concepts for bismaleimide-based addition polyimides are discussed. Attention is paid to the classical approaches and also to more recent developments.

1. Bismaleimides Based on Homopolymerization Reactions

The double bonds of bismaleimides fast polymerize by free radicals due to the electron-withdrawing character of the adjacent carbonyl groups giving thermostable polymers with high crosslinking degree (structure I)

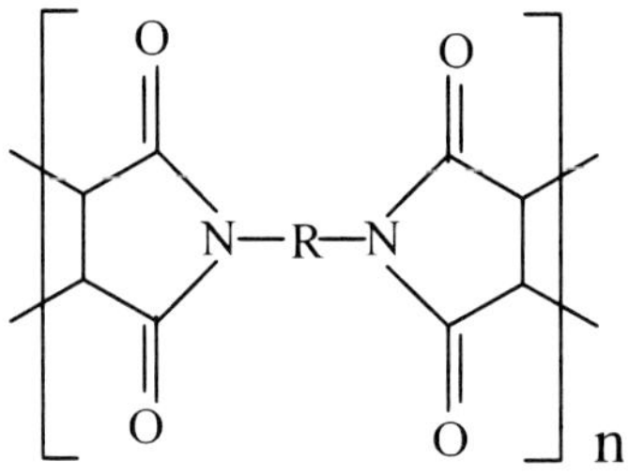

(I)

Stenzenberger [1] studied the thermostability of polybismaleimides containing aliphatic and aromatic bridges and evidenced that aromatic bridges are more stable that aliphatic bridges. In the case of aliphatic bridges, the thermal stability decreases with the increasing of aliphatic chain length. On the other hand, Gherasim and Stoicescu-Crivetz [2] studied thermal and photochemical polymerization of N,N'-bismaleimides and discovered a lower thermostability than those of polymers (I).

The researches of the last three decade were oriented on the improvement of thermostability and proccesability of polymers (I) through the synthesis of various polybismaleimide structures [3, 4a-b, 5, 6, 7, 8-17]. So, Mikroyannidis et al. described the synthesis of polybismaleimides possessing urea [5], amide [18, 19], imide and ester [18], or ketone [20] internal groups. The general structure of these polymers is presented in Scheme 1. The presence of azomethine, ketone groups, and heterocycles in the bridge chains between the imide groups leads to the increase of thermostability.

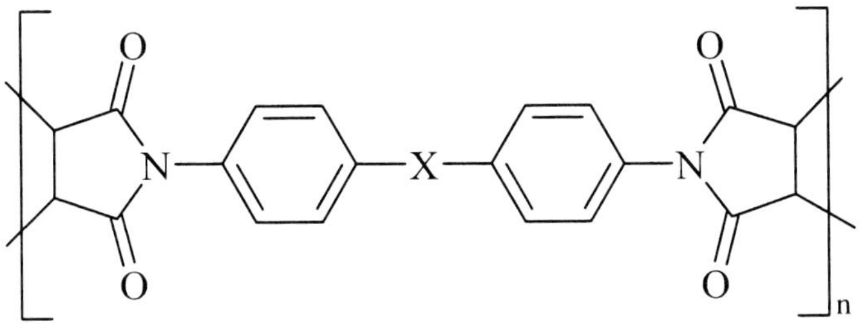

where X = NH-CO-NH; NH-CO-NH-Ph-NH-CO-NH; CO-NH; CO-Ph-O-Ph-CO; COO-

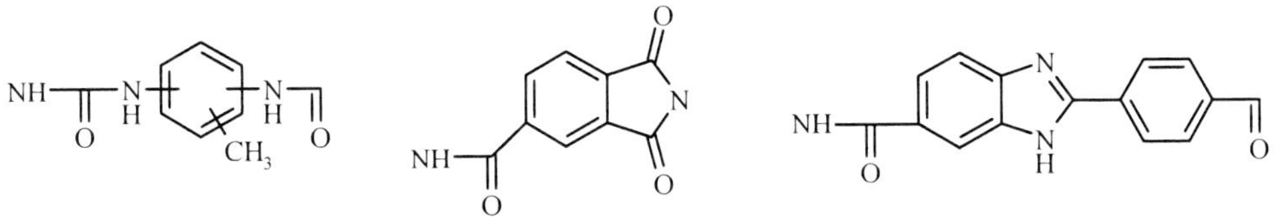

Scheme 1

Due to the considerable interest in resins with high thermostability used in the preparation of fiber reinforced composite materials, adhesives for aerospace applications, and laminate necessary for the isolation of electronic devices or for obtaining of printed boards, Mikroyannidis, Kumar and Ueyama [20, 10, 21] synthesized polytetramaleimides or phosphonitrile oligomers having more than four phenoxymaleimide groups according to Scheme 2. These structures were thermostable up to 350°C in nitrogen and air.

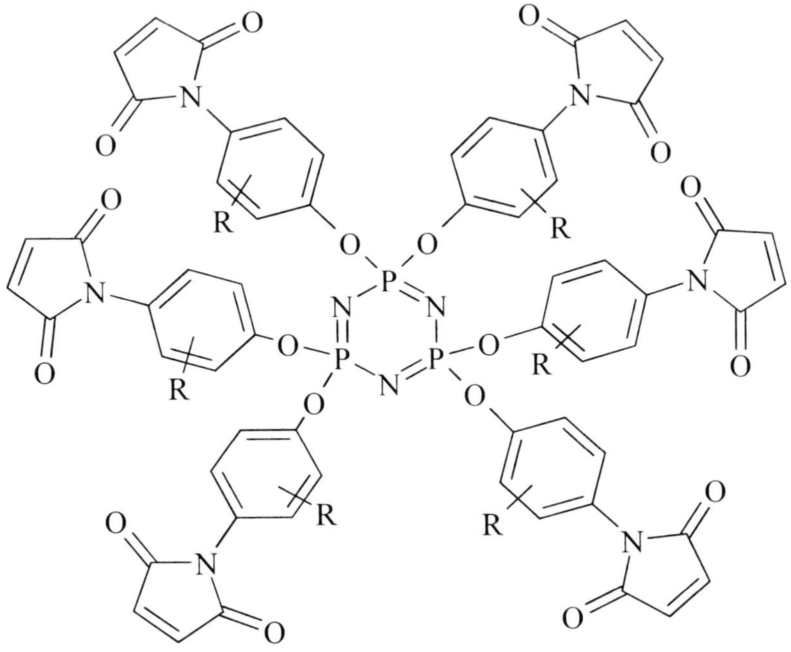

Scheme 2.

I.K.Varma et al. [22] synthesized new polybismaleimide structures for the preparation of laminate having good thermostability (250-360°C), Tg around 430°C and dynamic modulus around 17,505 GPa (Scheme 3).

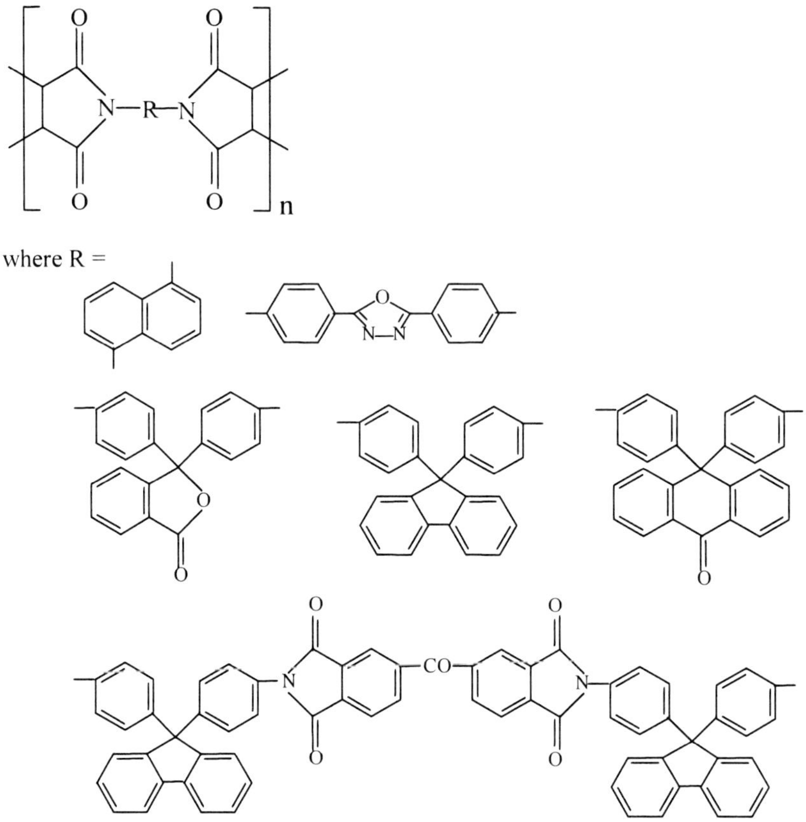

Scheme 3.

Takeda [17] obtained polybismaleimides with ether bonds that lead to the improvement of the polymer flexibility. The initial decomposition temperature (IDT) of bismaleimide resins containing aromatic ether bonds is higher than those with oxyethylene bonds. The T_g values increases with the increasing of crosslinking time. Soldfarb *et al.* synthesized bismaleimide resins containing oxyalkyl units with various lengths or flexible units with determined length having pendant aliphatic groups for the improvement of flexibility and in consequently reducing of crosslinking density [9]. By choosing the chain length and pendant aliphatic groups, the crosslinking interval width can be controlled without the modification of final thermostability or T_g.

Another way to decrease the crosslinking density and improvement of processability of polyimide resins was the synthesis of various oligomers or prepolymers structures with comparative long chains and end maleimide groups that thermally polymerized leading to the resins with good mechanical properties. So, Mikroyannidis *et al.* prepared pyromelitimide prepolymers (by the reaction between maleic/citraconic/dichloromaleic anhydride and diamines and pyromelitic anhydride) and azomethine prepolymers (by the reaction of aromatic diamines with maleic anhydride and terephthalic aldehyde) having maleimide end groups that polymerized leading to the resins with the thermostability ranging between 350-377°C in air and nitrogen and the char yield of solid residue at 800°C in anaerobic conditions around 50-63% [23-24]. By the reaction of trimellitic anhydride with diaminodiphenylmethane in NMP as solvent followed by the reaction with maleic anhydride, Kruh *et al.* synthesized a prepolymer containing amide-imide bonds and maleimide end units (Scheme 4) [25].

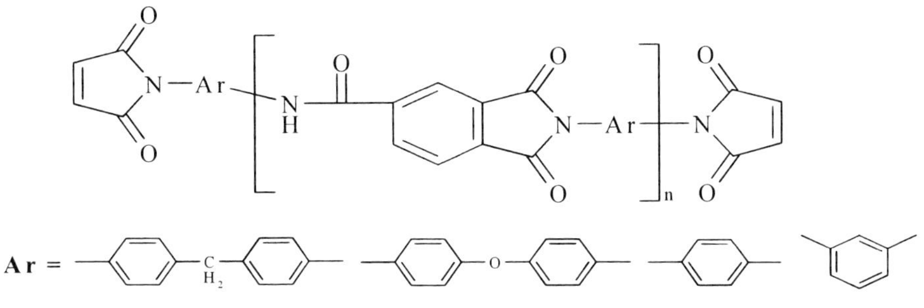

Scheme 4.

Such prepolymers obtained with various molecular weights present a poor processability. They have not melting transitions and they are only soluble in NMP.

The preparation of aryl-imide resins was also carried out by the using aromatic ether-sulfone oligomers with maleimide end groups that lead to the enhancement of solubility in the common organic solvents, incresing T_g and various modulus values depend on the molecular weights of ether-sulfone moiety (Scheme 5) [26-28]. A few properties of these resins are presented in Table 1 and Figure 1.

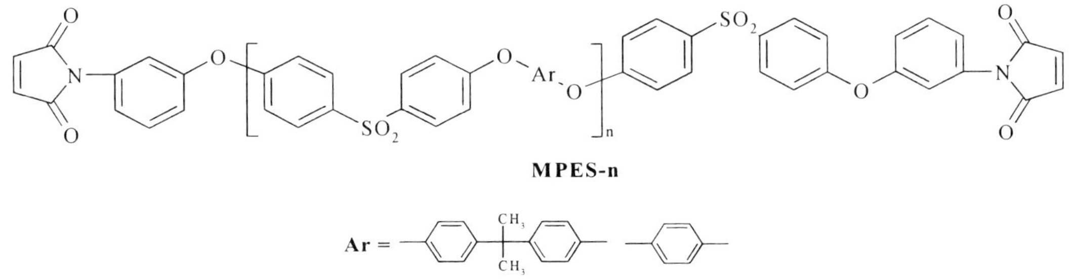

Scheme 5.

Table 1. The properties of resins based on ether-sulfone oligomers with maleimide end groups.

Oligomer	$\bar{M}_n$ [a]	T_g (°C)	E' (GPa)		Film
			25°C	200°C	
MPES-6	1400	277	2,60	2,30	brittle
MPES-8	1900	272	2,50	2,20	brittle
MPES-9	2400	253	2,30	2,00	brittle
MPES-12	>3000	253	2,05	1,90	flexible

[a]Number-average molecular weight of oligomer

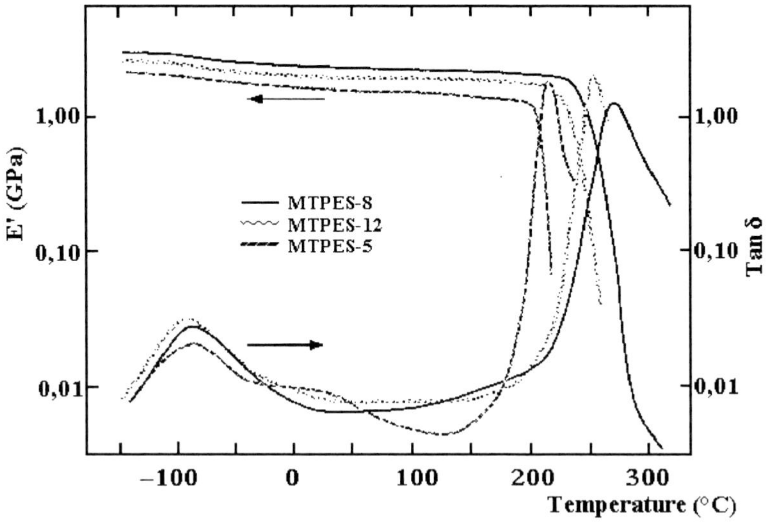

Figure 1. The thermomechanical curves of poly(ether-sulfone-imide) resins.

Another oligomer structures that presented real interest were poly(ether-ketone) oligomers with maleimide end that showed high fracture strength and insolubility in common solvents (Scheme 6) [29-36].

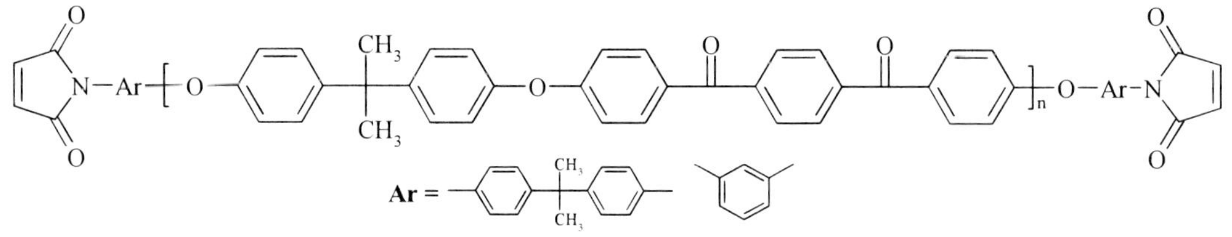

Scheme 6.

A few thermal properties of these polymer types are presented in Table 2.

Gaina *et al.* obtained urethane prepolymers by the reaction of urethane oligomers having end isocyan groups with maleic anhydride [31] or by the reaction of *N*-phenylmaleimidoisocyanate with diols (polybuthyleneadipate (PBA), polytetrahydrofurane (PTHF) with various molecular weights) (Scheme 7) [32]. The properties of these resins are presented in Table 3.

Table 2. Thermal properties of the resins base on poly(ether-ketone) oligomers with maleimide end [30].

$\bar{M}_n$ [a]	T_g (°C)	Tensile strength (MPa)	Elongation (%)	Elongation at break (%)
2500	177	70	9	10
5000	175	70	10	11
10000	169	70	11	21

[a]Number-average molecular weight of oligomer.

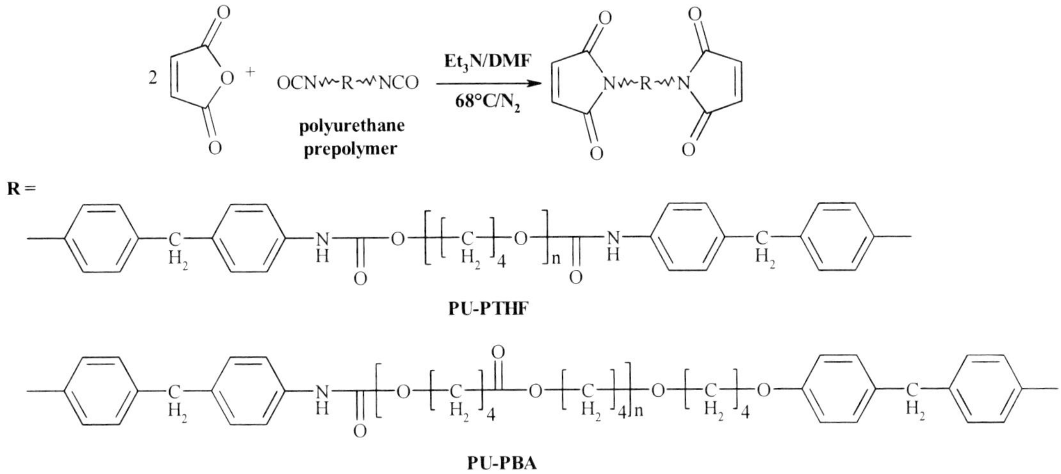

Scheme 7.

Table 3. The main properties of the urethane-bismaleimide resins, depending on the M_n value of the used polyol

Resin	$\bar{M}_n$ polyol	Thermostability (°C)				Tensile strength (MPa)
		T_g	T_m	$T_{5\%}$	$T_{10\%}$	
PU-PTHF1	650	14	-	346	364	41
PU-PTHF2	1000	-39	-	353	371	20
PU-PTHF3	2000	-52	-	365	385	11
PU-PBA1	650	12	-	352	370	36
PU-PBA2	1000	-13	-	361	382	19
PU-PBA3	2000	-18	56,5	371	391	27

Park and Jang reacted diglycidylbisphenol A (DGEBA) with 3(4)-maleimidobenzoic acid or 3(4)-maleimidophenol obtaining endcapping maleimide oligomers with various lengths (Scheme 8) [33].

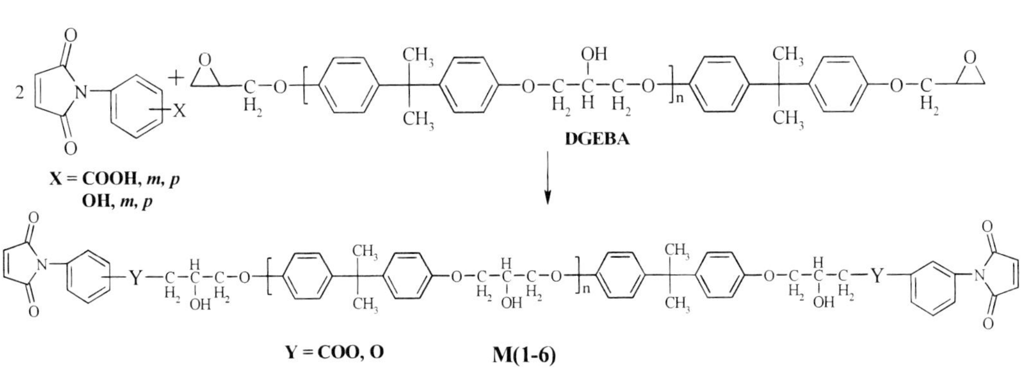

Scheme 8.

An increase of T_g and the improvement of thermal properties were observed by modification of epoxy resins (Table 4).

Table 4. The properties of modified epoxy resins beside of near epoxy resins and polyaminobismaleimides [33]

Resin	Y	Position of Y	T_g (°C)	Coefficient of thermal expansion (CTE) ($x10^{-6}$/°C)	Initial water absorption (%)	Temperature for 10% weight loss (°C)
Epoxy	-	-	100	60	0,20	220
Kerimid 601	-	-	260	45	0,37	440
M1	COO	*metha*	238	50	0,26	410
M2	COO	*para*	228	54	-	400
M3	O	*metha*	201	49	-	410
M4	O	*para*	209	54	-	410

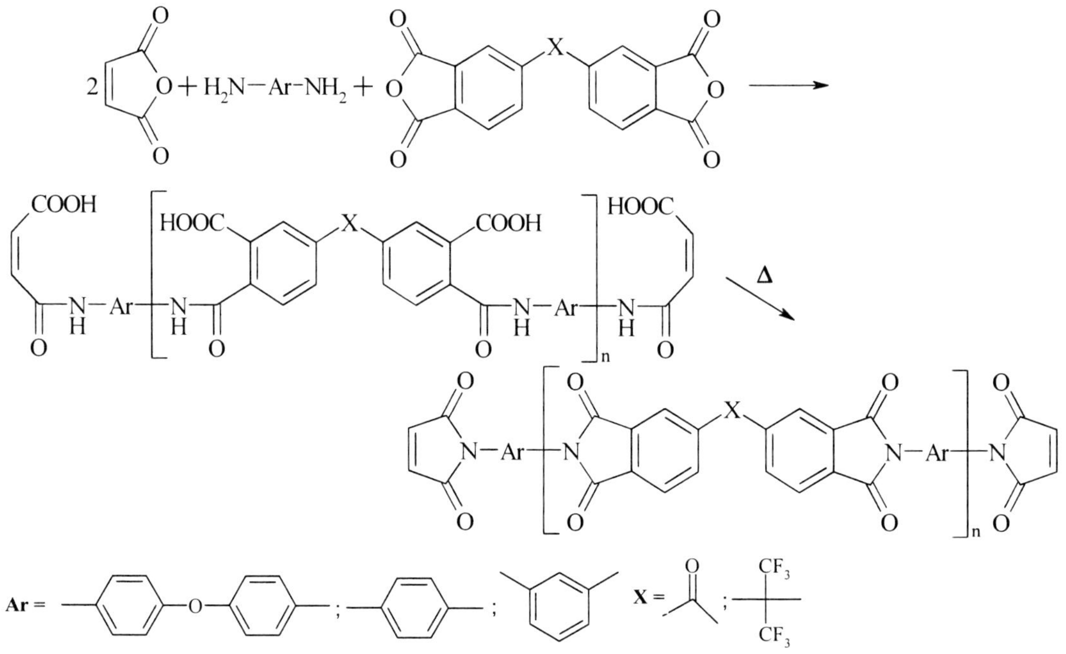

Scheme 9.

Hirano *et al.* reported the synthesis of new imide oligomers having maleimide ends through the reaction of aromatic diamines with diphenyltetracarboxilyc dianhydride and maleic anhydride, in dimethylacetamide and xylen mixture as solvent (Scheme 9) [34]. The thermal cyclization of maleamic acid carried out in presence of pyridine as a catalyst and molecular weights of oligomers depended on the monomer ratio.

W. J. Monacelly obtained oligomers having the structures presented in Scheme 10 [35].

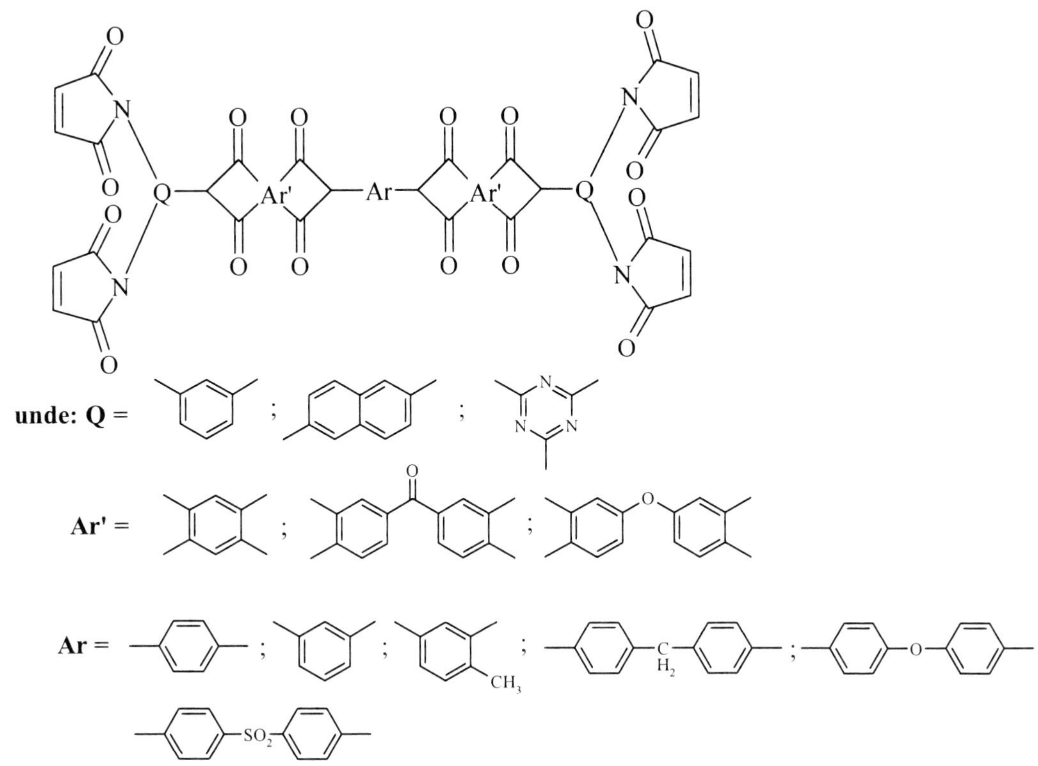

Scheme 10.

Hasagane *et al.* prepared polyamidoimide resins having thermostability ranging between 375-432°C by the reaction of an aromatic dicarboxylic acid with diisocyanate and maleic anhydride or 4-maleimidobenzoic acid (Scheme 11) [36].

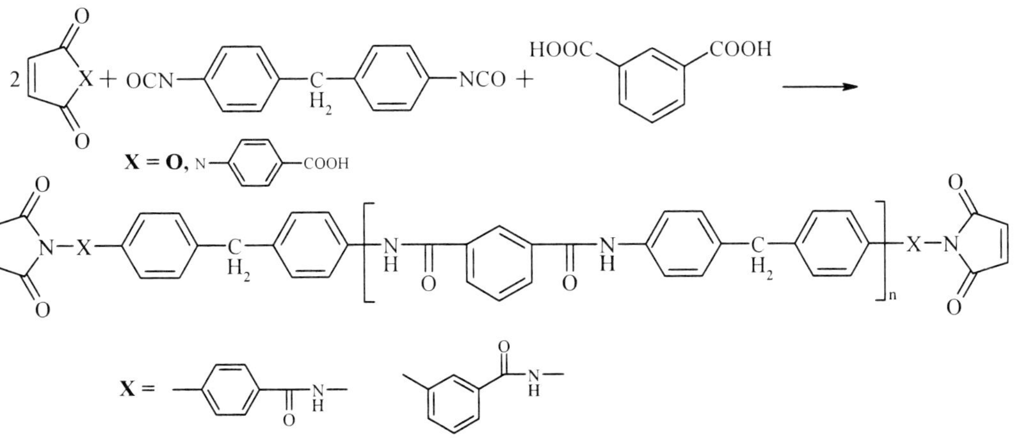

Scheme 11.

New polyimide resins based on polysiloxane oligomers were reported by Ph. Klemarczyk and P. Barthelemy (Scheme 12) [37-38]. The resins presented a tensile strength varying between 122-1010 psi, the lowest value corresponding to the sample based on siloxane monomer with molecular weight of 5000.

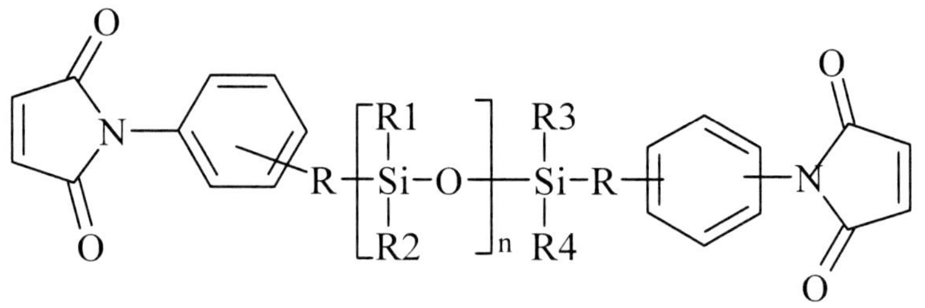

where:R = -X-$(CH_2)_x$-, X = O, S, SO, SO_2, R1-R4 = CH_3, C_6H_5

Scheme 12.

Polybismaleimides with liquid crystal properties were synthesized by Hoyt [4a, 8] (Scheme 13).

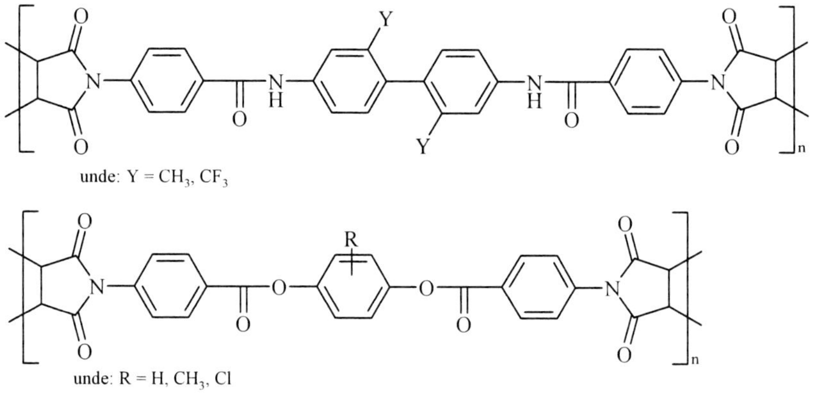

Scheme 13.

Recently, a new bismaleimide containing epoxy units and cyclic phosphinic oxide followed by its homopolymerization lead to a polymer with enhanced fireproof (Scheme 14) [39].

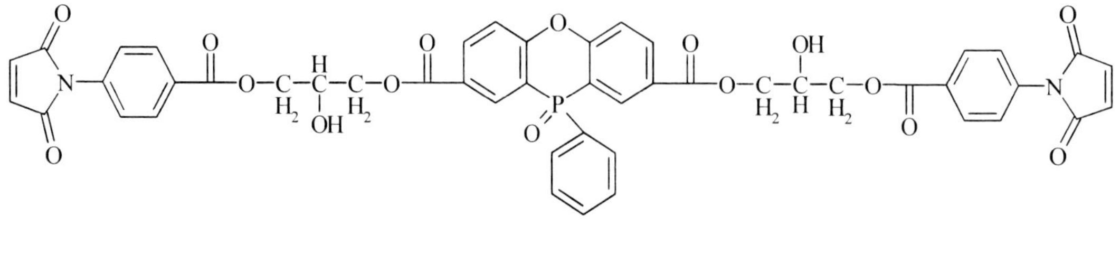

Scheme 14.

2. Polymers Based on Polyaddition Reactions

Bismaleimides allow easily nucleophylic addition with numerous nucleophylic reagents due to their bifunctionality and electron deficitary double bonds. The main nucleophyls used in this reaction are: diamines and their derivatives, bisalcohols and bisthiols, disulfinic acids, dialdoxime etc.

2.1. Polyaminobismaleimides (Polyaspartimides)

Poly(aspartamide)s exhibit properties similar to poly(imide)s but benefit from a facile, one-pot synthesis and increased solubility in organic solvents relative to aromatic poly(imide)s and poly(amide)s [40].

Polyaminobismaleimides (PABMI) were obtained by the nucleophylic polyaddition reaction of diamines to bismaleimides (Scheme 15)

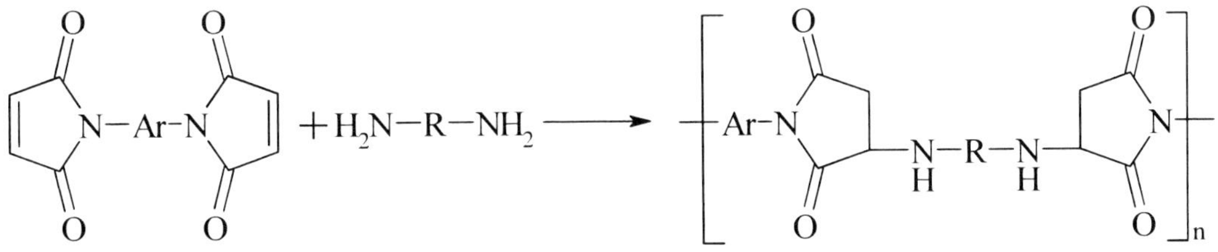

Scheme 15.

Poly(aspartamide)s constitute a class of addition polymers derived from bismaleimides and diamine monomers. Poly(aspartamide) synthesis relies on a facile Michael addition of the amine functional group to the maleimide double bond and does not require a high temperature poly(amic acid) cyclization reaction as for poly(imide)s.

The addition reactions of BMI, in solution, were reported since 1959 [41], but the polymers presented lower molecular weights. The similar results were obtained by the thermal polyaddition in the melt. The polyaddition reaction takes place in the base or acid catalysis and follows a Michael addition mechanism [42-45]. Crivello synthesized poly(aspartamide)s through the reaction of aromatic diamines with aromatic bismaleimides in cresol at 110°C for 3 h with catalytic amounts of acetic acid [42]. The lower basicity of the aromatic amines allowed the use of acetic acid, which activated the maleimide group, through protonation [46]. The model reactions suggested that the polymerization mechanism involved charged intermediates due to increased rates in polar solvents. The poly(aspartamide)s exhibited glass transition temperatures near 210°C and thermal decomposition temperatures near 350°C. This decomposition occurred at the same temperature (TGA) in both air and nitrogen atmospheres, suggesting that depolymerization occurred rather than oxidation. The poly(aspartamide)s possessed tensile strengths near 95 MPa and elongations of 5.5%. Flexural strengths and moduli as high as 140 MPa and 3.1 GPa, respectively, were reported.

The reaction between BMI and diamines was studied on the low molecular compound [42, 44] (Scheme 16).

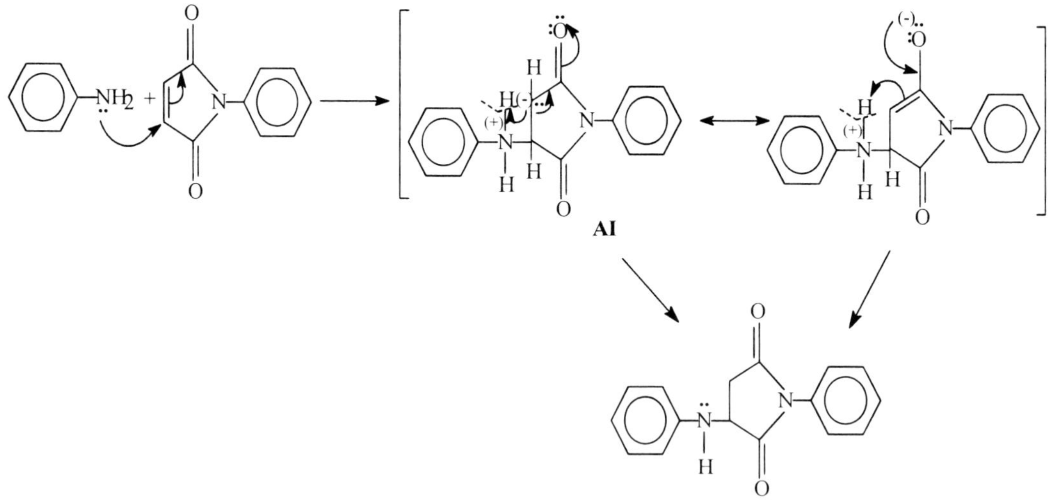

Scheme 16.

Because the reaction rate depends on the polarity of the reaction medium, the authors suggest that the reaction mechanism involves a charged intermediate or transition state (AI), which is stabilized by the medium. Although certain Lewis acids, strong protonic acids, and bases were found to exhibit no catalytic activity, the weak protonic acids (acetic, propionic, trifluoroacetic, formic, and maleic acid) have been shown to have a marked effect on the reaction (Scheme 16).

A study on the adsorption of maleimide on a polycrystalline iron electrode in neutral and acidic media by the impedance method showed the modification of partial charges in the maleimide molecule (Figure 2) [47].

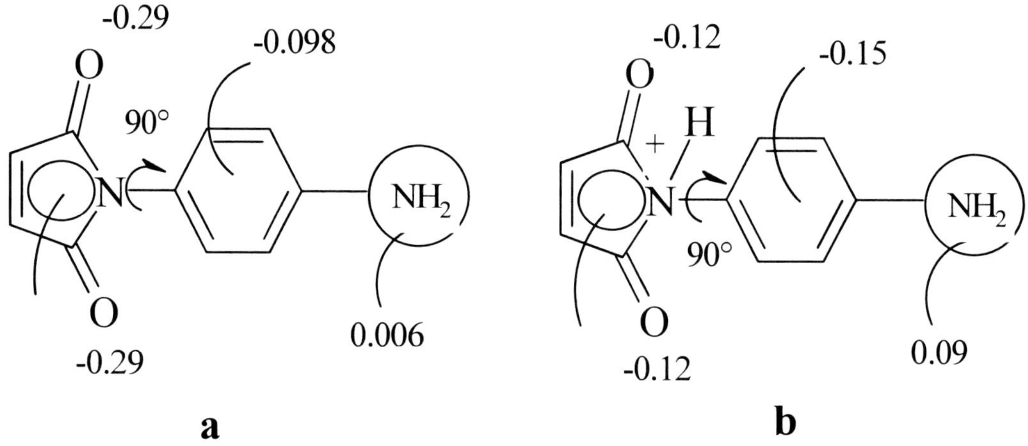

Figure 2. Molecular structure and distribution of partial charges in the molecules of (a) *N*-(*p*-aminophenyl) maleimide and, (b) protonated form of *N*-(*p*-aminopheny1)-maleimide.

Many patents and publications describe synthesis, properties, and applications of poly(aminobismaleimide)s (PABMI) (see Table 5). Different solvents are used in the polyaddition reaction: *m*-crezol, DMF, DMAc, NMP, and DMSO, *o*-chlorophenol or mixtures of them with xylene, toluene, methylethylketone, and THF [42-67]. As nucleophyle components are mentioned following amines: primary diamines [68, 42-45, 48-61], secondary diamines [62-64], telechelic oligomers with amine ends, such as: siloxane [65-67], butadiene-acrylonitrile rubbers [69] etc.

Table 5. Structures and some properties of PABMI.

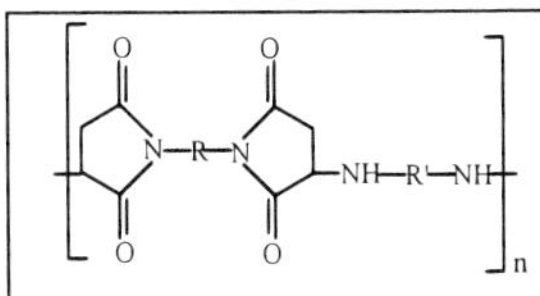 R R'		Properties	Literature
1	2	3	4
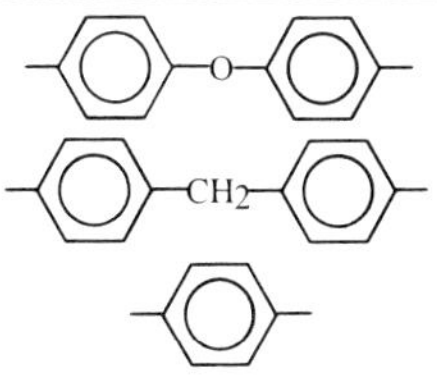	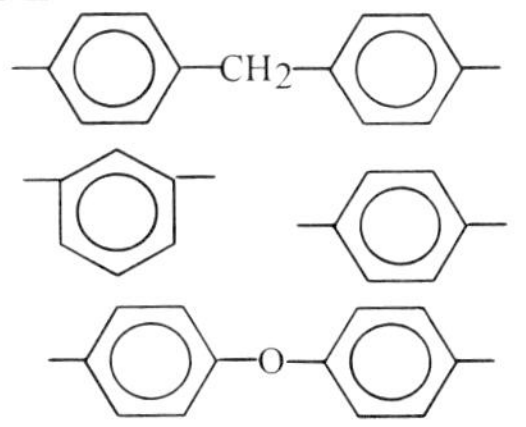	*T_d =350 ℃.* *T_g = 211 ℃* *Flexible films.*	[42] [48] [49]
-	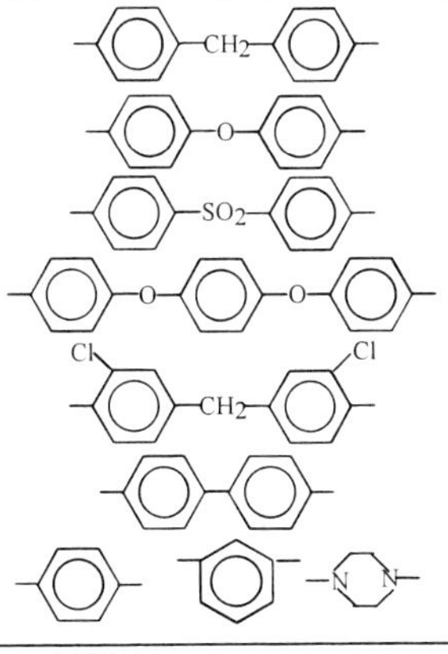	*Thermal stability ~300 ℃* *η=0,09-0,39 dl/g (free catalyst)* *η=0,2-0,56 dl/g (in the presence of catalyst)*	[50]
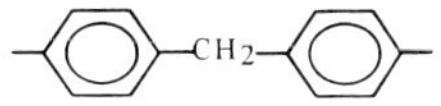	$—(CH_2)_n—$ n = 2,4,6	*Thermal stability ~180 ℃*	[51]
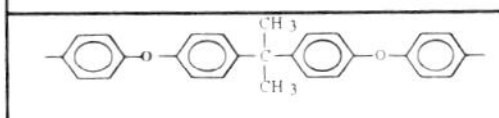	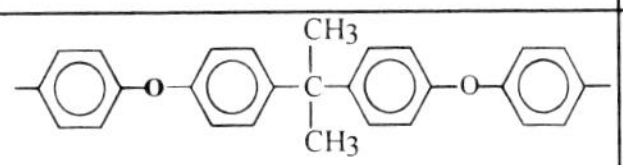	*Thermal stability ~300 ℃*	[52]
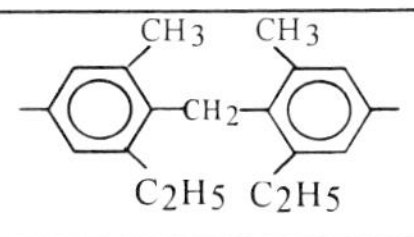	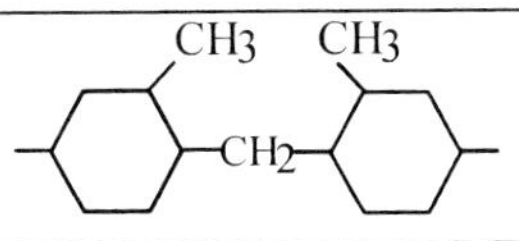	*Dielectric constant = 2,9* *η=0,85 dl/g*	[53]

Table 5. Continued.

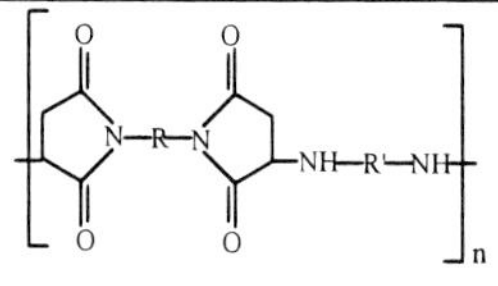 R R'		Properties	Literature
1	2	3	4
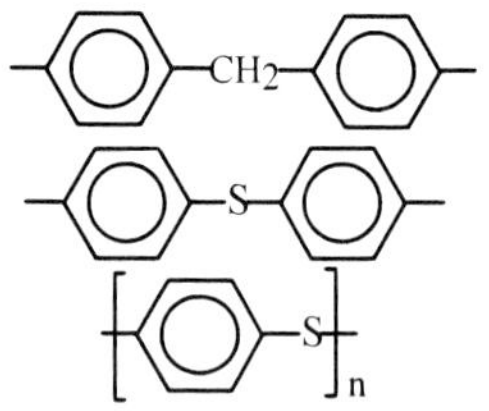	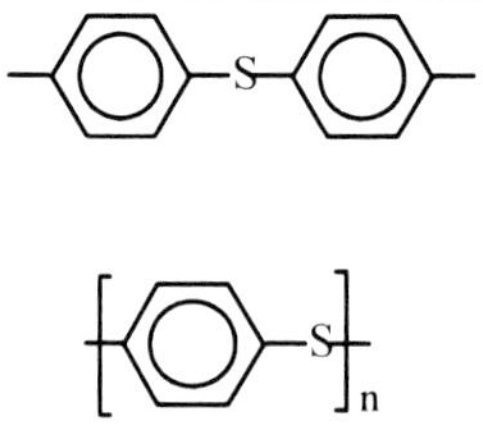	*Thermal stability ~400 °C*	[52]
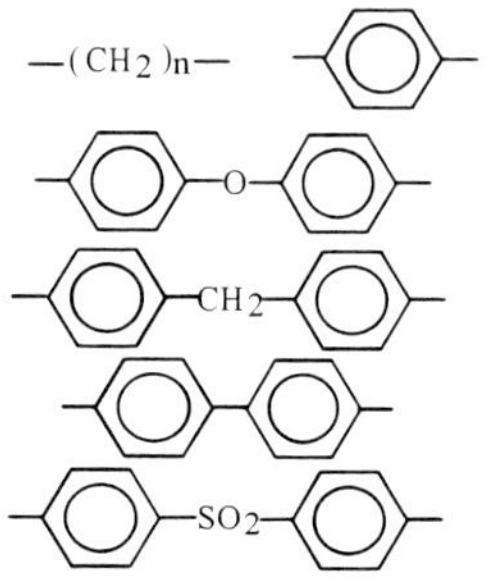	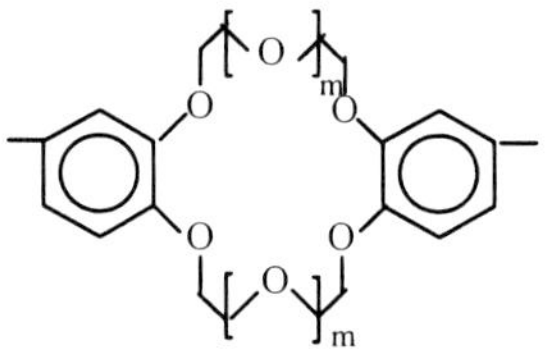	*High capacity to binding of alkaline metallic ions.*	[54-55]
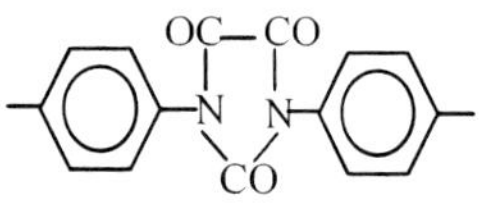	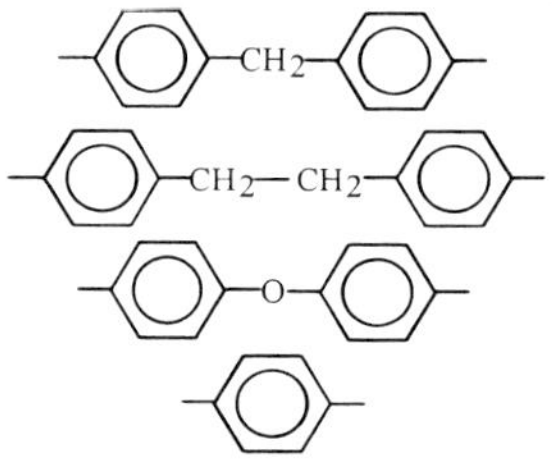	*Thermostable powders.* $T_g > T_d$	[56]
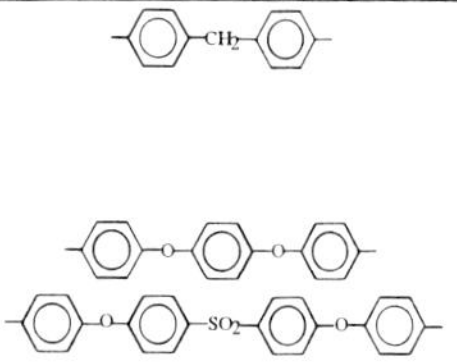	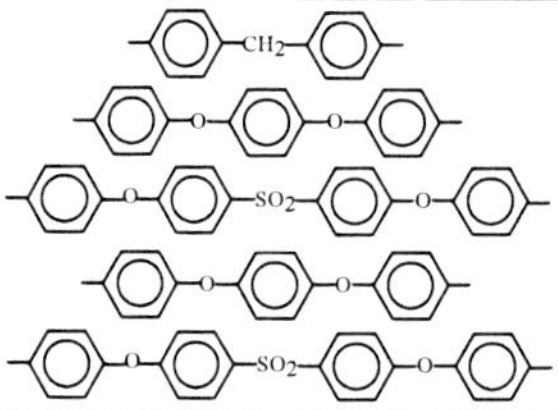	*Thermal stability ~300 °C*	[57]

T_d=temperature of decomposition

The polyadditon reaction can be performed in the melt and under pressure [70], and in solution at temperature ranging between 25 and 160°C. The molar ratio between bismaleimide and diamine can be widely varied. If excess BMI is employed, a prepolymer with mainly maleimide terminations is formed. This may be cured upon heating at temperatures around 200°C to form the poly(aminobismaleimide). The BMI is always employed in excess to obtain a resin susceptible to crosslinking. The kinetic study of the polyaddition of diamines to BMI performed by many methods: conductometry [71], spectrometry IR-FTIR [72], FT-NIR [73], calorimetry (DSC) [74-77], size exclusion chromatography (SEC) and FTIR [78] etc.

The polymerization kinetic of BMI with diamines is a second-order reaction and the activation energy varies between 50 and 60 kJ/mol [78]. The step of rate determining involves the reaction between primary amine and the complex due to the hydrogen binding between amine and maleimide groups [73]. *Ab initio* calculation of the addition reaction of a secondary amine (dimethylamine) to maleimide in the neutral medium proves that the reaction takes place by a zwitterionic intermediate (Figure 3) [79].

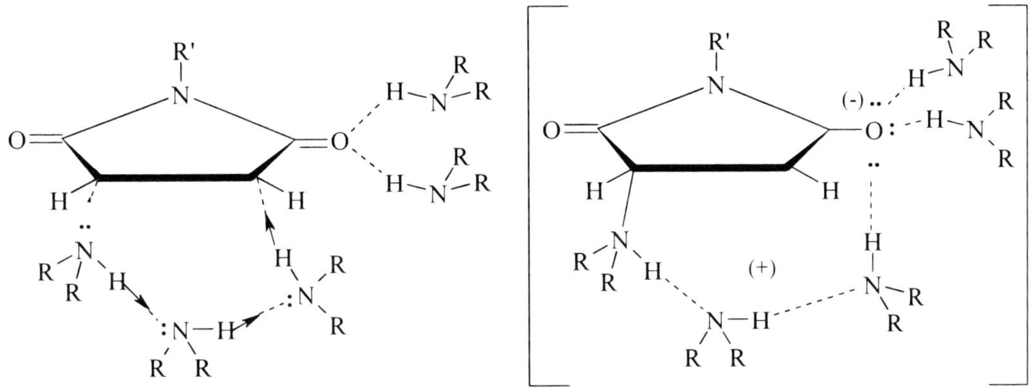

Figure 3. Nucleophilic attack and proton relays in the neutral Michael addition.

Linear polymers (A), oligomers with maleimide end (B) or oligomers with amine end (C) can be obtained depending on the ratio reactants (Scheme 17).

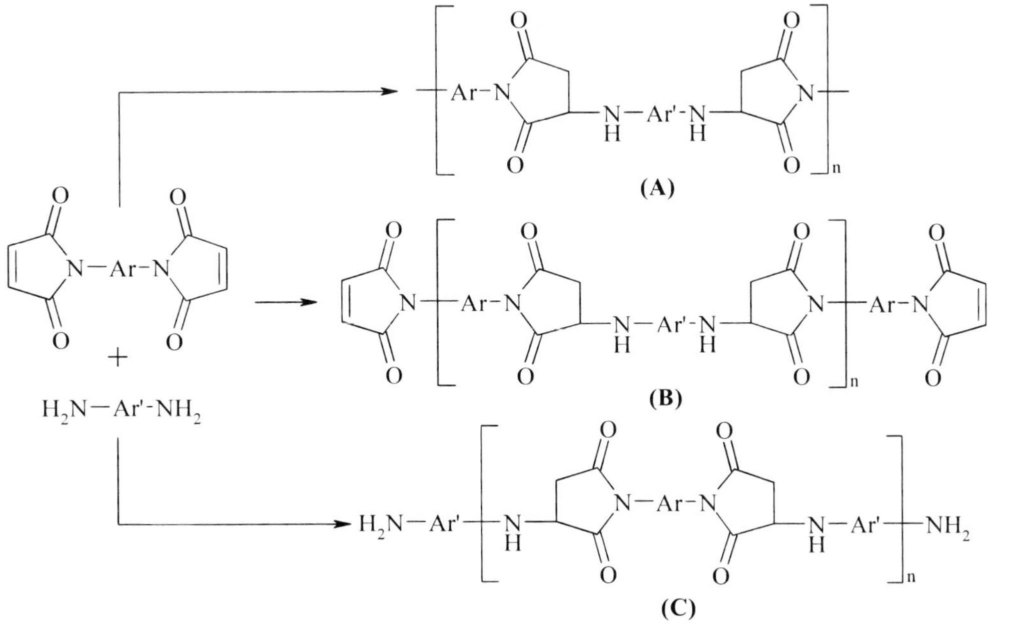

Scheme 17.

In general, the linear poly(aspartimide)s are soluble in aprotic dipolar solvents, have good thermostability ($T_{decomp} > 300°C$) and are used as materials for electronic applications.

Some ways were approached to improve the solubility and processability of polymer structures based on bismaleimides:

- incorporation of asymmetric groups (methylnaphthalene [80], methylanthracene [81], anisyl [82], phenylphosphine oxide [83], dichlorobenzophenone [84]);
- the using of bismaleimides having aromatic spacer with various substituents [85];
- bismaleimides containing different polar substituents and diamines having oxyethylene bridges [86-87].

Another method to enhance the solubility of polymers was the synthesis of poly(aminobismaleimide)s by the reaction of 5-aminoisophthalic acid with diamines (Scheme 18) [88].

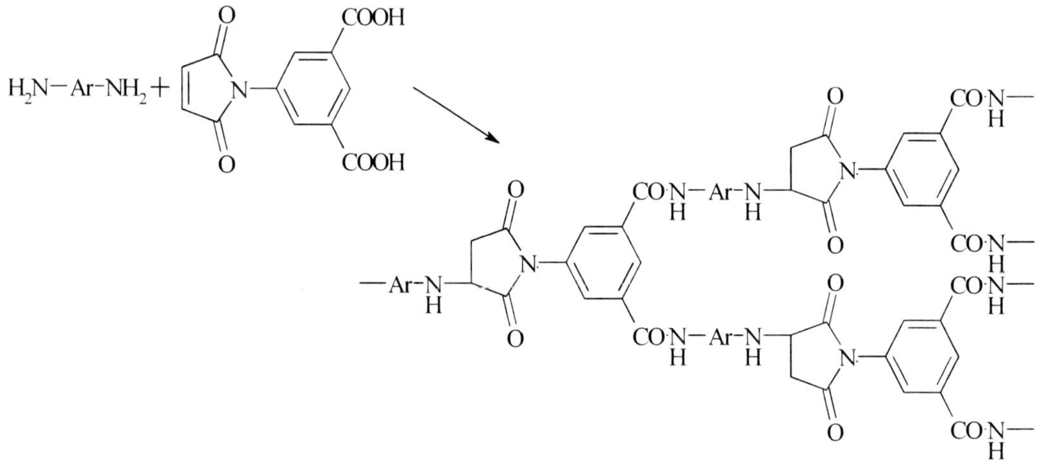

Scheme 18.

Poly(aspartamide)s have potential application in second order non-linear optical (NLO) materials due to their high glass transition temperatures (~260 to 290°C) which help to maintain the critical orientation of the NLO chromophores. In addition, poly(aspartamide)s do not exhibit poling interferences due to water evolution as seen with polyimides during cyclization of the precursor poly(amic acid). Wu et al. used Michael addition to obtain an NLO-containing poly(aspartamide) prepolymer with maleimide end groups that was thermally cured to obtain the final product [88]. At higher curing temperatures, the glass transition temperature of the cured poly(aspartamide)s appeared to increase, which demonstrated that curing continued to 250°C. By the using diamines containing chromophore groups (Figure 4), colored PABMI with NLO properties, good polarizability and high thermostability were synthesized [89-92].

Oligomers with the maleimide ends (Scheme 17 B) have most practical applications and are used in the synthesis of composite materials with higher thermomecanical properties than that of the epoxy and phenolic resins, such as:

- good dimensional stability in the range of -50 and 250°C;
- absence of microcracks during of the thermal treatment;
- resistantance to fire, oil, chlorinated solvents, dilution acids, and moisture;
- excellent radiation resistance (up to 10^{10} rad/s);

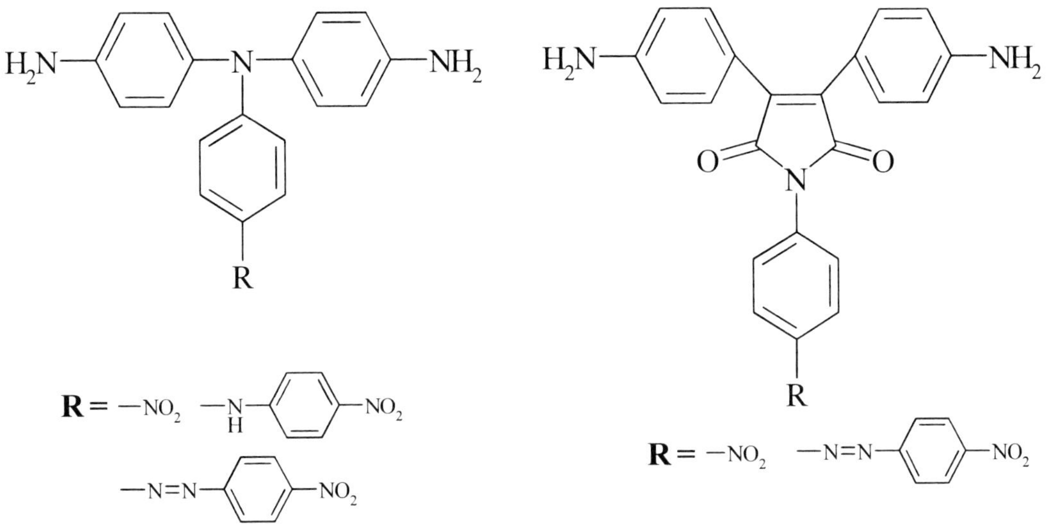

Figure 4. Structures of diamines useed in the synthesis of PABMI with nonlinear optical properties.

- low value of the flow at high temperatures under load due to the high rigidity at elevated temperatures;
- maintaining of the physicol-mechanical and electrical properties in the temperature range 150-250°C.

Based on above polymer structures, Rhone-Poulenc Company commercializes two types of composites under Kerimid 601 and Kinel trademark [93]. Kinel compounds contain structural and abrasive compounds that have self-lubricant fillers: graphite, molybdenum sulphide, and powder of polytetrafluoroethylene. Also, a composite material based on PABMI patented as Ponimid, was prepared in three types: L – for copper laminates [94], T – for molding compounds and P – for compressing compounds [95-97].

PABMI prepolymers having maleimide ends were used in the synthesis of the matrix for composite materials as such or in the mixture with other components [98-110]. The study on thermal behavior of PABMI oligomers was monitored by thermal methods (differential calorimetry, thermogravimetric analysis) [111]. DSC curves of the prepolymers obtained by melt polymerization exhibit multimodal exotherms that are characteristic to nucleophylic addition of secondary amine groups to maleimide groups at temperature above 150°C and polymerization of maleimide-chain-end at temperature above 200°C.

Prepolymers with amine end (Scheme 17 C) were obtained with excess of diamine and used as curing agents for epoxy resins [112-121]. Also, PABMI were used in the mixture with other polymers, such as: polyetherimide, [128], polyurethanes [123-124], vinylether polymers

[125], polybenzimidazole [126], polyamidoamine [127], etc. Many patents describe synthesis of the composite materials based on PABMI [128-138].

Thermostable polymers were obtained by using other amine derivatives (dihydrazide [139], hydrazide of 4-aminobenzoic acid [140-141], urea and thiourea [142], aminoacids [143], hidantoine [144], cyanamide [145], hydroxylamine [146], aminophenol [147-148]).

2.2. Polyaddition of Bisalcohols and Bisphenols

Poly(maleimido ether)s with good mechanical and thermal properties were obtained by the addition reaction of bisphenols to bismaleimides in basic catalysis, in the dipolar aprotic solvents (Scheme 19) [149-154]. The following bisphenols were used: azodiphenol [149], bisphenol A [155, 150], 1,4-butandiol [151], 1,2-ethylene glycol, 1,6-hexandiol, diethylene glycol and poly(ethylene glycol)adipate [152], tartaric acid and 5,5'-methylenedisalicylic acid [156], phenolic prepolymers with OH end groups [153, 157].

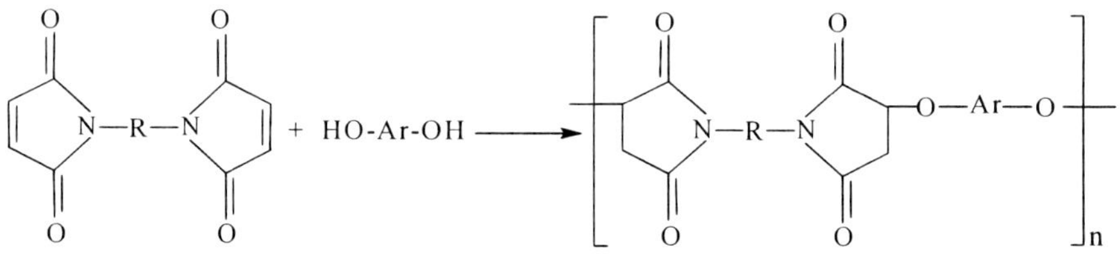

Scheme 19.

Oligomers containing mesogene groups with lower molecular weights exhibit thermotrope liquid crystal behavior [158].

2.3. Polyaddition of Bisthiols

The curing process of protein containing free SH groups with bismaleimides has known since 1956 [159]. Further studies [160-165] reveal the synthesis of poly(imidothioether)s by the addition reaction of bisthiophenols, bisthioalcohols or hydrogen sulfide to bismaleimides (Scheme 20).

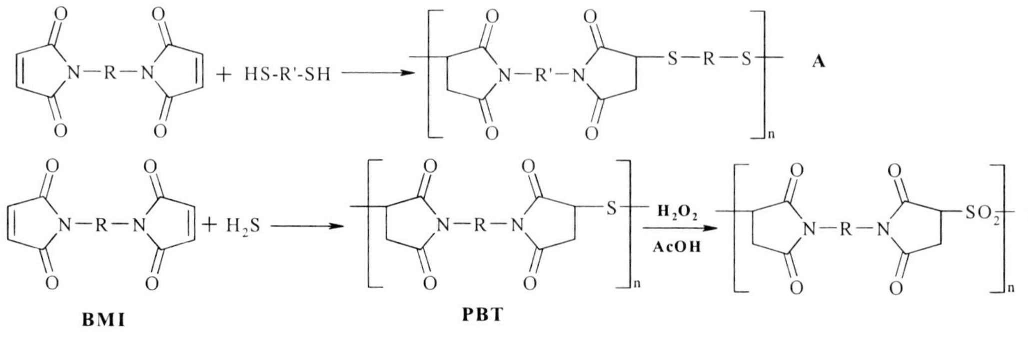

Scheme 20.

The reaction occurs in basic solvent (DMF, DMAc, pyridine, neutral or weak acids such as *m*-cresol, tricresols) and in the presence of basic catalyst (triethylamine, tri-*n*-buthylamine). Crivello performed the reaction between BMI and H_2S (Schema 20 B) followed by the curing in the base solvents [160-161] using DMF and acetic acid as proton donor, and *m*-cresol. The author also synthesized new poli(bismaleimidosulfone)s (PBS) by the oxidation of poly(bismaleimidothioeter)s with peracetic acid [161]. The polymers with similar structures were obtained by the reaction of *p*-aminothioether with citraconic anhydride [166].

By the polyaddition reaction of polysulfides with M_n=2000-8000 to BMI, block polymers containing poly(imidoether)s and polysulfide were obtained [165]. The reaction occurs at 40°C in *m*-cresol or dichloromethane in the presence of triethylamine. The polymer that contained up to 70% polysulfide is considered as thermoplastic. The stability of obtained block polymers decreases with the increasing of the polysulfides content but depends on their molecular weight and the structure of polysulfide block. Miao *et al.* synthesized polymers by the reaction of polyphenylsulfide oligomers containing SH end groups with bismaleimides [167]. New optically active poly(thioetherbismaleimide)s derived from chiral bismaleimides and 1,5-pentanedithiol were synthesized by Michael-type polyaddition [168]. The polyaddition proceeds with a quite low stereoselectivity in terms of asymmetric induction on the configuration of the newly formed chiral centers. The X-ray diffractograms revealed an amorphous structure of polymers due to a low degree of stereoregularity originated by the presence of randomly distributed constitutional repeating units having different diastereomeric configurations. Gaina reported the synthesis of new poly(thioetherimide)s obtained by the polyaddition reaction of bismaleimides having ester linkages and bisthioethylether [169] or by the reaction of biscitraconimide with same reagent [170]. The reaction between biscitraconimides with bisthiols and bis(2-oxazoline) also leads to the new polymers with elastic properties [171]. In the case of the reaction of biscitraconimides with bisthiols it was observed that the addition reaction of thiol to citraconimide group regioselectively takes place to the substituted carbon of bond double, yield being above 90% (Scheme 21).

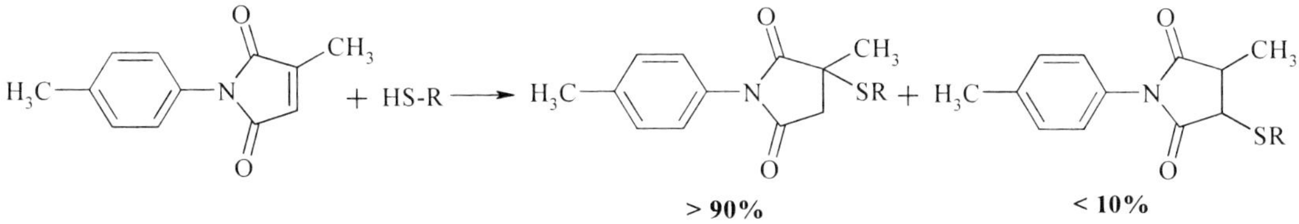

Scheme 21.

Poly(imidothioether)s have high mechanical resistance and thermostability and are recommended for electrical insulating films and compressing compounds.

2.4. Polyaddition of other Nuchleophylic Reactants

Besides nucleophylic agents already reported, the following agents were also used on large scale: disulfinic acids (AD) [171-173], dialdoxime (DA) [174], bis-(2,3-dimethyl-6-

chinoxolinimide) (BDC) [175-176]. The addition reactions of such agents were presented in Scheme 22.

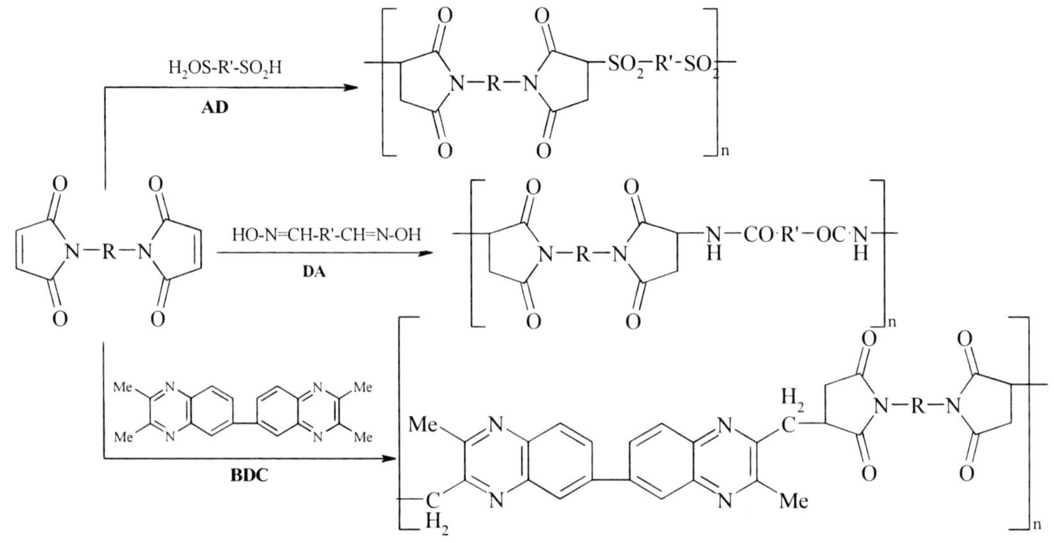

Scheme 22.

3. Polycycloaddition Reactions

Bismaleimides participated to the Diels-Alder-type addition reaction due to their unsaturated structure followed by the closure of new rings, alongside of other components having two, three or four active centre. The products resulted by polycycloaddition reactions present condensed rings along of the macromolecular chain that lead to the high thermal stability.

3.1. [2+2] Polycycloaddition

[2+2] Polycycloaddition reactions of bismaleimides are radicalic reactions and can occur by the photochemical way [177-183] or in the presence of peroxides [184] (Scheme 23). The kinetic studies on the polycycloaddition reactions of bismaleimides [177-180] emphasized two different mechanisms.

In case of the sensitized process the end-group chromophore is excited by the energy transfer at each step of chained reaction. For unsensitized process each chain-propagating step involves a photochemical reaction, the whole process being considered a multistep reaction.

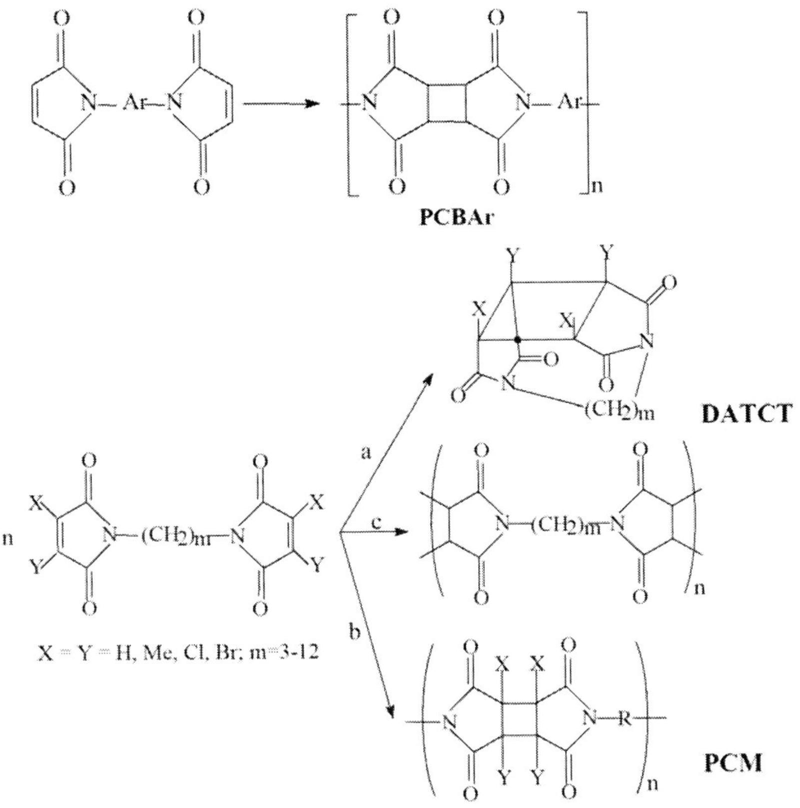

Scheme 23.

[2+2] Polycycloaddition of aromatic bismaleimides leads to the forming of polyimides from cyclobuthantetracarboxylic acid (PCBAr). In the case of using the aliphatic bismaleimides, three reaction types are produced depending on the length of aliphatic chain (Scheme 23, a-c) [132-134]. For m=3-7 and X=Y=H, diazatetracyclotetrone (DATCT) is obtained by the photocyclomerization reaction (Scheme 23, (a)). Polymers with high molecular weights (PCM) are obtained for just m=7-12. De Schryver et al. assigned the necessary conditions for the elimination of the possible competitive ways between a, b, and c, and a general reaction scheme for the photopolymerization is proposed [178]. Sonntag *et al.* studied the photoaddition reaction of maleimides and proved the presence of various reaction types: electron transfer, deprotonation, and cyclobutane dimer formation [185]. By the two-step reaction of cyclobutantetracarboxylic dianhydride with aromatic or aliphatic diamines, polymers PCBAr and PCM were synthesized [186].

3.2. [3+2] Polycycloaddition

Cycloaddition, [3+2] system or 1,3-dipolar cycloadditions represents a reaction between of 1,3-dipol (a-b-c) with unsaturated system (bismaleimides) that lead to thermostable heterocyclic polymers containing five-atoms rings (Figure 5) [187].

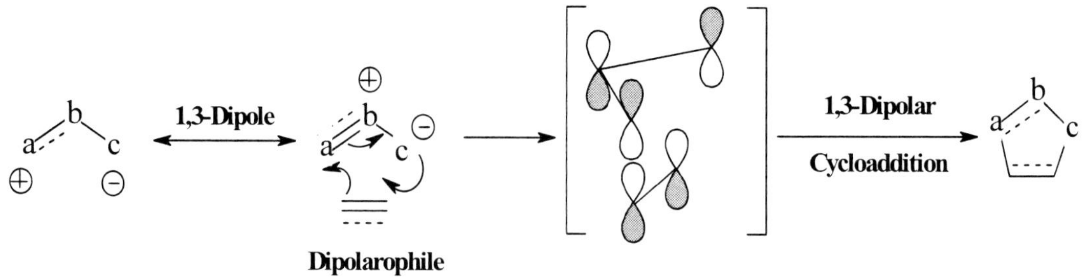

Figure 5.

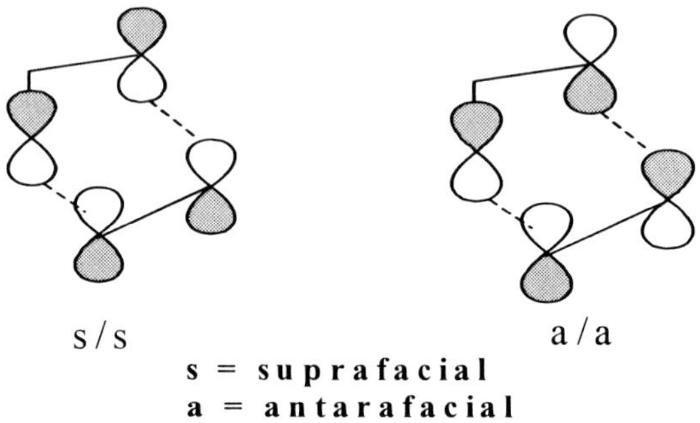

Figure 6. A summary of the *suprafacial* and *antarafacial* orbital interactions that are permitted or disallowed according to the Woodward and Hoffmann selection rules [207].

1,3-Dipolar cycloadditions share many features with the more well-known Diels-Alder reactions. An important difference, when studying medium effects, is that for 1,3-dipolar cycloadditions rate constants sometimes decrease and sometimes increase with solvent polarity, and in some cases are almost completely insensitive to the medium. In many cases, rate constants are influenced far less by the polarity of the medium than in the case of typical Diels-Alder reactions, and hardly ever exceed 1 order of magnitude over the full range of solvent polarities, i.e., from *n*-hexane to water [188-190]. Same exemples of 1,3-dienophiles and polymer structures are presented in Table 6.

Studies on the 1,3-dipolar cycloaddition mechanism evidenced a synchronic process with the formation of an activated rigid complex [196]. This mechanism is substantiated by both low value of activation enthalpy and high enough negative values of entropy. Cycloaddition reactions can occur either across the same face of the π-orbital system in the case of a *suprafacial* addition or across opposite faces of the π-orbital system—in the case of an *antarafacial* addition process (see Figure 6).

Woodward and Hoffmann proposed a series of selection rules to predict permissible cycloadditions and those that are forbidden when thermal conditions are employed [207]. Furthermore, these selection rules revealed that reactions which feature $4n + 2$ π-electrons (where n = integer value) are allowed when the starting materials are heated whereas those cycloaddition reactions which feature $4n$ π-electrons are forbidden thermally, yet are allowed under photochemical exposure.

Table 6. The structures of some 1,3-bisdienophiles and resulted polymers

Nr.	1,3-Dienophile	Structure of polymer	Reference
1.	Ph-CH=N-N=CH-Ph benzaldehyde [(1E)-phenylmethylene]hydrazone	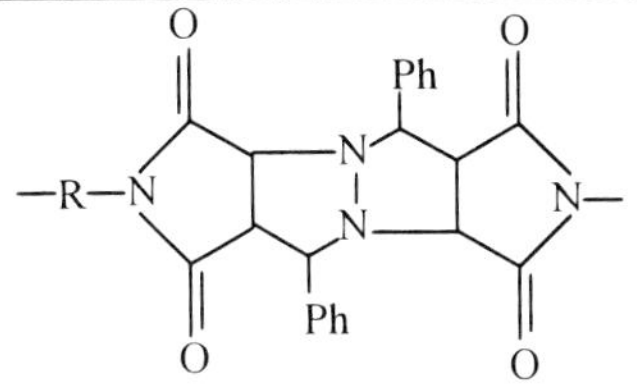	[191-193]
2.	Ph-N(→O)=CH-Ar-CH=N(→O)-Ph bis-nitrone	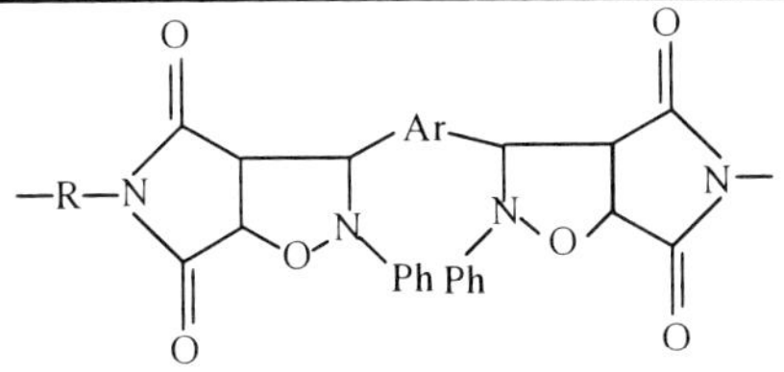	[191, 194-198]
3.	$^{-}O\text{-}N^{+}\equiv C\text{-}C_6H_4\text{-}C\equiv N^{+}\text{-}O^{-}$ dinitrileoxide	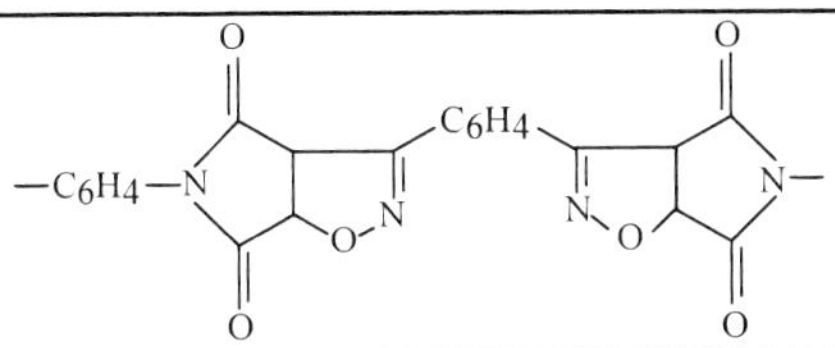	[192, 197, 199]
4.	$\bar{N}\text{-}N=\overset{+}{C}\text{-}C_6H_4\text{-}\overset{+}{C}=N\text{-}\bar{N}$ (N-Ph, N-Ph) dinitrilimine	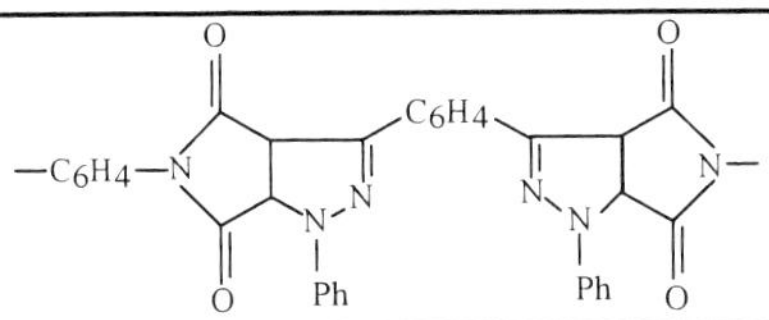	[200]
5.	$N_3\text{-}R'\text{-}N_3$ bis-azide	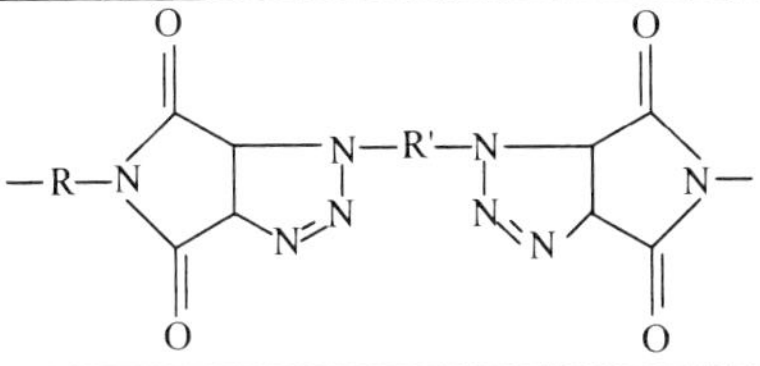	[197, 201-202]
6.	$\overset{+}{N}=N\text{-}\bar{C}H\text{-}R'\text{-}\bar{C}H\text{-}N=\overset{+}{N}$ diazoalkane	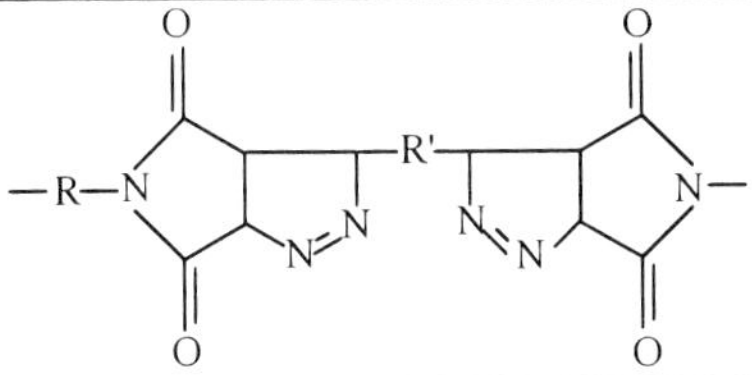	[197, 201]
7.	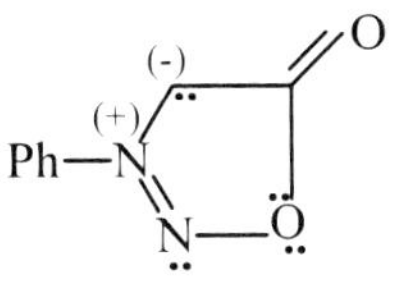 N-phenyl-sidnone	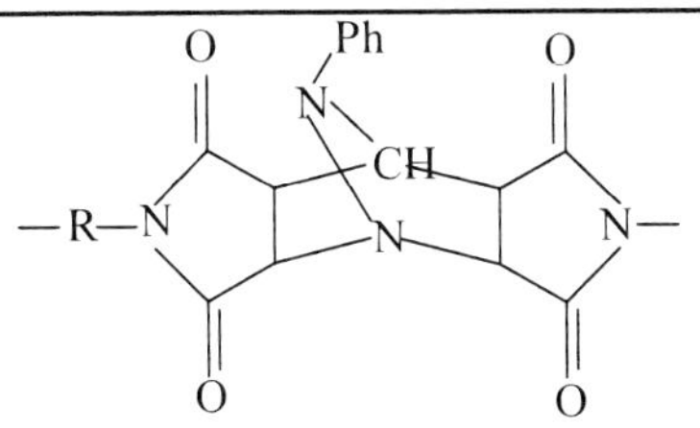	[197, 203-205]

Table 6. Continued.

Nr.	1,3-Dienophile	Structure of polymer	Reference
8.	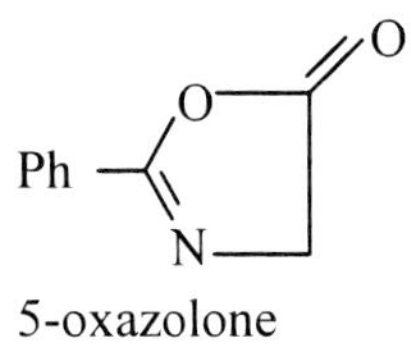 5-oxazolone	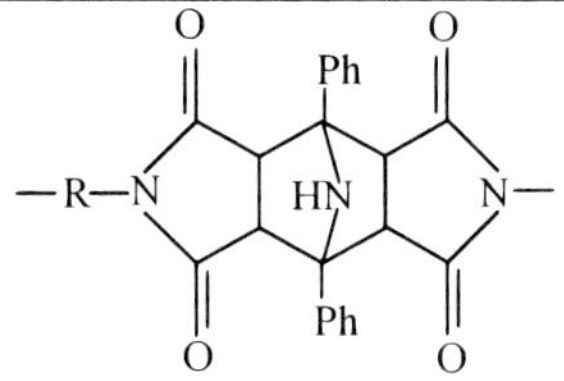	[206]

1.3-Dipolar cycloaddition reactions are catalyzed by dicarboxylic acids [208], Lewis acids [209] or organo-metallic catalysts [210].

3.3. [4+2] Polycycloaddition

The Diels-Alder reaction is one of the most important carbon-carbon bond forming processes available to organic chemists. This transformation belongs to the class of pericyclic reactions and is a [4π+2π] cycloaddition, usually between an electron-rich 1,3-diene **1** and an electrondeficient dienophile **2** (Figure 7). The reaction creates two carbon-carbon bonds and up to four new stereocenters. The mechanism of this reaction has been extensively studied and debated for several decades. The most accepted version of the reaction mechanism considers carbon-carbon bond formation by concerted mixing of the dienophile and diene π and π* orbitals. In the course of thc reaction, a six-member carbocycle is formed. In fact, two π bonds belonging to the substrates are converted into two sigma bonds in the six-membered ring. Because six π electrons are involved, it has been viewed as proceeding through an aromatic transition state, which is energetically more favored than a non-concerted reaction path. The Diels-Alder reaction was a successful candidate investigated in the field of catalytic antibodies, and obviously became an attractive target also for researchers in the ribozyme catalysis field.

Two methods are occurred in the synthesis of polyimides by the Diels-Alder reactions using bismaleimides as dienophil: (i) homopolycycloaddition reaction of AB monomers that contain both functions (diene and dienophil) [197, 211-220]; (ii) copolycycloaddition reaction of bisdienes to bismaleimides [213, 222-260].

The secondary method principle bases on the reaction between BMI and a dienic component (bispyrone [197, 211, 213], cyclopentadienone [212-213] or thiophen dioxides [213-214]) and leads to the formation of inactive compound followed by the elimination of the low molecular weight molecules (CO_2, CO or SO_2) and obtaining the AB monomers. Several types of AB monomers, dienes, as well as corresponding polymers are presented in Table 7 and 8, respectively.

1 2

Figure 7. General Diels-Alder reaction of diene (1) and dienophile (2).

Nowadays there is a considerable interest not only in the synthesis of new types of polymeric materials but also in the modification of existing polymers in order to alter their properties to meet requirements for new applications. One of the methods for modification is the grafting reaction, which provides an opportunity to vary physical and chemical properties of polymers, by the cycloaddition reaction of copolymers functionalized with anthracene units (either in the macromolecular chain [216], or in side-chain of polymer [217-219]) or with terpolymers [220].

It is known that the cycloaddition reaction of the diene (the furan) to dienophile (maleimide) is reversible. The Diels-Alder adduct is formed below 60°C. Above 90°C the adduct opens to form the original furan and maleimide moieties (Scheme 24).

Table 7. Structures of AB monomers and polymers resulted by [4+2] homopolycycloaddition

Nr.	Structure of AB monomer	Polymer structure	Reference
1.	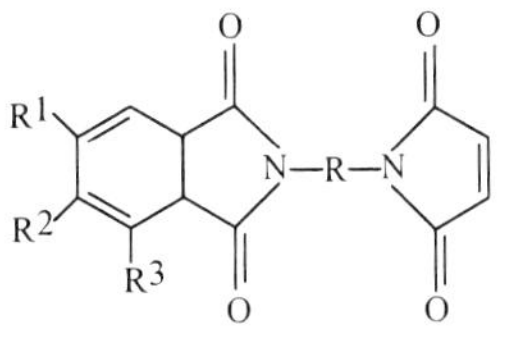	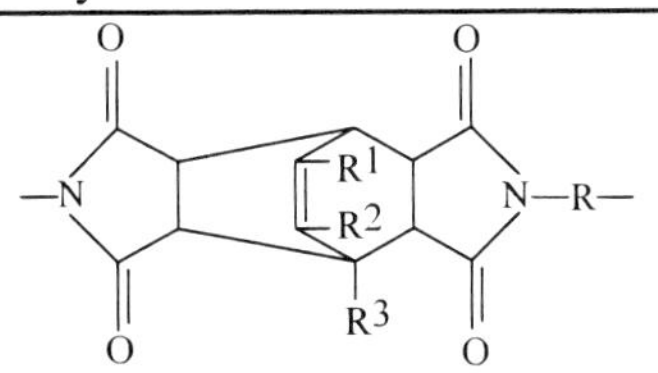	[197], [211-212]
2.	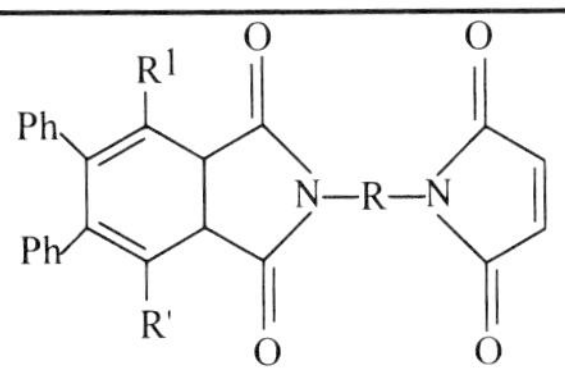	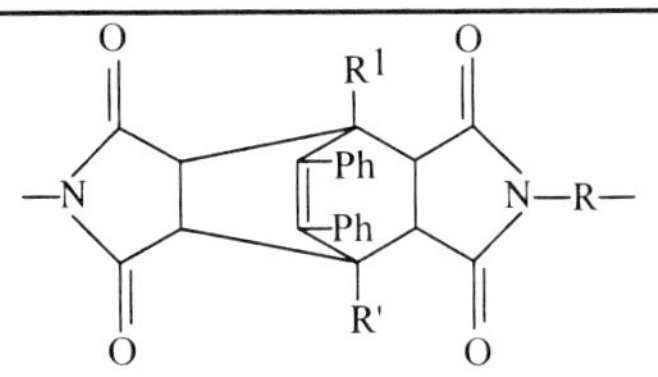	[213-214]
3.	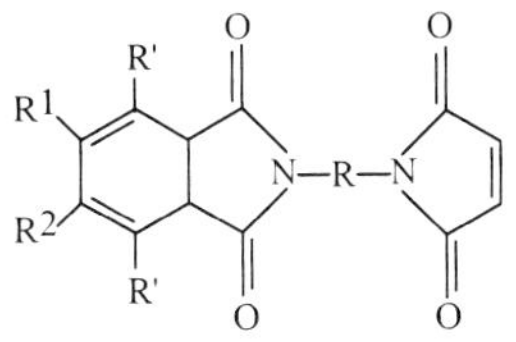	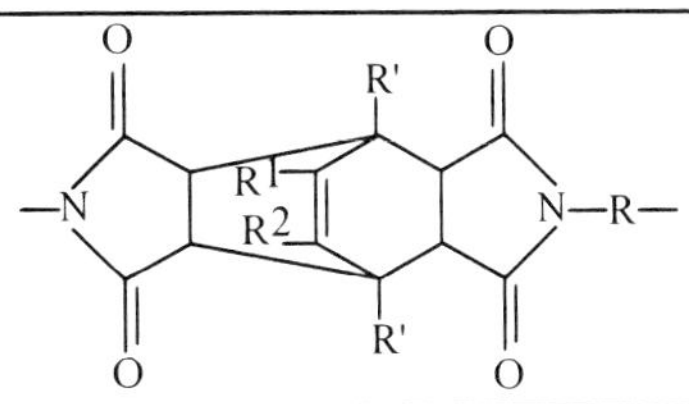	[212, 214]
4.	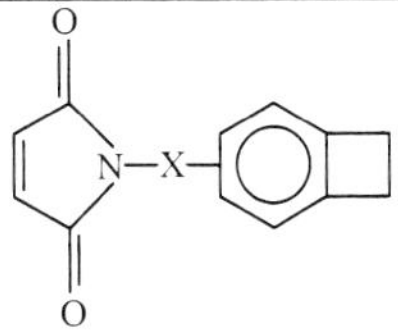	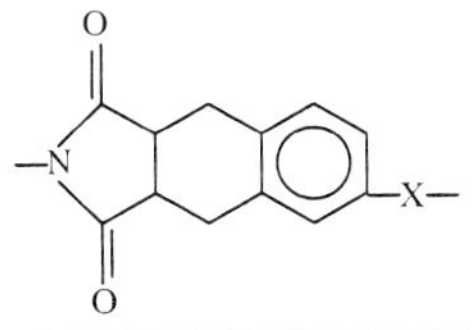	[215]

Table 7. Continued.

Nr.	Structure of AB monomer	Polymer structure	Reference
5.	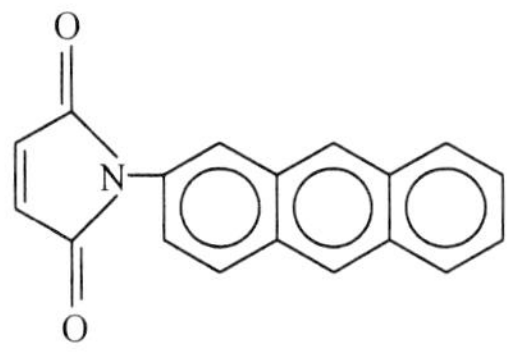	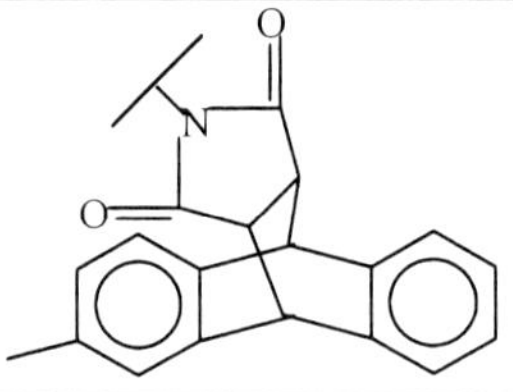	[216]
6.	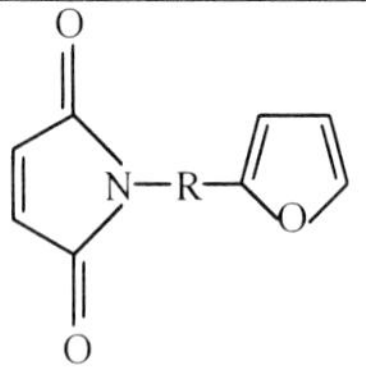	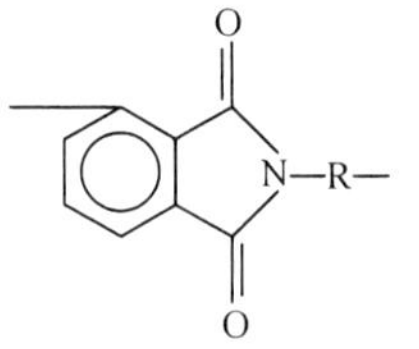	[217, 220]

Table 8. Structures of dienes and polymers resulted by [4+2] copolycycloaddition process

Nr.	Diene	Polymer structure	Reference
1.	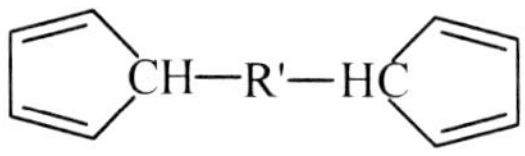	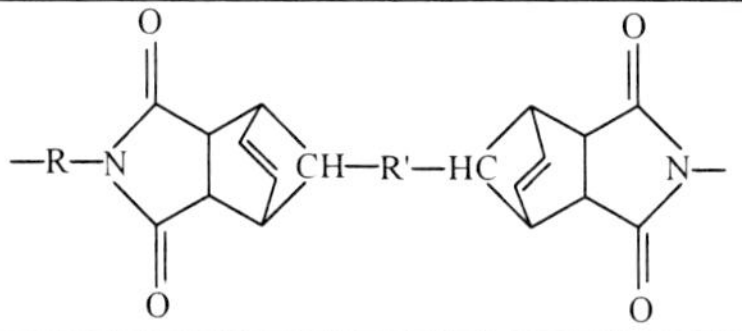	[213, 221]
2.	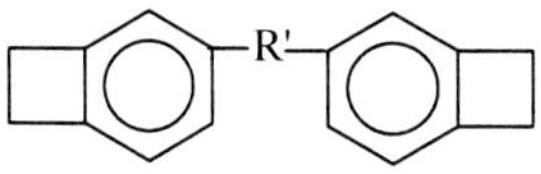	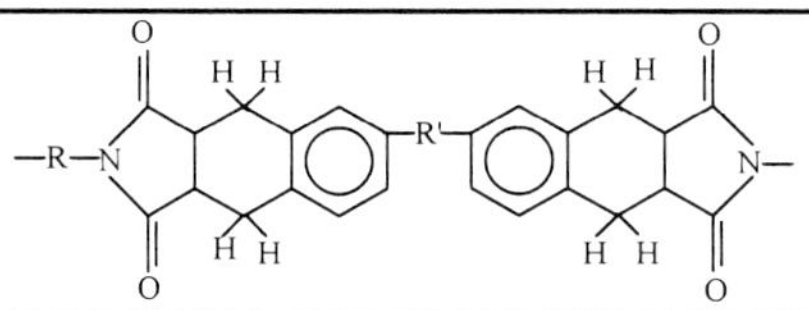	[222-225]
3.	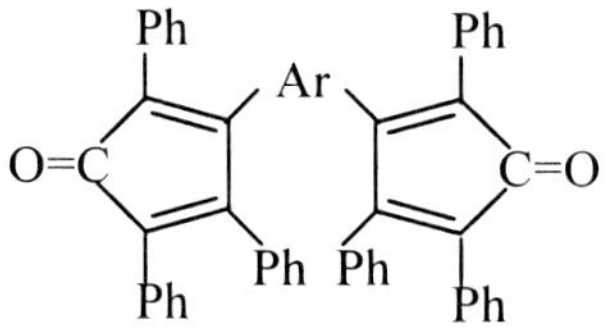	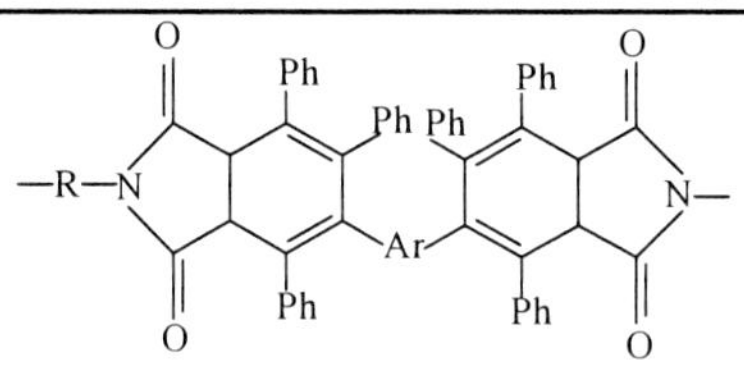	[226-227]
4.	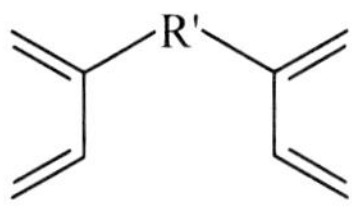	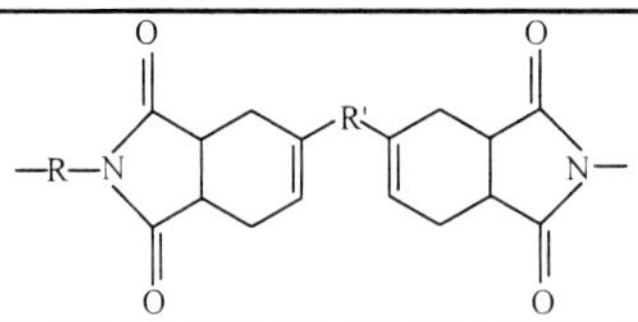	[228-231]
5.	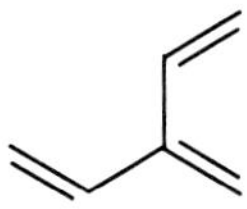	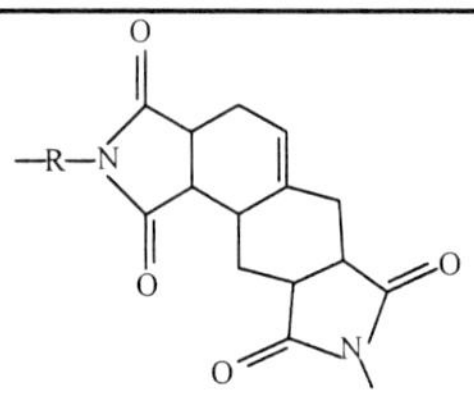	[192, 213] [228]

Nr.	Diene	Polymer structure	Reference
6.	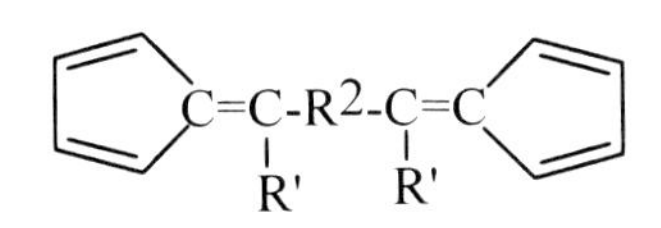	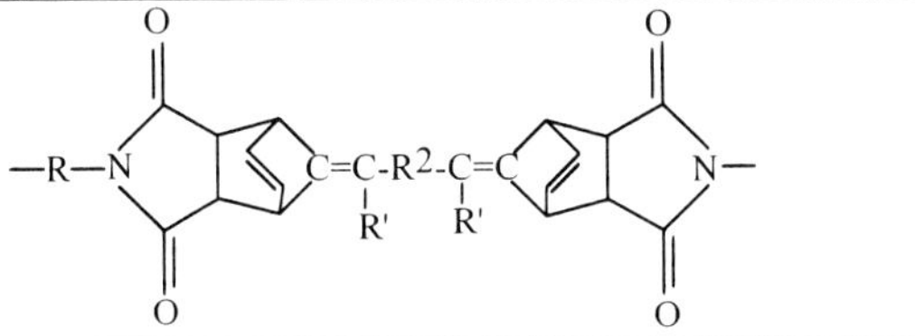	[232]
7.	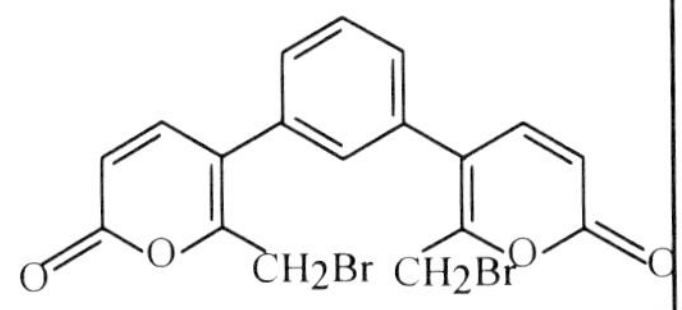	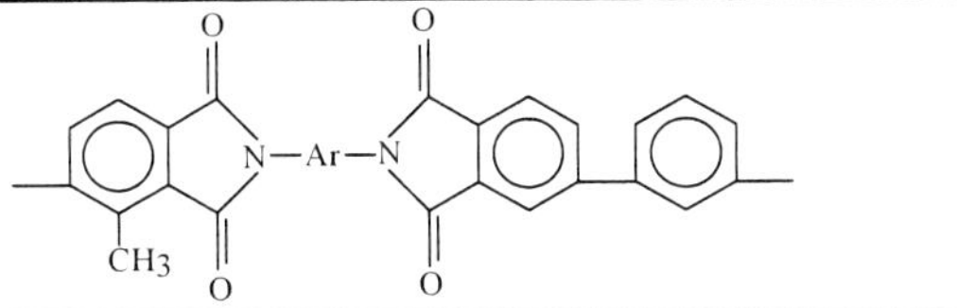	[211], [233-239]
8.	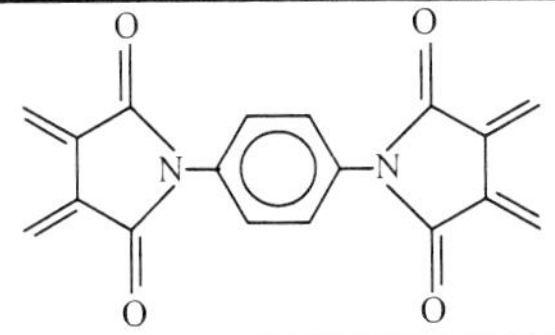	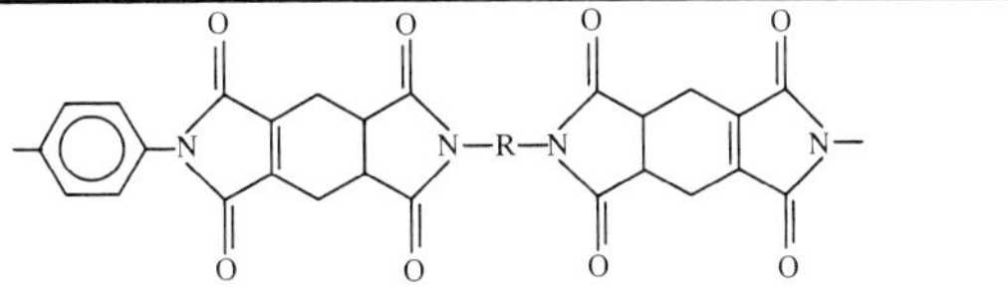	[240]
9.	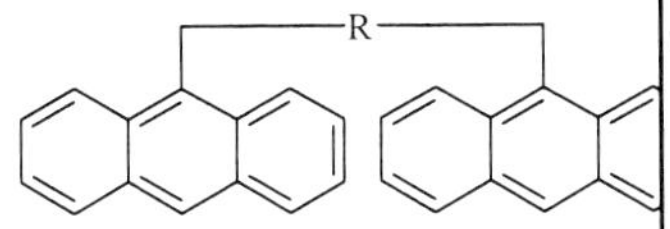	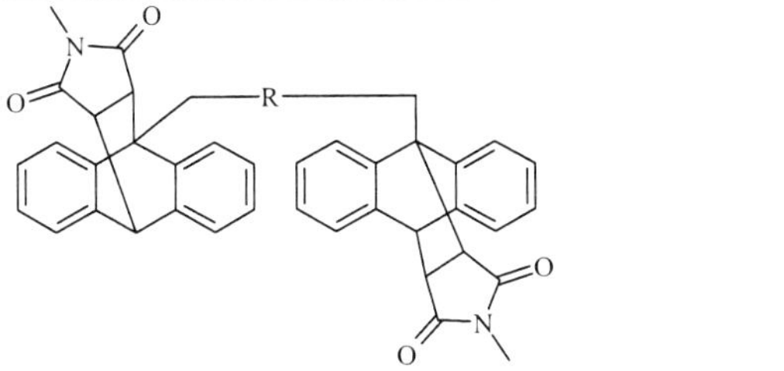	[241-242]
10.	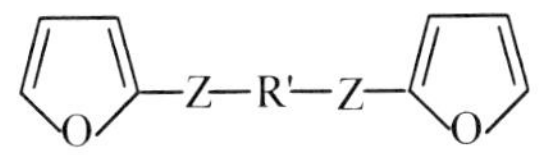	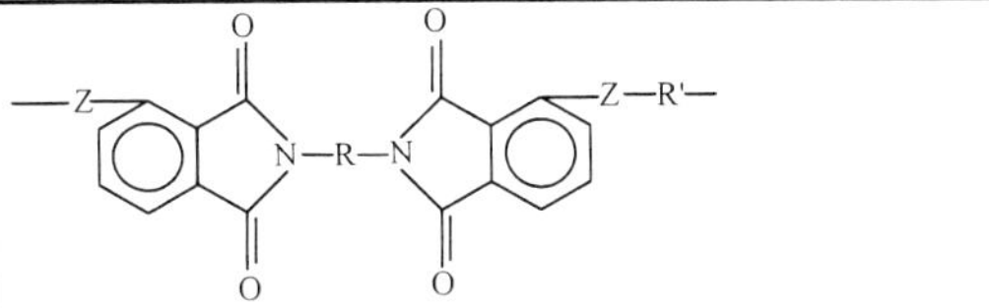	[216, 231], [242-259]
11.	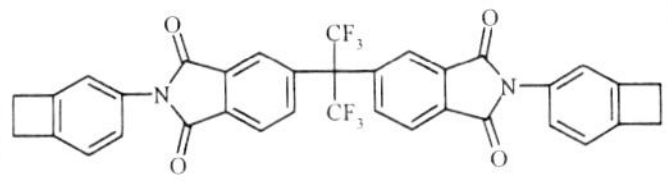	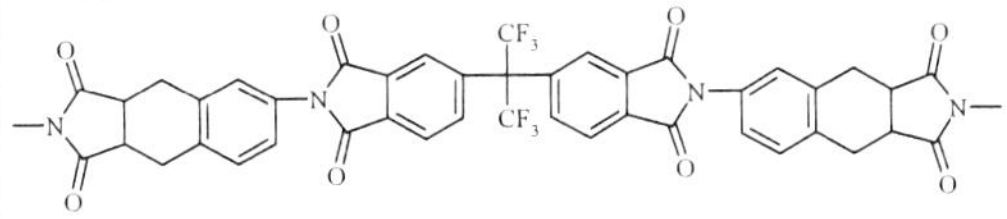	[260]

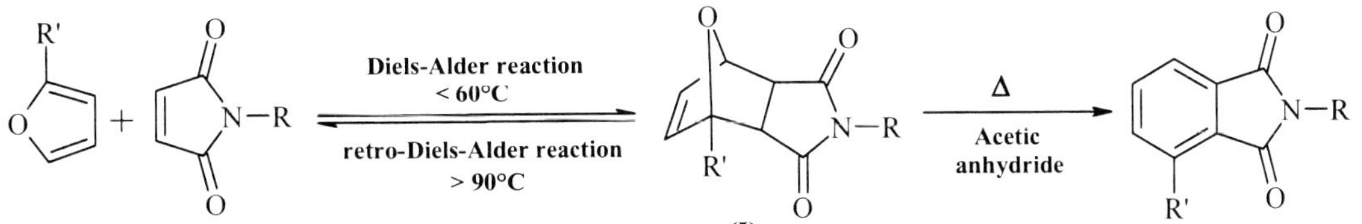

(I)

Scheme 24.

The products from Table 7 and 8, number 6 and 10, respectively, were obtained by the chemical cyclization of intermediary compounds (I) resulting from the cycloaddition reactions (Scheme 24).

Utilization of Diels-Alder reaction in polymer cross-linking reactions to build up the polymer networks results in a new class of thermally-reversible cross-linked polymers. In recent years these thermally-reversible cross-linked polymers are widely studied to explore applications in encapsulants, structural materials, coatings, etc. For example, semiconductor can be redeemed from bad encapsulation by using such polymers as encapsulants; self-repairing polymeric materials made by this kind of polymers can be used as smart materials.

By R choosing in the diene, which contain a reactive epoxy group (Scheme 24), Aubert succeeded to formulate removable elastomeric adhesives that are able to cure via conventional epoxy reactions [266-267]. This area was build upon recent results obtained by Loy *et al.* [268] and McElhanon *et al.* [268-269], which demonstrated that epoxy resins containing Diels-Alder adducts can be used in removable epoxy foams. Previously, McElhanon *et al.* [270] demonstrated the use of Diels-Alder adducts in the preparation of thermally responsive dendrimers. Chen *et al.* [271] recently demonstrated a self-healing polymer that contains Diels-Alder adducts.

Watanabe *et al.* [272] synthesized copolymers of furyl-telechelic poly(ethylene adipate) (PEA2F) with bismaleimide, and trismaleimide, and evidenced the recyclability of these copolymers. Though the reaction with bismaleimide proceeded partially to give a polymer whose molecular weight was only twice as large as PEA2F, the reaction with trismaleimide gave a product with a 3D-network structure. The advantage of the use of a macromonomer is that polymers with various properties can be designed by simply replacing the main chain of the macromonomer. The cycles of Diels-Alder and retro-Diels-Alder reactions could be repeated up to four times without the decomposition of PEA2F. The recyclability of this polymer was also confirmed up to eight times. The cycles of DA and retro-DA reactions induced neither loss of elasticity of the DA products nor reduction of the molecular weight of the retro-DA product.

Gheneim *et al.* synthesized new, thermally reversible, elastomeric networks via Diels-Alder and retro-Diels-Alder reactions [273]. Chen *et al.* reported new remendable cross-linked polymeric materials based on thermally reversible DA cycloaddition [274-275]. These materials are hard, colorless, and transparent at room temperature and do not require solvent for the polymerization processes.

Diels-Alder reaction is also used for obtaining thermoreversible polymer networks by the reaction between copolymers containing furan or maleimide groups and bismaleimides or furane compounds etc. So, by the reaction of styren-furfuryl methacrylic copolymer with bismaleimide, at room temperature, Goiti *et al.* obtained gel in the solution that forming the starting compounds by heating at 110°C (Figure 8) [276]. The studies are developed for obtaining thermoreversible gels [277-278], organic-inorganic polymer hybrids [279-280], high fluorescence materials [281], thermoreversible adhesives [282] etc.

3.4. Polyimides by Polycycloaddition Reaction of Benzeneor Alkyl Derivatives of Benzene to Bismaleimides

Polycycloaddition of benzene and its alkyl derivatives to bismaleimides is a very interesting photochemical process [283-288]. This reaction begins through a [2+2] cycloaddition, which leads to homoanular intermediate (DHA), and are followed by a Diels-Alder process to result in polyimides (PI) (Scheme 25). Xylene derivatives behave similarly [287].

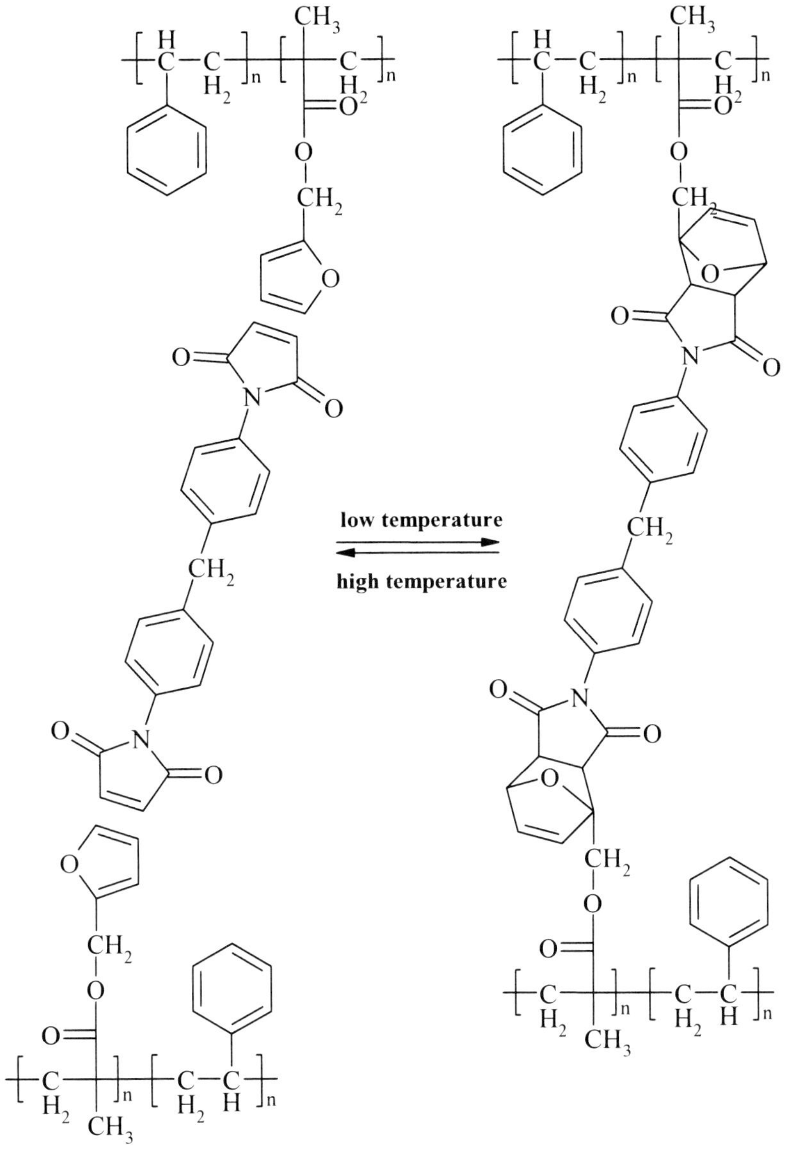

Figure 8. The crosslinking of styrene-furfuryl methacrylate copolymer with bismaleimide.

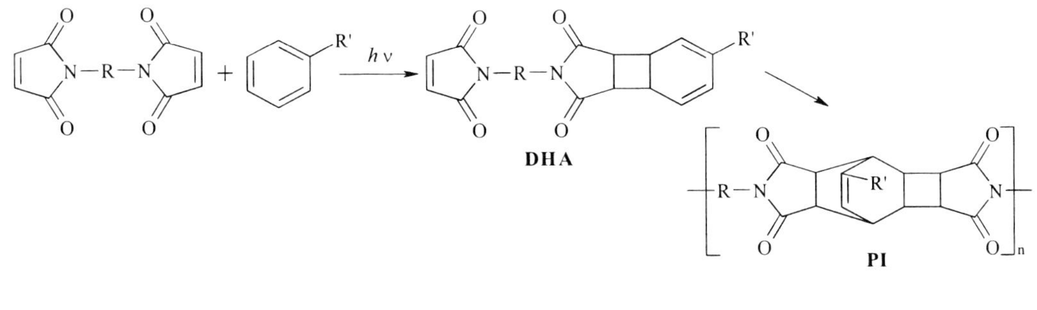

Scheme 25.

3.5. Polymers by Addition-Cycloaddition Diels-Alder Reactions

A special case of cycloaddition reactions is the reaction between maleimides and allyl phenolic compounds (Scheme 26).

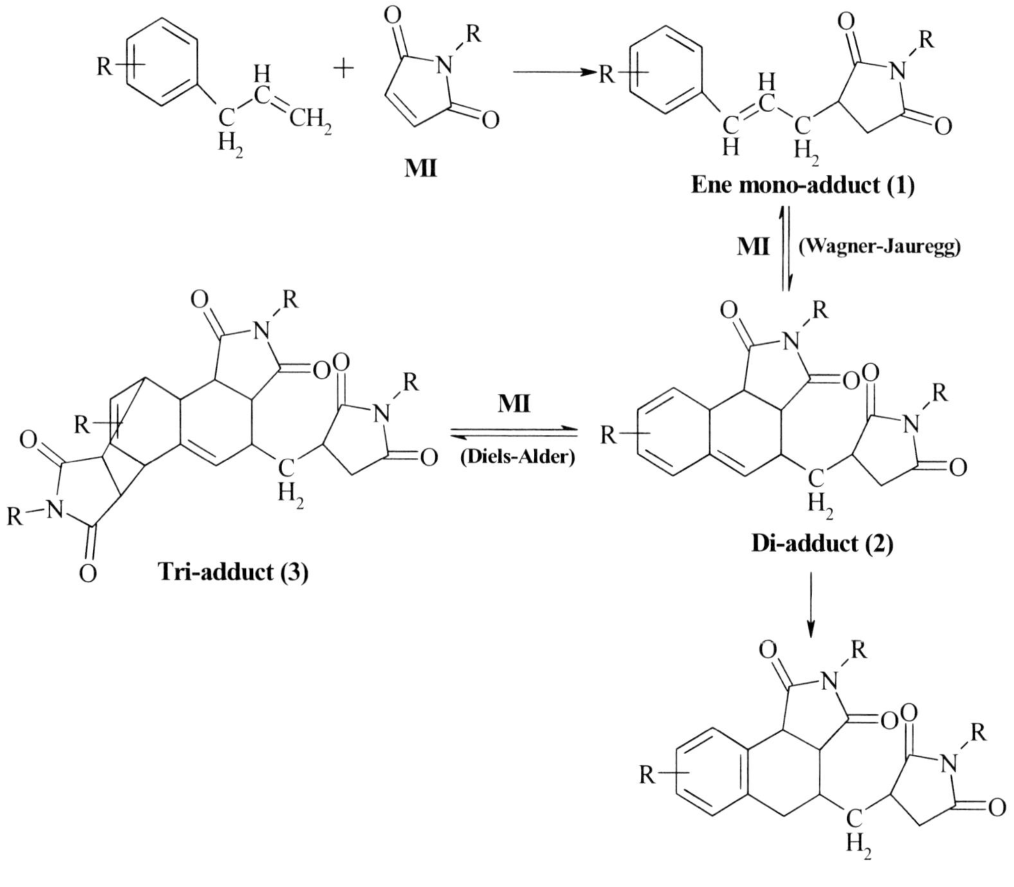

Scheme 26.

Earlier studies done by Enoki *et al.* [289] showed that the reaction between the two components proceeds *via* the Ene reaction. The unsaturated Ene adduct intermediate (1) undergoes a further Diels-Alder type reaction with BMI to give the bis- and tris adducts (2 and 3). The intermediate step (Diels-Alder) is sometimes referred to as Wagner–Jauregg reaction too. The total reaction sequence is also referred to as Alder-ene reaction. At very high temperatures, the occurring of a retro-Diels-Alder reaction is also suggested. Investigation of Reyx *et al.* [290-292] through reaction of model compounds, and identification of products confirmed the Enoki mechanism. The reaction sequences are shown in Scheme 27.

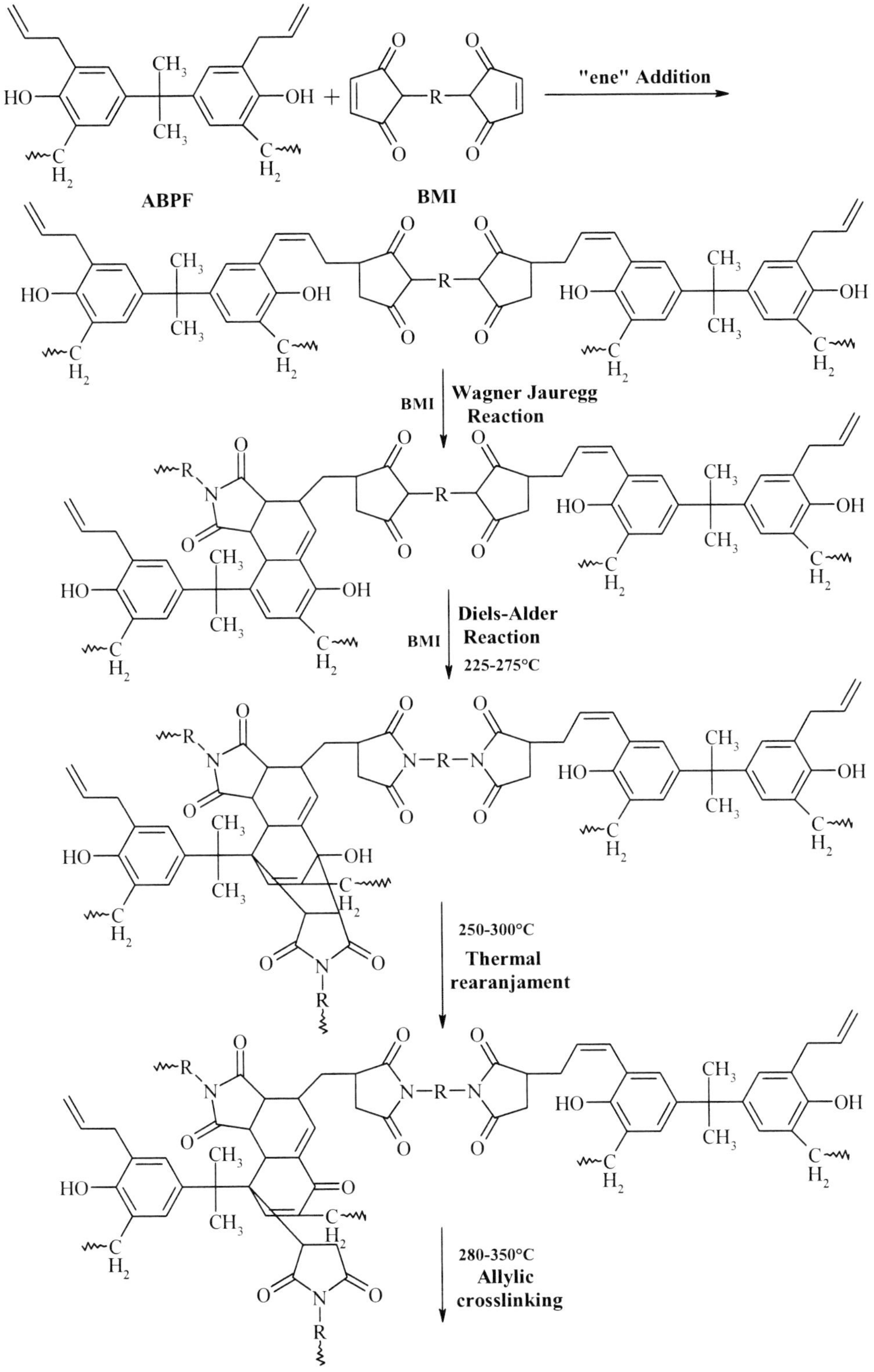

Scheme 27 (Continued on next page.)

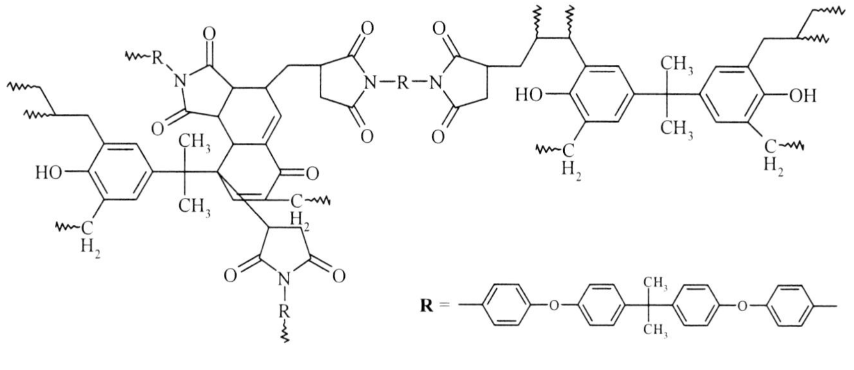

Scheme 27

A multitude of reactions occurs at different temperatures and reactant stoichiometries. All the three structures are possible, but their relative concentrations could depend on the stoichiometry. Ideally, a ratio 1:3 (allyl:maleimide) gives a maximum crosslinking and enhanced thermal stability, but this could lead to brittle matrices. Lower maleimide-content ensures the ene structure-dominated, tough matrix with good flexural properties, but having reduced Tg and thermal stability. In most of the cases, a compromise of various properties is achieved at an allyl/BMI ratio of 1:2.

The extrapolating the same chemistry, the cured structure of the resin can be depicted as in Scheme 27 [299-300]. Several allyl phenols-bismaleimides systems are presented in Table 9 [293-322]. These polymeric matrices in mixture with other components lead to the materials having various applications [323-331].

Table 9. Different allyl phenols/bismaleimides systems

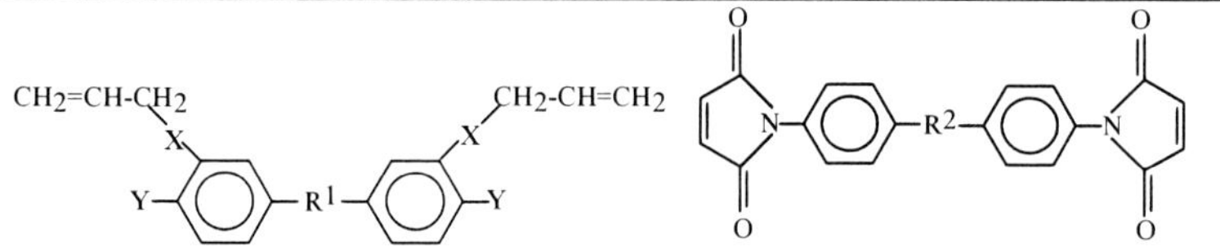

R^1	X	Y	R^2	T_g (°C)	Reference
$C(CH_3)_2$	-	OH	CH_2	200-360	[293-310]
$C(CH_3)_2$	OCO	-	-	-	[310]
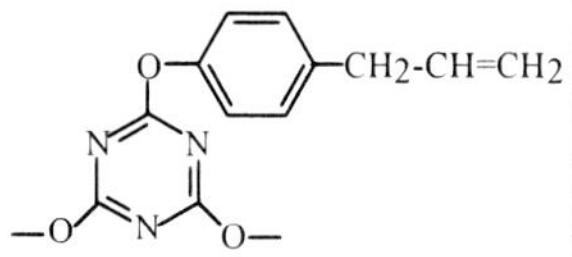	-	-	CH_2	265	[312]
CH_2	O	CH_3	CH_2	-	[313]
-OOC-$(CH_2)_8$-COO-	-	-	CH_2	-	[314, 318]
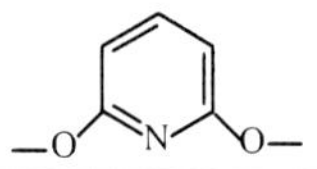	-	-	CH_2	-	[317]

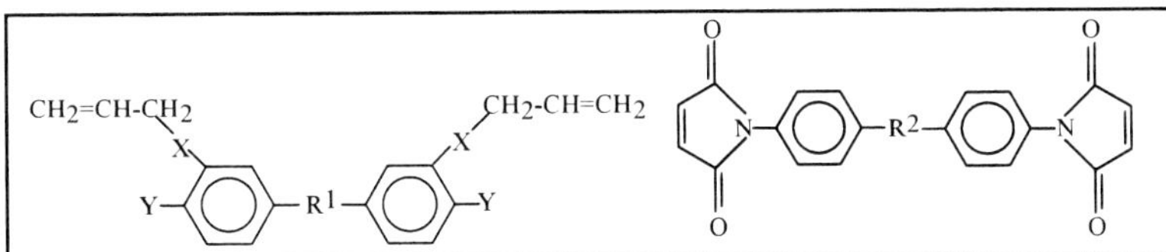

R^1	X	Y	R^2	T_g (°C)	Reference
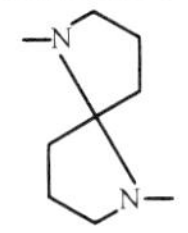	-	OH	CH_2	-	[316]
	O	-	CH_2	-	[316]
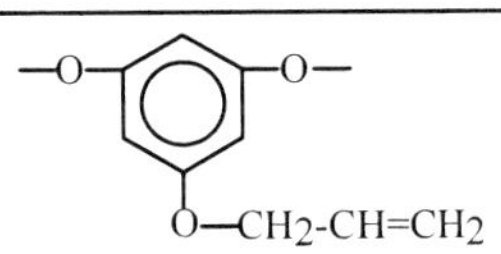	O	-	CH_2	202	[317]
-OOC-$(CH_2)_4$-COO-	-	OCH_3	CH_2	-	[318]
	-	OCH_3	CH_2	-	[318]
S	-	OH	CH_2	-	[319]
CH_2	-	OH	CH_2	-	[320]
triallylisocyanurat	-	-	CH_2	-	[321-322]

Recently, Reghunadhan Nair developed new materials based on the reaction between formaldehyde resins functionalized with maleimide or allyl groups and bismaleimides [332]. Gaina *et al.* obtained new compounds containing both allyl and bismaleimide moieties [333-334].

CONCLUSION

Throughout this chapter the chemical concepts employed to synthesize bismaleimide-based addition polyimides have been discussed. The properties has frequently been mentioned. The most important property of the imide backbone structure is the inherent thermal stability. The target of achieving the temperature performance of linear polyimide has not been reached, because of the aliphatic nature of the reactive endgroups, and because of the low molecular weight of the imide backbone required for processing. Future developments of addition polyimides will, as in the past, focus on the requirement of high thermal and thermal oxidative stability of the crosslinked polyimide and on a satisfactory processing behaviour of the uncured precursor. The Michael addition reaction serves important roles in polymer synthesis, allowing the development of novel polymers from diverse monomer feedstocks. Polymers of numerous different topologies including linear, dendritic, hyperbranched and network polymers are enabled. Michael addition also serves as a synthetic tool for coupling polymers with biological systems such as proteins and enzymes.

REFERENCES

[1] Stenzenberger, H. D.; Heinen, K. U.; Hummel, D.O. *J. Polym. Sci., Polym. Chem. Ed.* 1976, 14, 2911-2925.

[2] Gherasim, G. M.; Zugrăvescu, I. *Eur. Polym. J.* 1978, 14, 985-990.

[3] Nanjan, M. J.; Sivaraj, K. *J. Polym. Sci.: Part A: Polym. Chem.* 1989, 27, 375-388.

[4] a) Hoyt, A. E.; Benicewicz, B C. *J. Polym. Sci.: Part A: Polym. Chem.* 1990, 28, 3403-3415 and 3417-3427; b) Liu, F. J.; Munukutia, S.; Levon, K.; Tesoro, G. *J. Polym. Sci. Polym. Chem.* 1992, 30, 157-162

[5] Mikroyannidis, J. A.; Melissaris, A. P. *J. Appl. Polym. Sci.* 1987, 34, 2657-2672.

[6] Brown, I. M.; Sandreczki, T. C. *Macromol.,* 1990, 23, 94-100.

[7] Matsumoto, A.; Oki, Y.; Otsu, T. *Eur. Polym. J.* 1993, 29, 1225-1229.

[8] Hoyt, A. E.; Benicewicz, B. C. *Polym. Prepr.* 1989, 30(2), 536-537.

[9] Goldfarb, I. J.; Feld, W. A.; Saikumar, *J. Polym.* 1993, 34, 813-818.

[10] Kumar, D.; Gupta, A. D.; Khullar, M. *Polym. Prepr.* 1993, 34(1), 312-313.

[11] Mikroyannidis, J. A.; *J. Macromol. Sci.-Pure Appl. Chem.* 1992, 29, 137-152.

[12] Lee, B. H.; Chaudhari, M. A. *Polym. Prepr.* 1987, 28(1), 75-76.

[13] Packirisamy, S.; Gupta, M. *Eur. Polym. J.* 1990, 26, 703-706.

[14] Dolmatov, S. A.; Volkov, V. S.; Cegodaev, P. P.; Berdnikov, V. M.; Burbo, E. M.; Vasilenko, A. A.; Suvalov, B. F. *Visokomol. Soed.* 1990, 32-38.

[15] Rakoutz, M.; Balme, M. *Polym. J.* 1987, 19, 173-184.

[16] Kumar, D.; M.Fohlen, G.; Parker, J. A. *J. Polym. Sci., Polym. Chem. Ed.* 1983, 21, 565-578.

[17] Takeda, S.; Akiyama, H.; Kakiuchi, H. *J. Appl. Polym. Sci.* 1988, 35, 1341-1350.

[18] Melissaris, A. P.; Mikroyannidis, J. A. *J. Polym. Sci. Polym. Chem.* 1989, 27, 245-262.

[19] Mikroyannidis, J. A.; Melissaris, A. P. *J. Appl. Polym. Sci.* 1989, 37, 2587-2601.

[20] Mikroyannidis, J. *A. J. Polym. Sci.: Part A: Polym. Chem.* 1990, 28, 669-677 and 679-691.

[21] Ueyama, S.; Furukawa, M.; Nishikawa, T. Patent US 4855389/08.08.1989.

[22] Varma, I. K.; Fohlen, G. H.; Parker, *J. A. Polym. Prepr.* 1980, 21(1), 172-173.

[23] Mikroyannidis, J. A. *J. Polym. Sci. Polym. Chem.* 1989, 27, 3465-3479.

[24] Mikroyannidis, J. A.; Melissaris, A. P. *J. Appl. Polym. Sci.* 1988, 36, 691-702.

[25] Kruh, D.; Jablonski, R. J. *J. Polym. Sci., Polym. Chem. Ed.* 1979, 17, 1945-1951.

[26] Jin, S.; Yee, A. *F. Polym. Prepr.* 1991, 31(1), 453-454.

[27] Jin, S.; Yee, A. F. *J. Appl. Polym. Sci.* 1991, 43, 1865-1874.

[28] Lyle, G. D.; Jurek, M. J.; Mohanty, D. R.; Wu, S. D.; Hendrick, J. C.; McGrath, J. E. *Polym. Prepr.* 1987, 28(1), 77-78.

[29] Lyle, G. D.; Jurec, M. J.; Mohatny, D. K.; Wu, S. D.; Hendrick, J. C; McGrath, J. E. *Polym. Prepr*. 1987, 28(l), 77-78.

[30] Lyle, G. D.; Hendrick, J. C.; Lewis, D. A.; Senger, J. S.; Chen, D. H.; Wu, S. P.; McGrath, J. E. Polyimides: Materials, Chemistry and Characterization, Ed. Ferger, C.; Khajasteh, M. M.; McGrath, J. E. Amsterdam, 1989, 213.

[31] Liao, D. C.; Hsieh, K. H. *J. Polym. Sci.: Part A: Polym. Chem.* 1994, 32, 1665-1672.

[32] Gaina, V.; Gaina, C.; Sava, M.; Stoleriu, A.; Rusu, M. *J. Macromol. Sci.-Pure Appl. Chem.* 1997, 34, 2435-2449.

[33] Park, J. C.; Jang S. H. *J. Polym. Sci.: Part A: Polym. Chem.* 1992, 30, 723-729.

[34] Hirano, I.; Muramatsu, T.; Inoue, H. 1th Int.SAMPE I, 1989.

[35] Monacelly, W. J. US Patent 4418181, 1983.

[36] Hasagane, T.; Nakajima, H. US Patent 5247047

[37] Klemarczyk, Ph. EP Patent 0283204,1988 and US Patent 4923997, 1990.

[38] Barthelemy, P. EP Patent 0284519,1988.

[39] Shau, M.; Tsai, P.; Teng, W.; Hsu, W. *Eur. Polym. J.* 2006, 42, 1899-1907.

[40] Wu, C. S.; Liu, Y. L.; Chiu, Y. S. *Polymer* 2002, 43, 1773–1779.

[41] Kovacic, P.; Hein, R. W. *J. Am. Chem. Soc.* 1959, 81, 1187-1190.

[42] Crivello, J. V.; *J. Polym. Sci., Polym. Chem. Ed.* 1973, 11, 1185-1200 and DE Patent 2113063, 1971.

[43] Vygodski, Ya.S.; Aadigezalov, V. A.; Askadsky, A. A.; Slanimsky, G. L.; Bagirov, Sh. T.; Korshak, V. V.; Vinogradova, V. V.; Nagiev, Z. M. *Vysokomol. Soed.* 1979, 21A, 2672-.

[44] Crivello, J. V. *Polym. Prepr.* 1973, 14, 294-295.

[45] Tungare, A. V.; Martin, G. C. *J. Appl. Polym.Sci.* 1992, 46, 1125-1135.

[46] White, J.E.; Snider, D.A. *J. Appl. Polym. Sci.* 1984, 29, 891–899.

[47] Lazarova, E.; Yankova, T.; Aleksiev, B. *Langmuir*, 1995, 11, 2231-2236.

[48] Tungare, A.V.; Martin, G. C. *Polym. Eng. Sci.*, 1993, 33, 614-621.

[49] Volojin, A. J.; Osmalovscaia, L. Y.; Krutiko, E.T.; Rozmislov, A. A. Vesti Akad. Nauk. URSS, Ser. Him. 1989, 5, 74-77.

[50] Gherasim, M. G.; Zugrăvescu, I. *Eur. Polym. J.* 1978, 14, 985-990.

[51] Varma, I. K.; Gupta, S. P.; Varma, D. S. Angew. *Macromol. Chem.* 1987, 153, 15-31.

[52] Nanjan, M. J., Sivaraj, K. *J. Polym. Sci.: Part A: Polym. Chem.*, 1989, 27, 375-388.

[53] Toshuki, K.; Kenji, O. JP Patent 148360, 1994.

[54] Kurmanaliev, M.; Ergozhin, E. E.; Iztleuova, I.K. lzr. Nat. Akad. Nauck Resp. Kaz., Ser. Khim. 1993, 2, 46-, *Vysokomol. Soed.* 1988, B30, 814-821.

[55] Kurmanaliev, M.; Ergozhin, E. E.; Iztlenova, I.K. *Macromol. Chem.*, 1993, 194, 2655-2661.

[56] Gaina, C.; Gaina, V.; Sava, M.; Chiriac, C.; Cozan, V. *J. Macromol. Sci. - Pure Appl. Chem.* 1997, 34, 191-200.

[57] Mognonov, D. M.; Radnaeva, L. D.; Bazaron L. Y.; Erj, B.V. *Vysokomol. Soed.* Ser.B 1987, 29, 26-29.

[58] Dix, L. R; Ebdon, J. R.; Flint, N. J.; Hodge, Ph; O'Dells, R. *Eur. Polym. J.* 1995, 31, 647-652.

[59] Volkov, V. S.; Ganzha, V. I.; Dolmatov S. A.; Bojko, L. I. *Plast. Massy* 1986, 11, 38-39.

[60] Volkov, V. S.; Dolmatov S. A.; Khabenko, A. V. *Plast. Massy* 1986, 4, 62-63.

[61] Sergeev, V. A.; Nedelkin, N. I.; Juferov, E. A.; Pavlova, S. S. A.; Gribkova, P. N.; Vinogradova, X.; Komarova L. I.; Erz, B.V. *Acta Polym.* 1987, 38, 87-91.

[62] Batotsyrenova, A. I.; Radnaeva, L. D.; Mognonov, D. M.; Izyneev, A. A.; Yerzh B. V.; Yunnikov, V.V. *Vysokomol. Soed.* 1990, A32, 433-438.

[63] Suzuki, M.; Nagai, A.; Suzuki, M.; Takahashi, A. *J. Appl. Polym. Sci.* 1992, 44, 1807-1813; idem 1992, 45, 177-180.

[64] Laurienzo, P.; Malinconico, M.; Perenze, N.; Segre, A. L. *Macromol. Chem. Phys.* 1994, 195, 3057-3065.

[65] Bartheleny, P. FR Patent 2611724, 1987 and 2612195, 1988; EP Patent 284524, 1988.

[66] Tamami, B.; Betrabet, C.; Wilkes, G. L. *Polym. Bull.* 1993, 30, 393-399.

[67] Greco, R.; Laurienzo, P.; Malinconico, M.; Martuscelli, E.; Perenze, N.; Sorrentino, A. *Adv. Polym. Techn.* 1994, 13, 141-147.

[68] Mandric, G.; Chiriac, C.; Gherasim, G.; Sava, M.; Pecincu, S. *Mat. Plast.* 1980, 18, 133-138.

[69] Varma, I. K.; Fohlen, G. M.; Parker, I. A.; Varma, D.S. Phospforus-containing imide resins: Modification by elastomers, *Int. Chem.* 1982.

[70] Itoya, K.; Kumagai, Y.; Kanarnaru, M.; Sawada, H.; Kakinoto, M.; Imai, Y.; Fukunaga, O. *Polym. J.* 1993, 25, 883-890.

[71] Kotukhova, A. M.; Kandybko, A. M.; Novikova, I. G.; Ghegodaev P. P.; Dolmatov, S. A. *Vysokomol. Soed.* 1988, A30, 1398-1403.

[72] Tung, C. M. *Polym. Prepr.* 1987, 28(1), 7-8.

[73] Hopewell, J. L.; George G. A.; Hill, D. J. Polymer 2000, 41, 8221-8229 and 8231-8239.

[74] Khabenko A. V.; Dolmatov, S. A. *Plast. Massy* 1987, 9, 52-54.

[75] Varma, I. K.; Tiwari, R. *J. Therm. Anal. Calorimetry* 1987, 32, 1023-1038.

[76] Ninan, K. N.; Krishnan, K. *Thermochim. Acta* 1988, 134, 231-235.

[77] Tungare, A. V.; Martin, G. C. *J. Appl. Polym. Sci.* 1992, 46, 1125-1135.

[78] Fasce, D. P.; Williams, J. J. R. *Polym. Bull.* 1995, 34, 515-522.

[79] Okumoto, S.; Yamabe, S. *J. Org. Chem.* 2000, 65, 1544-1548.

[80] Wang, S.; Len, T. S. *J. Appl. Polym. Sci.* 1999, 73, 833-839.

[81] Hariharan, R.; Anitha, M.; Bhuvano, S.; Saroja Devi, M. *J. Macromol. Sci.-Pure Appl. Chem.* 2004, 41, 317-328.

[82] Hariharan, R.; Anuradha, G.; Bhuvana, S.; Saroja Devi M.; Anitha, M. *J. Polym. Research* 2004, 11, 239-245.

[83] Wu, C. S.; Liu, Y. L.; Chiu, Y. S. *Polymer*, 2002, 43, 1773-1779.

[84] Hariharan, R.; Bhuvana, S.; Sarojadevi, M. *High Perform. Polym.* 2006, 18, 163-184.

[85] Liaw, D. J.; Liaw, B. Y.; Chen, J. *J. Polymer* 2001, 42, 867-872.

[86] Sava, M. *J. Appl. Polym. Sci.* 2002, 84, 750-757.

[87] Sava, M.; Grigoraş, C. V. *J. Macromol. Sci.-Pure Appl. Chem.* 2005, 42, 1095-1108.

[88] Wu, C. S.; Tsai, S. H.; Liu, Y. L. *J. Polym. Sci.: Part A: Polym. Chem.* 2005, 43, 1923-1929.

[89] Wang, C.; Zhang, C.; Wang, P.; Zhu, P.; Wu, W.; Ye, C.; Dalton, L. R. *Polymer* 2000, 41, 2583-2590.

[90] Wu, W.; Wang, D.; Ye, C. *Polym. Bull.* 1998, 41, 401-408.

[91] Wu, W.; Zhang, Z.; Zhang, X. *Design. Monom. Polym.* 2004, 7, 423-434.

[92] Wu, W.; Wang, D.; Zhu, P.; Wang, P.; Ye, C. *J. Polym. Sci.: Part A: Polym. Chem.* 1999, 37, 3598-3605.

[93] Informations Chimie 112 (oct.), 5, 1972.

[94] Mandric, G.; Sava, M.; Pecincu, S.; Petrache, G.; Constantinescu, C.; Tatucu, E. S.; Irunca, M.; Pascane, C. RO Patent 87679, 1985.

[95] Mandric, G.; Sava, M.; Pecincu, S.; Constantinescu, C.; Lica, V. RO Patent 90889, 1986.

[96] Sava, M.; Mandric, G.; Pecincu, S.; Stere, E.; Lica, V.; Ionescu, V. RO Patent 100897, 1990.

[97] Sava, M.; Pecincu, S. RO Patent 116289, 1992.

[98] Witzel, J. M.; Jablonski, R. J.; Kruh D. *Org. Coat. Plast. Chem. Prep.* 1971, 31, 488-489.
[99] Yamamoto, H.; Ozawa S. JP Patent 63037126 1988.
[100] Narimasa, Y.; Masahiro, M.; Akihiro, Y. EP Patent 388251, 1988.
[101] Sastri, V. R.; Tesoro, G. C. J. Polym. Sci., Part C; *Polym. Lett.* 1989, 27, 153-159.
[102] Varma, I. K.; Sangita; Varma, D. S. *J. Polym. Sci. Polym. Chem. Ed.* 1984, 22, 1419-1433.
[103] Varma, I. K.; Gupta, S. P.; Varma, D. S. *J. Appl. Polym. Sci.* 1987, 33, 151-164.
[104] Lin, King-Fu; Lin, Jin-Sing *J. Appl. Polym. Sci.* 1993, 50, 1601-1610.
[105] Diakoumakos, C. D.; Microyannidis, J. A. *J. Appl. Polym. Sci.* 1994, 54, 2063-2073.
[106] Mahesh, K. P. O.; Alagar, M.; Kumar, S. *A. Polym. Adv. Technol.* 2003, 14, 137-146.
[107] Singh, R. P.; Nayak, N.; Kandpal, L. D. *J. Polym. Mater.* 1996, 13, 163-167.
[108] Yerlikaya, Z.; Öktem, Z.; Bayramli, E. *J. Appl. Polym. Sci.* 1996, 59, 165-171 and *J. Appl. Polym. Sci.* 1996, 59, 537-542.
[109] Wang, C. S.; Hwang, H. Y. *J. Polym. Sci. Part A: Polym. Chem.* 1996, 34, 1493-1500.
[110] Lin, K. F.; Lin, J. S.; Cheng, C. H. *Polymer* 1996, 37, 4729-4737.
[111] Gaina, V.; Gaina, C.; Sava, M. *Polym. Plast. Technol.* Eng. 2001, 40, 89-102.
[112] Pan, Jing-Pin; Shiau, Gwo-Yuh; Chen, Ker-Ming *J. Appl. Polym. Sci.* 1992, 44, 467-476.
[113] Kumar, D.; Fohlen, G. M.; Parker, J. A. *J. Polym. Sci. Polym. Chem. E*d. 1983, 21, 245-267.
[114] Nagai, A.; Takahashi, A.; Wajima, M.; Tsukanishi, K. *Polym. J.* 1988, 20, 125-130.
[115] Chandra, R.; Rajabi, L.; Soni, R. K. *J. Appl. Polym. Sci.* 1996, 62, 661-671.
[116] Patel, H. S.; Patel, N. P. *Polym. Polym. Compos.* 1996, 4,173-180.
[117] Patel, H. S.; Shah, S. N. *High Perform .Polym.* 1996, 8, 233-242.
[118] Patel, H. S.; Patel, N. P. *High Perform. Polym.* 1997, 9, 33-39.
[119] Shah, V. J.; Partch, R. *Int. J. Polym. Mater.* 1998, 39, 151-159.
[120] Patel, H. S.; Dixit, B. C.; Patel, N. P. *Polym. Plast. Technol. Eng.* 2000, 39, 351-362.
[121] Patel, M. C.; Patel, G. A. *Polym. Plast. Technol. Eng.* 2003, 42, 947-958.
[122] Keiko, K.; Kenji, O.; Yoshinobu, *M. JP Patent* 7316427, 1995.
[123] Gaina, V.; Gaina, C. *Mat. Plast.* 2004, 41, 16-19.
[124] Rhone-Poulenc, BE Patent 834680, 1976.
[125] Yasuhiko, Y.; Tetsuo, I.; Mamoru, K. JP Patent 63152645, 1988.
[126] Magnonov, D. M.; Batlaev, K. E. *Vysokomol. Soed. Ser.B* 1993, 35, 177-183.
[127] Patel, H. S.; Patel, N. P. *Polym. Polym. Compos.* 1997, 5, 159-164.
[128] Masahiro, M.; Kensaku, M.; Kenji, O.; Yoshihisa, S. JP Patent 62208939, 1987.
[129] Keiichiro, I.; Kenichi, S. JP Patent 62181364, 1987.
[130] Yasuo, F.; Yasushi, Y.; Sadao, S.; Shohei, E. JP Patent 62190226, 1987.
[131] Norimasa, Y.; Nabubito, K.; Masahiro, O.; Akihiro, Y. EP Patent 253586, 1988.
[132] Kaoru, K.; Yoshinobu, O. JP Patent 63077938, 1988.
[133] Kavacikova, M.; Vasiljev, R.; Benkov, P. CZ Patent 247273, 1988.
[134] Heinz, H.; Gerd, G. EP Patent 316775, 1990.
[135] Ogasawara. K.; Kashihara, K.; Yoshioka, S.; US Patent 5279895, 1994.
[136] Kamio, Y.; Masahiro, M.; Hidetaka, K.; Hiroaki, F.; Takao, H.; Satoshi, O. JP Patent 5310930, 1993.
[137] Yasuhiro, H.; Kazuov, T.; Mitsuhiro, S.; Shuichi, K. JP Patent 6049025, 1994.

[138] Sava, M. Mat. Plast. 1995, 32, 26-30.
[139] Asahara, T. JP Patent 14745, 1972.
[140] Stenzenberger, H. D. Pat.USA 4211861, 1980.
[141] Stenzenberger, H. D.; Herzog, M.; Romer, W. R.; Pierce, S.; Canning, M.; Fear, K. Proc. ANTEC, 1246, 1985.
[142] Asahara, T.; Tanaka, C. JP Patent 14140, 1972.
[143] Hang, T.; Kiefer, J. DE Patent 2529092, 1976.
[144] Fukami, T.; Yamainoto, Y. JP Patent 141900, 1977.
[145] Gruffan, M.; Lofebre, G. DE Patent 2517303, 1975.
[146] Balme, M.; Gruffan, M. FR Patent 2318886, 1977.
[147] Makino, M.; Tomoga, K. JP Patent 01217, 1978.
[148] Kiyoje, M.; Tsutomu, O. JP Patent 51499, 1977.
[149] Rhone-Poulenc, DE Patent 2506264, 1975.
[150] Fargo, I.; Schreiber, B.; Renner, A.; Hang, Th. DE Patent 2459925, 1975.
[151] Mueller, A.; Huag, T.; Renner, A. DE Patent 2459961, 1975.
[152] Taranu, V.; Pecincu, S. *Macromol. Reports*, 1994, 31, 45-52.
[153] Suzuki, H.; Furukawa, T.; Saito, M. JP Patent 34646, 1975.
[154] Kajiura, S.; Suzuki, S. JP Patent 26896, 1978.
[155] Renner, A.; Fargo, I.; Hofmann, W.; Ramsteiner, K. *Helv. Chim. Acta* 1978, 61, 1443-1453.
[156] Hulubei, C.; Cojocariu, C.; Pecincu, S.; Popescu, F. *J. Macromol. Sci. -Pure Appl. Chem.* 1997, 34, 1085-1095.
[157] Hulubei, C.; Cozan, V. *High Perform. Polym.* 2004, 16, 149-158.
[158] Hulubei, C.; Cozan, V.; Bruma, M. *High Perform. Polym.* 2004, 16, 405-418.
[159] Moore, J. E.; Ward, W. H. *J. Am. Chem. Soc.* 1956, 78, 2414-2418.
[160] Crivello, J.; Juliano, P. C. *J. Polym. Sci.: Part A: Polym. Chem.* 1975, 13, 1819-1842.
[161] Crivello, J. V. *J. Polym. Sci. Polym. Chem. Ed.* 1976, 14, 159-182.
[162] Sergeev, V. A.; Nedelkin, V.; Arustainyan, A. *Vysokomol. Soe*d. B21 1979, 391.
[163] Sergeev, V. A.; Nedelkin, V.; Juferov, E. A.; Pavlova, S. A.; Gribkova, P. N.; Vinogradova, N. K.; Komarova, L. I.; Erz, B. V.*Acta Polym.* 1987, 38, 87-91.
[164] White, J. E.; Snider, D. A.; Scaia, M. D. *J. Polym. Sci. Polym. Chem. Ed.* 1984, 22, 589-596.
[165] Dix, L. R.; Ebdon, J. R.; Hodge, *Ph. Eur. Polym. J.* 1995, 31. 653-658.
[166] Asahara, T. *Kobunshi Ronbunshu* 1975, 32, 452-454.
[167] Miao, G.; Yu, Z.; Wu, Q.; Chen, Y. *Eur. Polym. J.* 1997, 33, 1401-1402.
[168] Cianga, L. *High Perform. Polym.* 2005, 17, 117-134.
[169] Gaina, V.; Gaina, C. *Polym. Plast. Technol. Eng.* 2004, 43, 539-553.
[170] Gaina, C.; Gaina, V. *Design. Monom. Polym.* 2005, 8, 145-158.
[171] Gaina, C. *Chem. Bull. "Politehnica"Univ. Timisoara* 2006, 51, 1-2.
[172] Akiyama, K.; Matsuda, K. Japan Patent 7417437, 1975.
[173] Matsuda, I.; Akiyama, K.; Mizuta, M. *Kobunshi Ronbunshu* 1976, 33, 47.
[174] Matsuda, I.; Yoshizumi, A.; Akiyama, K. *Kobunshi Ronbunshu* 1974, 31, 47.
[175] Berlin, A. A.; Zelenetskaya, T. V.; Aseeva, R. M. Zh. Vses. Khimi Obshchest 1966, 15, 591.
[176] Delmon, A. D. *J. Macromol. Sci.*, C-2 1968, 165.

[177] De Schryver, F. C.; Feast, W.; Smets, G. *J. Polym. Sci. Part A1 Polym. Chem.* 1970, 8, 1939-1948.
[178] De Schryver, F. C.; Boems, N.; Smets, G. *J. Polym. Sci. Part A1 Polym. Chem.* 1972, 10, 1687-1699.
[179] Put, J.; De Schryver, F. C. *J. Am. Chem. Soc.* 1973, 95, 137-145.
[180] De Schryver, M. F. C. IUPAC, *Pure Appl. Chem.* 1973, 34, 213-225.
[181] Shim, C. S.; Bong, P. H. *Bull. Korean Chem. Soc.* 1982, 3, 115-119.
[182] De Schryver, F. C.; Boens, N.; Smith, G. *J. Amer. Chem. Soc.* 1974, 96, 6463-6469.
[183] De Schryver, F. C.; Boens, N.; Smith, G. *Macromolecules* 1974, 7, 399-406.
[184] Iwata, K.; Stille, J. K. *J. Polym. Sci. Polym. Chem. Ed.* 1976, 14, 2841-2843.
[185] von Sonntag, J.; Knolle, W.; Naumov, S.; Mehnert, R. *Chem. Eur. J.* 2002, 8, 4199-4209.
[186] Nakanishi, F.; Hasegawa, M.; Takahashi, *H. Polymer* 1973, 14, 440-444.
[187] Huisgen, R. *Bull. Soc. Chim. Fr.* 1965, 3431.
[188] Rispens, T.; Engberts, J. B. F. N. *J. Org. Chem.* 2003, 68, 8520-8528.
[189] Rispens, T.; Engberts, J. B. F. N. *J. Phys. Org. Chem.* 2005, 18, 908-917.
[190] Engberts, J. B. F. N.; Fernández, E.; Garcia-Rio, L.; Leis, J. R. *J. Org. Chem.* 2006, 71, 4111-4117.
[191] Stille, J. K.; Anyos, T. *J. Polym. Sci. Part A2 General Papers* 1964, 2, 1487-1491.
[192] Smets, G. *Rev. Plast. Mod* (Madrid) 1969, 140, 79-80.
[193] Maier, G.; Zech, G.; Voit, B.; Komber, H. *Macromol. Chem. Phys.* 1998, 199, 2655-2664.
[194] Zugrăvescu, I.; Stoicesu-Crivetz, L. Polimeri Heterociclici, Ed.Acad. RSR, Buc. 1971.
[195] Iwakuro, Y.; Akiyama, M.; Shiraishi, S. *Bull. Chem. Soc. Jap.* 1965, 38, 513.
[196] Manecke, G.; Klawitter, *J. Makromol. Chem.* 1967, 108, 292-295.
[197] Marquez, C. *Rev. Plast. Mod.* (Madrid) 1971, 22, 706-719.
[198] Vretik, L.; Ritter, H. *Macromol.* 2003, 36, 6340-6345.
[199] Iwakura, Y.; Shiraishi, S.; Akiyama, M.; Yuyama, M. *Bull. Chem. Soc.* Japan 1968, 41, 1648.
[200] Iwakura, Y.; Shiraishi, S.; Akiyarna, M. *Makromol. Chem.* 1966, 97, 278-281.
[201] Gilliams, J.; Smets, G. *Makromol. Chem.* 1968, 117, 1-11; idem 1969, 128, 263-271.
[202] Varma, I. K.; Choudhary, V.; Gaur, B.; Lochab, B.; Oberoi, S.; Chauhan, R. *J. Appl. Polym. Sci.* 2006, 101, 779-786.
[203] Oda, R.; Takagi, K.; Shona, T. *J. Polym. Sci. Part B, Polym. Lett.* 1969, 7, 11-15.
[204] Takagi, K.; Gakujutsu, K. JP Patent 37831, 1993.
[205] Takagi, T.; Shiro, M.; Takeda, S.; Nokamura, N. *J. Amer. Chem. Soc.* 1992, 114, 8414-8416.
[206] Rigby, G. W.; Stark, B. P. DE Patent 2036405, 1971.
[207] Woodward, R. B.; Hoffmann, R. The Conservation of Orbital Symmetry, Verlag Chemie GmbH, Weinheim/Bergstr, 1970.
[208] Howell, S. J.; Spencer, N.; Philp, D. *Org. Lett.* 2002, 4, 273-276.
[209] Nour El-Din, A. M.; Mourad, A. E. S.; Abdel-Nabi, H. *J. Chem. Eng. Dat*a 1986, 31, 259-260.
[210] Oderaotoshi, Y.; Cheng, W.; Fujitomi, S.; Kasano, Y.; Minakata, S.; Komatsu, M. *Org. Lett.* 2003, 5, 5043-5046.
[211] Alhakimi, G.; Görls, H.; Klemm, E. *Macromol. Chem. Phys.* 1994, 195, 1569-1576.

[212] Kraiman, E. A. Patent US 2890207, 1959.
[213] Stille, J. K. Fortschr. *Hochpolym.-Forsch* 1961, 3, 48.
[214] Chow, S. W.; Whelam, J. M. Patent US 2971944, 1961.
[215] Bishop, M. T.; Bruza, K. J.; Laman, S. A.; Lee, W. M.; Woo, E. *P. Polym. Prepr.* 1992, 33(1), 362-363.
[216] Stevens, M. P. *J. Polym. Sci. Polym. Lett. Ed.* 1984, 22, 467-471.
[217] Diakoumakos, C. D.; Mikroyannidis, J. A. *J. Polym. Sci.: Part A: Polym. Chem.* 1992, 30, 2559-2567.
[218] Mikroyannidis, J. A. *J. Polym. Sci.: Part A: Polym. Chem.* 1992, 30, 2017-2024.
[219] Gaina, V.; Gaina, C. *Rev. Roum. Chim.* 2003, 48, 881-890.
[220] Gousse, C.; Gandini, A. *Polym. Bull.* 1998, 40, 389-394.
[221] Stille, J. K.; Plummer, L. *J. Org. Chem.* 1961, 26, 4026-4029.
[222] Tan, L. S.; Arnold, F. E.; Soloski, E. J. *J. Polym. Sci. Polym. Chem. Ed.* 1988, 26, 3103-3117.
[223] Tan, L. S.; Soloski, E. J.; Arnold, F. E. ACS *Symp. Series* 1988, 367, 349-365.
[224] Reifsnider, K.; Giacco, A. *Polym. Prepr.* 1992, 33(1), 355-356.
[225] Denny, L. R.; Goldfarb, I. J.; Farr, M. P. *ACS Symp. Series* 1988, 367, 366-.
[226] Harris, F. W.; Norris, S. O. *J. Polym. Sci. Polym. Chem. Ed.* 1973, 11, 2143-2151.
[227] Harris, F. W.; Stille, J. K. *Macromol.* 1968, 1, 463-464.
[228] Bailey, W. J.; Economy, J.; Hermes, M. E. Amer. *Chem. Soc. Div. Polym. Chem. Prepr.*, 1960, 1, 1-2.
[229] Bailey, W. J.; Economy, J.; Hermes, M. E. *J. Org. Chem.* 1962, 27, 3295-3299.
[230] Smith, J. G.; Sun, J. F.; Ottenbrite, R. M. *Macromol.* 1996, 29, 1123-1130.
[231] Kamahori, K.; Tada, S.; Ito, K.; Itsuno, S. *Macromol.* 1999, 32, 541-547.
[232] Reeder, J. A. Patent US 3344071, 1967.
[233] Ryntz, R. A.; Kohl, R. *T. Polym. Prepr.* 1983, 24(1), 322-323.
[234] Klemm, E.; Alhakimi, G.; Schiitz, H. *Makromol. Chem.* 1993, 194, 353-361.
[235] Klemm, E.; Koltner, N.; Alhakimi, G. *Angew. Makromol. Chem.* 1994, 223, 193-201.
[236] Katlner, N.; Klemm, E. *Angew. Makromol. Chem.* 1998, 254, 39-45.
[237] Alhakimi, G.; Klemm, E. *J. Polym. Sci.: Part A: Polym. Chem.* 1995, 33, 767-770.
[238] Alhakimi, G.; Klemm, E.; Görls, H. *J. Polym. Sci.: Part A: Polym. Chem.* 1995, 33, 1133-1142.
[239] Mzani, O.; Romdhane, H. B.; Baklouti, M.; Mercier, R. *Design. Monom. Polym.* 2006, 9, 13-27.
[240] He, Z.; Tolbert, L. *Polym. Prepr.* 1992, 33(l), 1000-1001.
[241] Grigoras, M.; Colotin, *G. Polym. Int.* 2001, 50, 1375-1378.
[242] Kriegel, R. M.; Saliba, K. L.; Jones, G.; Schiraldi, D. A.; Collard, D. M. *Macromol. Chem. Phys.* 2005, 206, 1479-1487.
[243] Hao, J.; Jiang, L.; Zou, Y.; Jiang, B.; Gu, Y.; Cai, X. MACROAKRON 94, 174, 1994.
[244] Jiang, B.; Jiang, L.; Gu, Y; Cai, X. MACROAKRON 94, 643, 1994.
[245] Kuramoto, N.; Hayashi, K.; Nagai, K. *Polym. Sci.: Part A: Polym. Chem.* 1994, 32, 2501-2504.
[246] Patel, H. S.; Mathur, A. B.; Bhardwaj, I. S. *J. Macromol. Sci.-Pure Appl. Chem.* 1995, 32, 2025-2034.
[247] Patel, H. S.; Patel, N. R. *High Perform. Polym.* 1994, 6, 13-19.
[248] Patel, H. S.; Patel, H. D. *Macromol. Reports* 1993, 30, 417-424.

[249] Gaina, V.; Gaina, C. *Polym. Plast. Technol. Eng.* 2002, 41, 523-538.
[250] Diakoumakos, C. D.; Mikroyannidis, *J. A. Eur. Polym. J.* 1994, 30, 465-472.
[251] Patel, H. S.; Patel, N. D. Phosphorous, Sulfur, Siliran Relot.Elem., 79(1-4), 69, 1993.
[252] Patel, H. S.; Vyas, H. S. *Eur. Polym. J.* 1991, 27, 93-96.
[253] Jiang, B.; Hao, J.; Wang, W.; Jiang, L.; Cai, X. *Eur. Polym. J.* 2001, 37, 463-470.
[254] Patel, H. S.; Patel, V. C. *Design. Monom. Polym.* 2000, 3, 191-198.
[255] Patel, H. S.; Patel, V. C. Design. *Monom. Polym.* 2001, 4, 369-380.
[256] Patel, H. S.; Patel, V. C. *Polym. Int.* 2003, 52, 304-309.
[257] Hawthorne, D. G.; Hodgkin, J. H.; Jackson, M. B.; Morton, T. C. *High Perform. Polym.* 1994, 6, 249-256.
[258] Patel, H. D.; Patel, H. S. *High Perform. Polym.* 1992, 4, 237-246.
[259] Patel, H. D.; Patel, H. S. *High Perform. Polym.* 1992, 4, 19-33.
[260] Tan, L. S.; Arnold, F. E.; Soloski, E. J. *J. Polym. Sci.: Part A: Polym. Chem.* 1988, 26, 3103-3117.
[261] Vargas, M.; Kriegel, R. M.; Collard, D. M.; Schiraldi, D. A. *J. Polym. Sci.: Part A: Polym. Chem.* 2002, 40, 3256-3263.
[262] Durmaz, H.; Colakoglu, R.; Tunca, U.; Hizal, G. *J. Polym. Sci.: Part A: Polym. Chem.* 2006, 44, 1667-1675.
[263] Gacal, B.; Durmaz, H.; Tasdelen, M. A.; Hizal, G.; Tunca, U.; Yagci, Y.; Demirel, Al. *Macromol.* 2006, 39, 5330-5336.
[264] Kang, J. W.; Kim, T. D.; Luo, J.; Haller, M.; Jen, A. K.Y. *Appl. Phys. Lett.* 2005, 87, 071109.
[265] Durmaz, H.; Karatas, F.; Tunca, U.; Hizal, G. *J. Polym. Sci.: Part A: Polym. Chem.* 2006, 44, 499-509.
[266] Aubert, J. H. *Polym. Prepr.* 2002, 43, 950-951.
[267] Aubert, J.H. *J. Adhes.* 2003, 79, 609-616.
[268] Loy, D. A.; Wheeler, D. R.; Russick, E. M.; McElhanon, J. R.; Saunders, R. S. US Patent 6,337,384 (2002).
[269] McElhanon, J. R.; Russick, E. M.; Wheeler, D. R.; Loy, D. A.; Aubert, J. H. *J. Appl. Polym. Sci.* 2002, 85, 1496-1502.
[270] McElhanon, J. R.; Wheeler, D. R. *Organic Letters* 2001, 3, 2681-2683.
[271] Chen, X.; Dam, M. A.; Ono, K.; Mal, A.; Shen, H.; Nutt, S. R.; Sheran, K.; Wudl, F. *Science* 2002, 295, 1698-1702.
[272] Watanabe, M.; Yoshie, N. *Polymer* 2006, 47, 4946-4952.
[273] Gheneim, R.; Perez-Berumen, C.; Gandini, *A. Macromol.* 2002, 35, 7246-7253.
[274] Chen, X.; Wudl, F. *Polym. Prepr.* 2003, 44, 87-88.
[275] Chen, X.; Wudl, F.; Mal, A. K.; Shen, H.; Nutt, S. R. *Macromol.* 2003, 36, 1802-1807.
[276] Goiti, E.; Huglin, M. B.; Rego, J. M. *Polymer* 2001, 42, 10187-10193.
[277] Chujo, Y.; Sada, K.; Saegusa, T. *Macromol.* 1990, 23, 2636-2641.
[278] Goussé, C.; Gandini, A.; Hodge, Ph. *Macromol.* 1998, 31, 314-321.
[279] Adachi, K.; Achimuthu, A. K.; Chujo, Y. *Macromol.* 2004, 37, 9793-9797.
[280] Imai, Y.; Itoh, H.; Naka, K.; Chujo, Y. *Macromol.* 2000, 33, 4343-4346.
[281] He, F.; Tang, Y.; Yu, M.; Wang, S.; Li, Y.; Zhu, D. *Adv. Funct. Mater.* 2007, 17, 996-1002.
[282] Kobayashi, M.; Mizuno, E.; Sakurai, A. US0188706, 2006.
[283] Arhipova, N. A.; Jybanov, B. A.; Rafikov, S.R. *Usp. Himii* 1978, 47, 705-738.

[284] Stevens, M. P.; Musa, Y. Div. *Polym. Chem. Prepr.* 1971, 12, 615.
[285] Musa, Y.; Stevens, M. P. *J. Polym. Sci. Part A-1: Polym. Chem.* 1972, 10, 319-327.
[286] Kardush, N.; Stevens, M. P. *J. Polym. Sci. Part A-1: Polym. Chem.* 1972, 10, 1093-1096.
[287] Zhubanov, B. A.; Akkulova, Z. G. *Vysokomol. Soed.* 1973, B15, 473-480.
[288] Zhubanov, B. A.; Nikitina, I. I. *Izv. Akad. Nauk. SSR, Ser. Khim.*, 1974, 24, 46.
[289] Enoki, T.; Okubo, H.; Ishii, K.; Shibahara, S. *Netsuku Kasei Jushi* 1991, 12, 9–18.
[290] Reyx, D.; Campistron, I.; Cailland, Ch.; Villatte, M.; Cavedon, A. *Macromol. Chem. Phys.* 1995, 196, 775-785.
[291] Rozenberg, B. A.; Dzhavadyan, E. A.; Morgan, R. J.; Shin, E. E. *Macromol. Symp.* 2001, 171, 87-96.
[292] Sunitha, A.; Reghunadhan Nair, C. P.; Krishnan, K.; Ninan, K. N. *Thermochim.Acta* 2001, 374, 159-169.
[293] King, J.; Caudhari, M.; Zahir, S. 29th Nat.SAMPE Symp. 1984, 29, 394.
[294] Stenzenberger, H. D.; Konig, P.; Romer, W.; Pierce, S.; Canning, M. 29th Nat.SAMPE Symp. 1984, 29, 1043.
[295] Carduner, K. R.; Chattha, M. S. *ACS Symp. Ser.* 1988, 367, 379-394.
[296] Abraham, T. J. Polym. Sci. C. *Polym. Lett.* 1988, 26, 521-528.
[297] Galvin, T. J.; Chaudhari, M. A.; King, J. *J. Chem. Eng. Progress* 1985, 81, 45-48.
[298] Chattha, M. S.; Dickie, R. A. *J. Appl. Polym. Sci.* 1990, 40, 411-416.
[299] Zahir, S.; Chaudhari, M. A.; King, *J. Makromol. Chem. Macromol. Symp.* 1989, 25, 141-154.
[300] Mijovic, J.; Andjelic, S. *Macromol.* 1996, 29, 239-246.
[301] Reyx, D.; Campistron, J.; Cailland, Ch.; Villatte, M.; Cavedon, A. *Macromol. Chem. Phys.* 1995, 196, 775-785.
[302] Morgan, R. J.; Shin, E. E.; Rosenberg, B.; Jurek, A. *Polymer* 1997, 38, 639-646.
[303] Morgan, R. J.; Jurec, R. J.; Yen, A. *Polym. Prepr.* 1992, 33(1), 426-427.
[304] Wu, Y.; Tesoro, G.; Engelberg, P. I. *Polym. Prepr.* 1991, 32(2), 140-141.
[305] Phelan, J. C.; Paik Sung, C. S. *Macromol.* 1997, 30, 6837-6844.
[306] Phelan, J. C.; Sung Paik, C. S. *Polym. Prepr.* 1997, 38(1), 227-228.
[307] Sumio, S.; Tokahiro, Y.; Shiya, M.; Sadao, J. H. *Polym. J.* 1998, 30, 404-409 and 410-413.
[308] Farong, H.; Jiuggang, W.; Yangsheng, J.; Zhangwen, L. *Reguxing Shuzhi* 1993, 8, 19-23.
[309] Liu, X.; Mi, J.; Gu, Y.; Yang, L.; Cai, X. *Reguxing Shuzhi* 1994, 9, 17-21.
[310] Gu, A.; Liang, G.; Lan, L. *J. Appl. Polym. Sci.* 1996, 62, 799-803.
[311] Loctite Corp. Neth. Patent 7511743, 1976.
[312] Yasuhisa, S.; Kunimasa, K.; Shuichi, K.; Yutaka S. EP Patent 0263915, 1988.
[313] Shigetsugu, H.; Hisashi, T.; Tokashi, M. US Patent 5173545, 1992.
[314] Varma, I. K.; Gupt, S. P.; Varma, D. S. *Polym. Mater. Sci. Eng.* 1989, 60, 639-641.
[315] Eisenbarth, Ph.; Linden, G.; Altstofdt, V.; Peter, R. US Patent 5003017, 1991.
[316] Chuan, H. H.; Wang, P. C. *Polym. Eng. Sci.* 1994, 34, 12-16.
[317] Akira, W.; Kaoru, K.; Masao, T. JP Patent 6100631, 1994.
[318] Varma, I. K.; Gupta, S. P.; Varma, D. S. *Angew. Macromol. Chem.* 1991, 184, 7-18.
[319] Guozheng, L.; Aijuan, S.; Liwen, L. *J. Adv. Mater.* 1996, 27, 61-64.
[320] Qiaolong, J.; Farong, H.; Yangsheng, J. *J. Appl. Polym. Sci.* 1996, 62, 459-464.

[321] Reghunadhan Nair, C. P.; Ninan, K. N. *J. Appl. Polym. Sci.* 2003, 88, 908-914.
[322] Varma, I. K.; Gupta, S. P.; Varma, D. S. *J. Appl. Polym. Sci.* 1987, 33, 151-164.
[323] Stenzenberger, H.; Herzog, M.; Franz, N.; Sarbacher, F. CA 109, 7374c, 1988.
[324] Wilkinson, S. P.; Liptak, S. C.; McGrath, J. E.; Ward, T. C. *Polym. Prepr.* 1992, 33(1), 424-425.
[325] Morgan, R. J.; Jurek, R. J.; Yen, A.; Donnellan, T. *Polymer* 1993, 34, 835-842.
[326] Iijima, T. Bismaleimide Resins in Polymeric Materials Encyclopedia, Ed. J.C.Salamone, vol. 1, 1996, p.683.
[327] Mai, K.; Huang, J.; Zeng, H. *J. Appl. Polym. Sci.* 1997, 66, 1965-1970.
[328] Liang, G.; Gu, A. *J. Appl. Polym. Sci.* 1997, 64, 273-279.
[329] Iijima, T.; Shiono, H.; Fukuda, W.; Tomoi, M. *J. Appl. Polym. Sci.* 1997, 65, 1349-1357.
[330] Iijima, T.; Nishina, T.; Fukuda, W.; Tomoi, M. *J. Appl. Polyrn. Sci.* 1998, 67, 769-780.
[331] Iijima, T.; Ohnishi, K.; Fukuda, W.; Tomoi, M. *Polym. Int.* 1998, 45, 403-413.
[332] Reghunadhan Nair, C. P. *Progr. Polym. Sci.* 2004, 29, 401-498.
[333] Gaina, C.; Gaina, V. *High Perform. Polym.* 2007, 19, 160-174.
[334] Gaina, C.; Gaina, V. *Design. Monom. Polym.* 2007, 10, 91-104.

INDEX

A

B

C

D

E

F

G

H

J

K

L

M

N

O

P

V

W

X

Y

Z